ENERGETICS OF ORGANIC FREE RADICALS

Structure Energetics and Reactivity in Chemistry Series (SEARCH series)

Series Editors

JOEL F. LIEBMAN
Department of Chemistry and Biochemistry
University of Maryland Baltimore County
Baltimore, MD 21228

ARTHUR GREENBERG
Department of Chemistry
University of North Carolina at Charlotte
Charlotte, NC 28223

The volumes in this series are comprised of state-of-the-art reviews, explicitly pedagogical in nature, in which specific topics are treated in depth. The series acronym SEARCH reflects the interplay between *S*tructure, *E*nergy *A*nd *R*eactivity in *CH*emistry and how these are also manifested in physical properties and biological activities.

Others titles in the Series

1. **Mesomolecules: From Molecules to Materials**
   Edited by G. David Mendenhall, Arthur Greenberg and Joel F. Liebman

2. **Active Oxygen in Chemistry**
   Edited by Christopher S. Foote, Joan Selverstone Valentine, Arthur Greenberg and Joel F. Liebman

3. **Active Oxygen in Biochemistry**
   Edited by Joan Selverstone Valentine, Christopher S. Foote, Arthur Greenberg and Joel F. Liebman

ENERGETICS OF ORGANIC FREE RADICALS

EDITORS:
JOSÉ ARTUR MARTINHO SIMÕES
Universidade de Lisboa

ARTHUR GREENBERG
University of North Carolina at Charlotte

JOEL F. LIEBMAN
University of Maryland Baltimore County

BLACKIE ACADEMIC & PROFESSIONAL
An Imprint of Chapman & Hall
London · Glasgow · Weinheim · New York · Tokyo · Melbourne · Madras

Published by
Blackie Academic & Professional, an imprint of Chapman & Hall, Wester Cleddens Road, Bishopbriggs, Glasgow G64 2NZ

Chapman & Hall, 2–6 Boundary Row, London SE1 8HN, UK

Blackie Academic & Professional, Wester Cleddens Road, Bishopbriggs, Glasgow G64 2NZ, UK

Chapman & Hall GmbH, Pappelallee 3, 69469 Weinheim, Germany

Chapman & Hall USA, 115 Fifth Avenue, Fourth Floor, New York NY 10003, USA

Chapman & Hall Japan, ITP-Japan, Kyowa Building, 3F, 2-2-1 Hirakawacho, Chiyoda-ku, Tokyo 102, Japan

DA Book (Aust.) Pty Ltd, 648 Whitehorse Road, Mitcham 3132, Victoria, Australia

Chapman & Hall India, R. Seshadri, 32 Second Main Road, CIT East, Madras 600 035, India

First edition 1996

Softcover reprint of the hardcover 1st edition 1996

Typeset in 10/12pt Palatino by Words & Graphics Ltd, Anstey, UK

ISBN-13: 978-94-010-6532-0 e-ISBN-13: 978-94-009-0099-8
DOI: 10.1007/978-94-009-0099-8

A catalogue record for this book is available from the British Library
Library of Congress Catalog Card Number: 94–81420

Printed on permanent acid-free text paper, manufactured in accordance with ANSI/NISO Z39.48-1992 (Permanence of Paper).

Contents

Preface

The current volume consists of eight chapters which interweave various aspects of the structure, energetics and reactivity of organic free radicals, all combining pedagogical insights with current research. The first is by Walling in which a personalized overview is given by one of the modern pioneers of the discipline. In the next two chapters, Tsang, and Traeger and Kompe, present key thermochemical and kinetic quantities from the complementary vantage points of the studies of neutral and cationic species. The fourth chapter by Francisco and Montgomery discusses the armamentarium of modern theory as applied to species with unpaired electrons, while the next chapter by Goodman presents the theory, methodology and results from photoacoustic calorimetry, a novel and powerful experimental technique. Martinho Simões and Minas da Piedade interrelate organometallic and free radical chemistry, while Greenberg and Liebman consider resonance energy and rearrangements as applied to small molecules and enzyme cofactors alike. The volume ends with the chapter by Tanko and Suleman, which describes the surprising and diverse solvent effects which modulate free radical chemistry. This volume will make it apparent to entry-level graduate students and senior researchers alike that much is known and much remains to be done in both the qualitative understanding and quantitative insights of the chemistry of organic free radicals.

José Artur Martinho Simões
Arthur Greenberg
Joel F. Liebman

Editorial Advisory Board

Series Preface

The purpose of this series is the presentation of the most significant research areas in organic chemistry from the perspective of the interplay and inseparability of structure, energetics, and reactivity. Each volume will be modeled as a text for a one-semester graduate course and will thus provide groundwork, coherence, and reasonable completeness. In this context, we have made the editorial decision to defer to the authors the choice of the desired blend of theory and experiment, rigor and intuition, practice and perception. However, we asked them to engage in the spirit of this venture and to explain to the reader the basis of their understanding and not just the highlights of their findings. For each volume, and each chapter therein, we have aimed for both a review and a tutorial of a major research area. Each volume will have a single theme, unified by the common threads of structure, energy, and reactivity for the understanding of chemical phenomena.

Structure, energetics, and *reactivity* are three of the most fundamental, ubiquitous, and therefore seminal concepts in organic chemistry. The concept of structure arises as soon as even two atoms are said to be bonded, since it is there that the concept of bond length and interatomic separation begins. Three-atom molecules already introduce bond angle into our functioning vocabulary, while four atoms are needed for the introduction of the terms *planarity, nonplanarity,* and *dihedral angles.* Of course, most organic chemists are interested in molecules of more than four atoms, so new shapes (tetrahedra, cubes, dodecahedra, prisms, and numerous exotic polyhedra) and new degrees of complexity arise. These new molecular shapes, in turn, function as templates for the next molecular generation. Still, the basic assumption remains: molecular structure determines energy and reactivity, and even though Van't Hoff, LeBel, and Sachse explained chemical reality with palpable molecular models over a century ago, we still do much the same thing on the screens of personal computers.

The concept of energetics arises in chemistry as soon as there is a proton and an electron and remains with us throughout our discipline. There are the fundamental, experimentally measurable quantities of bond energies, proton affinities, ionization potentials, pK_a values, and heats of formation. There are the derived quantities such as strain and resonance energies, acidity, and basicity. There are also the widely used, generally understood, and rather amorphous concepts such as *delocalization, conjugation,* and *aromaticity*. Indeed, the shape, conformation, and therefore function of a protein are determined by a balance of energetics contributions – resonance in the peptide linkage, hydrogen bonding, hindered rotation of a disulfide, Van der Waals forces, steric repulsion, Coulombic interactions and salt bridges, and solvent interactions.

The concept of reactivity inseparably combines structure and energetics and introduces more concepts and words: *stereospecificity, intramolecularity, nucleophilicity, catalysis, entropy of activation, steric hindrance, polarizability, hard and soft acids and bases, Hammett/Taft parameters*. Reactivity is a more difficult concept than structure and energetics. One must specify reaction conditions and usually accompanying reagents, since reactivity generally refers to two species or at least two seemingly disjoint parts of the same molecule.

Structure, energetics, and reactivity in chemistry have been probed by a plethora of experimental and theoretical methods. These tools have different degrees of accuracy and applicability, and consensus is rare as to when our understanding is deemed adequate. Indeed, diverse approaches – heats of hydrogenation and Hartree-Fock calculations, line intensities and LD_{50} values, ease of substitution and of sublimation, coupling constants and color – all contribute to the special blend of rigor and intuition that characterizes modern organic chemistry.

As people, and not just as scientists and editors, we wish to acknowledge the unity of the intellect and the emotions. We are grateful for the love and support of our families and for the inspiration, agitation, and stimulation from our mentors, colleagues, and students, and so we dedicate these volumes *"To Research and to Reason, To Family and to Friendship."*

JOEL F. LIEBMAN
Baltimore, Maryland

ARTHUR GREENBERG
Charlotte, North Carolina

Contributors

Joseph S. Francisco
Department of Chemistry and Department of Earth and Atmospheric Sciences, Purdue University, West Lafayette, IN 47907, USA

Joshua L. Goodman
Department of Chemistry, University of Rochester, Rochester, NY 14627, USA

Arthur Greenberg
Department of Chemistry, University of North Carolina at Charlotte, Charlotte, NC 28223, USA

Barbara M. Kompe
School of Chemistry, La Trobe University, Bundoora, Victoria 3083, Australia

Joel F. Liebman
Department of Chemistry and Biochemistry, University of Maryland Baltimore County, Baltimore, MD 21228, USA

José Artur Martinho Simões
Departamento de Química e Bioquímica, Faculdade de Ciências, Universidade de Lisboa, 1700 Lisboa, Portugal

Manuel E. Minas da Piedade
Centro de Química Estrutural, Complexo I, Instituto Superior Técnico, 1096 Lisboa Codex, Portugal

John A. Montgomery, Jr.
Lorentzian, Inc., 140 Washington Avenue, North Haven, CT 06473, USA

N. Kamrudin Suleman
Department of Chemistry, University of Guam, UOG Station, Mangilao, Guam 96923

James M. Tanko
Department of Chemistry, Virginia Polytechnic Institute and State University, Blacksburg, VA 24061, USA

John C. Traeger
School of Chemistry, La Trobe University, Bundoora, Victoria 3083, Australia

Wing Tsang
Chemical Kinetics and Thermodynamics Division, National Institute of Standards and Technology, Gaithersburg, MD 20899, USA

Cheves Walling
Department of Chemistry, University of Utah, Logan, UT 84322, USA

1

Free Radical Reactions

CHEVES WALLING

INTRODUCTION

Free radicals, molecules (or atoms) having one or more unpaired electrons are recognized as important short-lived, highly reactive intermediates in a variety of organic, and some inorganic, reactions. These include most of the processes involving molecular oxygen, e.g. combustion, autoxidation, and respiration, and much atmospheric chemistry including ozone formation and destruction and smog production. Many technical processes rely on radical reactions. Oxygen is our cheapest, most ubiquitous and cleanest oxidant and it is used in the manufacture of major chemicals ranging from phenol (from cumene) to terephthalic acid (from xylene). Further, most vinyl polymerizations and hydrocarbon halogenations also involve radicals. Much of organic photochemistry involves the formation of transient free radicals. Finally, the role of free radicals in biochemical processes is becoming increasingly evident. They play a harmful role in radiation damage and aging phenomena, but are also intermediates in enzyme mediated processes, many oxidation–reductions and rearrangements, e.g. those involving coenzyme B_{12}.

Because, as discussed further below, free radicals are usually very short-lived, their reactions, to be of any significance, must be very rapid and have low activation energies. This, in turn, requires that they be exothermic, or, at least, only slightly endothermic. Prediction of whether a radical reaction will be energetically favorable, and accordingly at least possible, can be made from a knowledge of bond dissociation energies, e.g. for a radical displacement

$$R\bullet + X{-}Y \rightarrow R{-}X + Y\bullet \qquad (1.1)$$

$\Delta H = D(X{-}Y) - D(R{-}X)$. Alternatively, ΔH can be calculated from the difference between the heats of formation of products and reactants, but since determining the heat of formation of a radical usually involves the knowledge of some bond dissociation energy, a bond dissociation energy is required here as well.

Calculating the energetics of a radical addition to a double bond is a bit more complicated. Consider the thermodynamic cycle:

$$H_2 \rightarrow 2\,H\bullet \qquad \Delta H = 104\ \text{kcal mol}^{-1} \qquad (1.2a)$$

$$CH_3{-}CH_2\bullet + H\bullet \rightarrow CH_3{-}CH_3 \qquad \Delta H = -98\ \text{kcal mol}^{-1} \qquad (1.2b)$$

$$CH_2{=}CH_2 + H\bullet \rightarrow CH_3{-}CH_2\bullet \qquad \Delta H = -38\ \text{kcal mol}^{-1} \qquad (1.2c)$$

$$CH_2{=}CH_2 + H_2 \rightarrow CH_3{-}CH_3 \qquad \Delta H = -32\ \text{kcal mol}^{-1} \qquad (1.2d)$$

where all but (1.2c) are experimentally measured quantities. Reactions (1.2b) and (1.2c) both form primary C–H bonds, but (1.2c) involves the disruption of a π-bond. Accordingly it is plausible to take the difference between (1.2b) and (1.2c), 60 kcal mol^{-1}, as the π-bond energy $D(\pi)$ associated with opening the ethylene double bond. One may quibble about whether the name is entirely accurate, but ΔH for the addition of any other species X• to ethylene is well approximated by $D(C_2H_5{-}X) - D(\pi)$. Similar calculations for substituted and conjugated olefins give slightly smaller $D(\pi)$ values, e.g. for styrene ~ 54 kcal mol^{-1}.

Knowledge of bond dissociation energies is particularly useful because, while they are usually referred to the gas phase, radicals are generally non-polar and not highly solvated so that the same values can be used in solution, even in highly-polar solvents. There are, however, some important exceptions to this, e.g. for the case of radical ions, and some examples are discussed in the last section of this chapter.

Free radical reactions have by now developed an enormous literature, which cannot be covered. Rather, this chapter summarizes the major classes of elementary steps, says something about the manner in which they may be strung together, and looks briefly at how individual rate constants may be measured and considers some generalizations about structure and reactivity.

CLASSES OF ELEMENTARY RADICAL REACTIONS

Since radicals are generally only transient intermediates, an overall radical reaction as observed consists of some radical-forming process, any transformations that the radicals undergo, and then the reactions by which they are destroyed. These elementary steps can be classified into a number of groups.

Radical Generating Reactions

Thermal bond scissions. At high temperatures almost any covalent bond may dissociate into free radicals. If a molecule contains a weak bond with a strength of 25–35 kcal mol^{-1}, dissociation may occur at a convenient rate near room temperature. Organic peroxides and azo compounds are good examples.

$$RO{-}OR \rightarrow RO\bullet + \bullet OR \qquad (1.3)$$

$$R{-}N{=}N{-}R \rightarrow R\bullet + N_2 + \bullet R \qquad (1.4)$$

Activation energies are close to the strengths of the bonds being broken and a variety of structures are available that can be chosen to yield radicals at desired rates at different temperatures. Table 1.1 lists a few examples, but it should be noted that the chemistry of organic peroxides is extremely complicated and many decompose by paths giving few, if any, radical products. Metal–carbon bonds are also usually weak. Tetra-alkyl lead compounds can serve as radical sources in high-temperature gas-phase reactions. Of particular interest currently are molecules with carbon–cobalt(III) bonds, which are very weak (E_a 18–30 kcal). Coenzyme B_{12} is such a species and its biological activity involves dissociation.

Photochemical radical production. The quanta of UV light at 268 nm have an energy equivalent to 100 kcal mol^{-1}, more than enough to dissociate most chemical bonds. Usually the excited states of molecules produced by light absorption decay without dissociation, but, if the molecule contains a weak bond as in the above cases, dissociation into radicals occurs (in fact, even peroxides that are poor thermal radical sources give high radical yields on photolysis). Such direct photolysis is the simplest case. Absorption of light by one molecule may lead to radical production in another either by energy transfer, electron transfer, or chemical reaction. As an example of the first, the triplet excited states of ketones will transfer energy to a diacyl peroxide, leading to its photosensitized dissociation. As an example of the second, an excited electron-rich donor, e.g. phenanthrene, may donate an electron to an electron-poor acceptor, e.g. dicyanobenzene to yield two radical ions

$$\text{Phen}^* + \text{DCB} \rightarrow \text{Phen}\bullet^+ + \text{DCB}\bullet^- \tag{1.5}$$

Of the third, the triplet states of suitable ketones readily abstract hydrogen from reactive substrates to generate a radical pair

$$R_2C{=}O^* + H{-}R \rightarrow R_2C\bullet{-}OH + \bullet R \tag{1.6}$$

TABLE 1.1 Decomposition of Some Convenient Radical Sources

Radical source	E_a (kcal M^{-1})	$t_{1/2}$(°C)
$(CH_3)_3CO{-}OC(CH_3)_3$ di-*t*-butyl peroxide	39	4.5 h (135°)
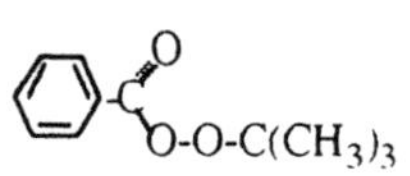 *t*-butyl peroxybenzoate	34.5	5.8 h (110°)
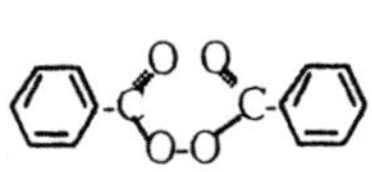 benzoyl peroxide	30	5 h (80°)
$(CH_3)_3CO\text{-}OC(O)C(O)O\text{-}OC(CH_3)_3$ di-*t*-butyl peroxyoxalate	25.5	5 h (27°)
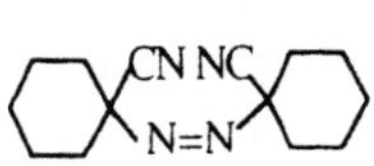 azo-cyclohexanecarbonitrile	34.5	2 h (100°)
$H_3C\text{-}C(CN)(CH_3)\text{-}N{=}N\text{-}C(CN)(CH_3)\text{-}CH_3$ azo-bis-isobutyronitrile (AIBN)	31	1.2 h (80°)

High energy radiation. The energy from γ-rays, electron beams, etc. also dissociates molecules. Thus irradiation of water yields hydroxyl radicals and hydrogen atoms (the latter in equilibrium with protons and solvated electrons)

$$H_2O \rightarrow HO\bullet + H\bullet \rightleftharpoons H^+ + e \tag{1.7}$$

With more complicated molecules, dissociation may be quite indiscriminate giving complicated fragmentation patterns, similar to those observed in a mass spectrometer.

Oxidation–reductions. Any one electron oxidation or reduction of a non-radical substrate generates a free radical, and many oxidations and reductions go through such a path. Reagents may be a metal ion as in the Fenton's reagent reaction

$$Fe^{2+} + H_2O_2 \rightarrow FeOH^{2+} + \bullet OH \tag{1.8}$$

or a metal as in ketyl formation

$$Ph_2C{=}O + Na \rightarrow PhC\bullet{-}O^- + Na^+ \tag{1.9}$$

or an electrolysis as in the Kolbe reaction

$$RCO_2^- - e \rightarrow RCO_2\bullet \rightarrow R\bullet + CO_2 \tag{1.10}$$

An example of a photo-oxidation reduction was given in equation (1.5).

Bimolecular reactions. ΔH for the reaction

$$(CH_3)_3CH + (CH_3)_2C{=}CH_2 \rightarrow 2\,(CH_3)_3C\bullet \tag{1.11}$$

is about 50 kcal mol^{-1}, much less than any of the C–C or C–H bond strengths in the molecules involved. There is good evidence that this kind of radical-forming process occurs at high temperatures, and it can be observed nearer room temperature in suitably designed systems where ΔH should be much smaller. It is the reverse of the disproportionation reaction (equation (1.13)) discussed below. Other more complicated polymolecular radical-forming reactions are known and have been named 'molecule-induced homolyses'.

Mechanical processes. If a polymeric plastic is cut or fractured C–C bonds in the polymer chains are broken and free radicals can be detected. Radicals are thus produced in the milling of rubber or other plastic, as a result of osmotic swelling of cross-linked polymers, and in polymer solutions by vigorous stirring or exposure to ultrasonic vibrations. In this last case, the collapse of cavitation bubbles produces violent local shear, pulling the chains apart, but radical production can also be detected in some non-polymeric solutions due to the intense adiabatic heating of solvent vapor in the bubbles during collapse.

Radical Destroying Reactions

Once formed, radicals generally have short lives because of very fast radical-destroying reactions, of which the most general is:

Radical coupling and disproportionation. When radicals encounter each other they usually couple to give a dimeric product

$$R\bullet + \bullet R \rightarrow R{-}R \qquad (1.12)$$

although, if coupling is sterically hindered, they may disproportionate, e.g.

$$2(CH_3)_3C\bullet \rightarrow (CH_3)_3CH + (CH_3)_2C{=}CH_2 \qquad (1.13)$$

Thus ethyl radicals give almost entirely *n*-butane while *t*-butyl radicals give chiefly the products indicated. Whichever path is followed the reactions in solution are usually close to diffusion controlled with rate constants of $\geqslant 10^9$ sec^{-1} M^{-1}. The chief exceptions are highly hindered radicals that also cannot easily disproportionate. Here steric hindrance to coupling is so large that the radicals may persist for seconds, minutes, or longer or exist in measurable equilibrium with their dimers. The classic example is triphenylmethyl, which exists in equilibrium with a quinoid dimer. The expected dimer, hexaphenylethane, is evidently too sterically hindered to form

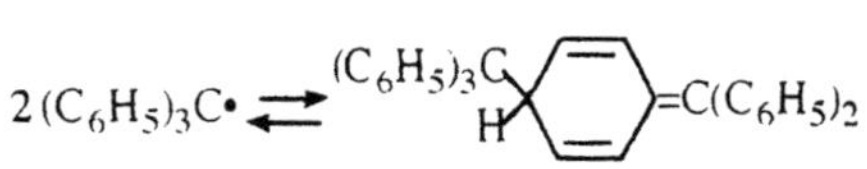

Such long-lived species are best called *persistent radicals,* reserving the older term *stable radicals* for species that are stabilized energetically by delocalization of the odd electron with resulting decreased values of $D(R{-}H)$. Such species are usually not *persistent.*

1.2.2.2 Oxidation reduction reactions. Just as radicals can be formed by one-electron oxidation–reductions, they can be destroyed by the same means. Many examples involve metal ions and may be either a simple outer-sphere electron transfer, as with Fe^{3+}

$$Fe^{3+} + \bullet CH_2OH \rightarrow Fe^{2+} + CH_2{=}OH^+ \rightarrow CH_2{=}O + H^+ \qquad (1.14)$$

or a more complex inner-sphere process as with Cu

$$R\bullet + Cu(II) \rightarrow RCu(III) \rightarrow Cu(I) + \text{non-radical products} \quad (1.15)$$

Electron transfers require that metal ion and radical have suitable oxidation–reduction potentials while inner-sphere processes assume the possibility of rapid ligand exchange.

Radical–Substrate Reaction

Reaction of a radical with a non-radical substrate generates a new radical and thus greatly increases the variety of overall radical processes.

1.2.3.1 Radical additions. Most radicals add readily to carbon–carbon double bonds

$$Br\bullet + CH_2{=}CHR \rightarrow BrCH_2{-}C\bullet HR \quad (1.16)$$

addition occurring predominantly to the less substituted, less sterically hindered carbon. A considerable entropy loss accompanies the addition, so it may be significantly reversible even when it is moderately exothermic. Additions may also occur to unsaturated centers involving heteroatoms, e.g. $>C{=}O$, $-C{\equiv}N$, $-N{=}N-$, $>C{=}S$. However, since the π-bond energies of such systems appear to be significantly larger (except for $>C{=}S$), the reactions are more reversible and harder to observe. For example, $D(\pi)$ for carbonyl groups, $>C{=}O$, can be estimated via cycles paralleling equation (1.2) and lies in the range of 70–80 kcal mol^{-1} and varies with structure in a manner paralleling the ease of ionic additions.

Radical displacements. Reactions of the type

$$R\bullet + X{-}Y \rightarrow R{-}X + Y\bullet \quad (1.17)$$

play an important part in free radical chemistry. In the commonest examples X=H or halogen (in the order of reactivity $F \ll Cl < Br < I$), but displacements are also known on the oxygen of peroxide bonds and, more readily, on disulfides. Displacements on carbon are restricted to highly strained ring systems such as cyclopropanes, in notable contrast to the behavior of nucleophilic and electrophilic displacements. Thus radical halogenation of cyclopropane leads to ring opening as well as substitution.

Reactions involving valence expansion. Higher-row elements may add radicals to give metastable products, which then fragment. The most-studied reactions involve trivalent phosphorus, which yield intermediate phosphoranyl radicals.

$$RO\bullet + PR_3 \rightarrow RO-\dot{P}R_3 \begin{cases} \nearrow R\bullet + OPR_3 \\ \searrow RO - PR_2 + R\bullet \end{cases} \quad (1.18)$$

The phosphoranyl radical may undergo either *α*- or *β*-scission. In the case cited, both occur. The relatively easy displacements on sulfur, selenium, and higher-row halogens may also involve such transient intermediates, although they have not, in general, been detected.

Electron deficient elements on the left of the periodic table, e.g. boron, behave similarly

$$ROO\bullet + BR_3 \rightarrow ROO-\dot{B}R_3 \rightarrow ROOBR_2 + R\bullet \quad (1.19)$$

and there are many other examples.

1.2.3.4 Electron transfer and reactions with ions. Electron transfer may occur between a radical and another species generating a new radical. Thus an important step in the facile autoxidation of carbanions is the transfer

$$ROO\bullet + R^- \rightarrow ROO^- + R\bullet \quad (1.20)$$

Radicals and radical ions may be interconverted by a number of reactions involving ions. For example, the so-called $S_{RN}1$ reaction involves a reversible step

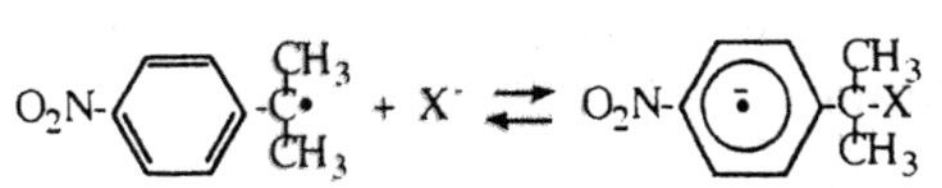

(1.21)

while radical cations may give radicals by proton loss

$$[ArCH_3]^{+\bullet} \rightarrow ArCH_2\bullet + H^+ \quad (1.22)$$

Unimolecular and Intramolecular Reactions

β-Scissions. These are the reverse of addition reactions (1.16). While they are entropically favored, those involving C—C bond scission are generally endothermic by 20–25 kcal M^{-1} and require high temperatures. Ethylene may be obtained by the thermal cracking of paraffins via the sequence

$$RCH_2-CH_2\bullet \rightarrow R\bullet + CH_2=CH_2 \rightarrow \text{etc.} \quad (1.23)$$

and some polymers, e.g. polymethylmethylacrylate, decompose similarly to monomer in good yield. Many other groups with weaker bonds to carbon, e.g. Br—, RS—, R_3Sn— are lost more readily near room temperature, so that addition reactions that form such *β*-substituted radicals are significantly reversible. The *β*-scission of alkoxy radicals also competes with their other reactions

$$R_3C{-}O\bullet \rightarrow R_2C{=}O + R\bullet \qquad (1.24)$$

Generally the most resonance stabilized radical, R•, is lost

1,2-Shifts. Unlike the behavior observed with carbocations, there is little evidence for 1,2-shifts of H or alkyl groups in radicals. However 1,2-shifts of unsaturated groups, e.g. vinyl or aryl, are readily observed and go through intermediate 3-membered rings

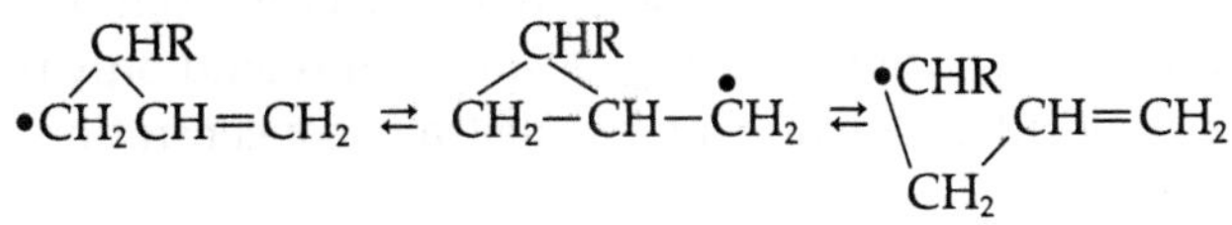

(1.25)

which can be trapped in some cases. 1,2-shifts of halogens and some other groups are also known and apparently do not simply involve *β*-scission and readdition but a symmetric bridged species that may well only be a transition state rather than a metastable intermediate as above.

More remote intramolecular additions and displacements. These are also well known and are increasingly being used in synthesis. Additions involving various substituted 5-hexenyl radicals occur with particular ease, and may yield either 5- or 6-membered rings depending on structure and reaction conditions.

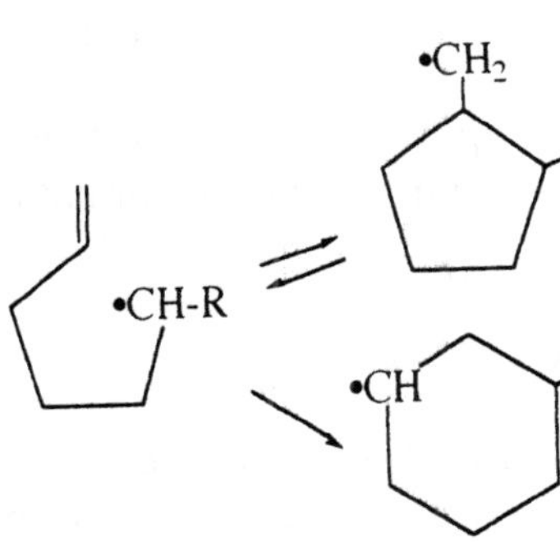

(1.26)

Although the 5-membered ring closure is less exothermic, in aliphatic systems it is kinetically favored both energetically and entropically. However, if the initial radical is highly resonance-stabilized, 5-ring closure becomes reversible and either product may be observed depending on conditions. Interestingly, ring closure on a carbonyl group may occur at a comparable rate but is much more reversible so the cyclized product is observed only if the intermediate alkoxy radical is trapped by some very fast H-abstraction reaction.

Intramolecular hydrogen abstraction reactions occur most readily through cyclic 6-membered transition states, e.g. for alkoxy radicals

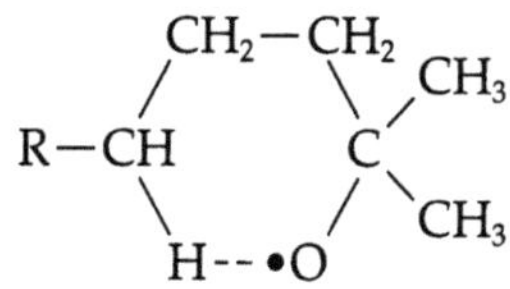

(1.27)

The analogous reaction involving a carbon radical is important in producing chain branching in ethylene polymerization. Both additions and abstractions are very facile, occuring at rates that would only be observed for an intermolecular reaction at a substrate concentration of several hundred molar.

CHAIN REACTIONS

Reaction sequences that transform one radical into another en route to regenerating the original radical yield cycles that may be repeated many times in a chain reaction, giving many molecules of product for every radical initially produced. The steps can be arranged in many ways. In vinyl polymerizations, *successive additions* produce long thread-like polymers via the repeating step

$$\sim CH_2{-}CHR\bullet + CH_2{=}CHR \rightarrow \sim CH_2{-}CHR\bullet \quad (1.28)$$

and are of enormous technical importance. *Alternating successive displacements* are involved in reactions like free radical chlorination

$$Cl\bullet + H{-}R \rightarrow HCl + R\bullet \quad (1.29)$$

$$R\bullet + Cl{-}Cl \rightarrow RCl + Cl\bullet \quad (1.30)$$

Alternating additions and displacements lead to the addition of one molecule across the double bond of another as in the classic 'abnormal addition' of HBr to olefins

$$Br\bullet + CH_2{=}CHR \rightarrow BrCH_2{-}CHR\bullet \quad (1.31)$$

$$BrCH_2{-}CHR\bullet + H{-}Br \rightarrow BrCH_2{-}CH_2R + Br\bullet \quad (1.32)$$

The typical steps in hydrocarbon autoxidations make up a similar cycle:

$$R\bullet + O_2 \rightarrow R{-}OO\bullet \qquad (1.33)$$

$$R{-}OO\bullet + H{-}R \rightarrow ROOH + R\bullet \qquad (1.34)$$

Many other combinations are possible, including ones involving several successive steps in each cycle. Thus the rather unlikely looking process

$$RSSR + P(OR)_3 + CO \rightarrow RCOSR + SP(OR)_3 \qquad (1.35)$$

occurs readily and appears to involve a four-step cycle

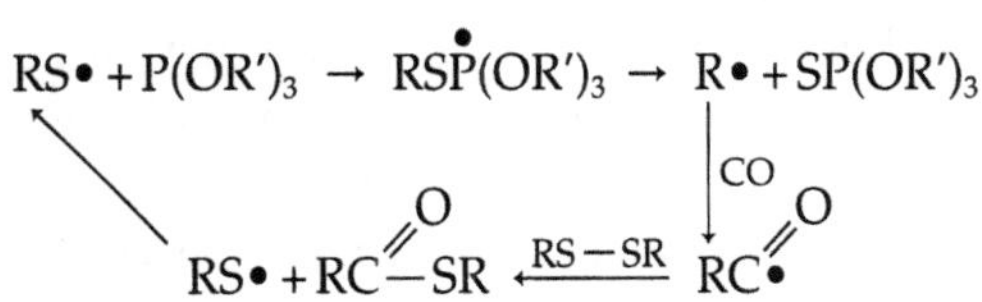

(1.36)

Another example is the deoxygenation of alcohols developed by Barton. Here the alcohol, R_2CHOH, is first converted to a thioester, $R_2CHOCSR$, and then reacted with tributyl tin hydride in the presence of a radical source

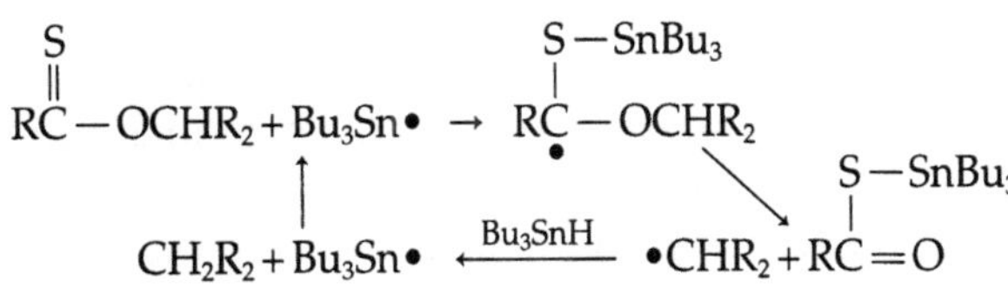

(1.37)

Many complex cycles involving ring-closures, often with the tributyl tin radical, have been devised by synthetic chemists to build up elaborate fused-ring systems in a single overall reaction.

Appreciation of the *'time schedule'* of radical reactions is important in understanding radical chain reactions. To take a typical example, if radicals are being generated in a system at a rate $R_i = 4 \times 10^{-7}\ M^{-1}\ sec^{-1}$ (about that expected from the decomposition of 0.01 M benzoyl peroxide at 80°C) and disappear by bimolecular recombination with a rate constant k_t of $2 \times 10^9\ M^{-1}\ sec^{-1}$, their concentration will be 10^{-8} M and their average lifetime (their concentration divided by their rate of disappearence) will be 0.025 s. For a chain reaction to occur with a kinetic chain-length of 250, each cycle would have to take place in 10^{-4} s. Decreasing R_i would give longer lifetimes, but with lower radical concentrations and slower overall rates. It is clear that only very fast, low activation energy steps can be involved and even a non-chain reaction would have to fit into the 0.025 s time slot.

Although chain reactions appear complicated, their overall kinetics can be relatively simple. Considering a polymerization as in equation (1.28) and letting [M] represent the monomer concentration and [~M•] the concentration of growing chains, the latter's concentration by the argument in the paragraph above will be $(R_i/2k_t)^{1/2}$. If they add monomer with a rate constant k_p the overall rate of monomer consumption becomes $-\mathrm{d}[\mathrm{M}]/\mathrm{d}t = k_p\,[\mathrm{M}]\,(R_i/2k_t)^{1/2}$.

A notable characteristic of and diagnostic test for radical chain reactions is that experimentally their rates may be highly variable. They are accelerated by radical sources (sometimes incorrectly called catalysts; chain initiators is a better term) and retarded or stopped altogether by species which have been called retarders, inhibitors, or negative catalysts. The acceleration is easily understood as a simple increase in R_i. Retarders and inhibitors work by reacting with chain carrying radicals and converting them to other species that react further too slowly to carry on the chain. Interestingly, an inhibitor that effectively stops one type of chain may participate perfectly well in another. Thus traces of molecular O_2 strongly inhibit most radical chains involving carbon-centered radicals because their reactions with O_2 (reaction (1.33)) occur at close to a diffusion-controlled rate, while the resulting peroxy radicals, ROO•, attack hydrocarbon substrates only very slowly via reaction (1.34) and simply float about until they encounter another radical. On the other hand, many organic substrates in O_2-saturated solutions autoxidize quite rapidly by the (1.33)–(1.34) cycle. The explanation is that under these conditions ROO• species are essentially the only radicals present and they undergo mutual destruction only by rather complicated processes involving reversible formation of a tetroxide detectible at very low temperatures.

$$2\ \mathrm{ROO}\bullet \rightarrow \mathrm{ROO{-}OOR} \rightarrow O_2 + \text{non-radical products} \qquad (1.38)$$

which then goes on to non-radical products. This path has an appreciable overall activation energy and may have an overall rate constant of $<10^4$, so that the low rate constant for (1.34) is compensated for by a much higher radical concentration.

RATES OF RADICAL REACTIONS

Although the overall rates of radical reactions are of practical importance, they are not of much help in understanding what is going on. As the discussion of polymerization above shows, the overall rate depends upon how rapidly radicals are being produced in the system and the rate constant for chain termination as well as that for chain propagation, and the latter two terms cannot be separated. Further, many reactions are more complicated than this and may involve minor side reactions that lead to unreactive species that stop chain growth. Notable examples involve double bond additions to substrates that also possess allylic hydrogens. Ethyl acrylate, $CH_2{=}CHCOOEt$, polymerizes readily to give high molecular weight polymer with hundreds or thousands of monomer units in each chain. Its isomer, allyl acetate, $CH_2{=}CHCH_2OAc$, reacts much more slowly to give a polymer of only about 20 units. The explanation is in the following scheme.

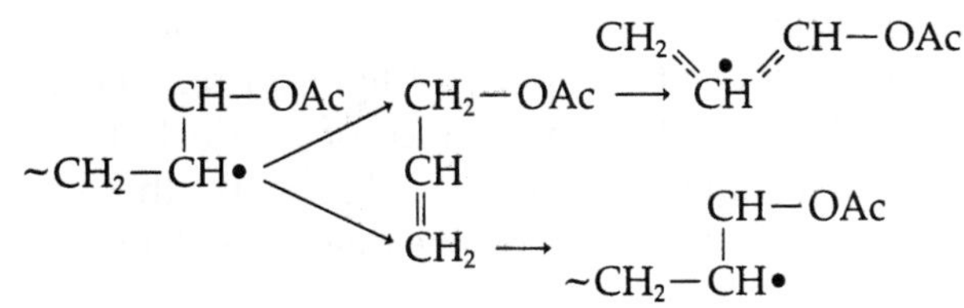

(1.39)

About one time in twenty the growing chain radical abstracts allylic hydrogen rather than adding to the double bond, and the resulting allylic radical is too unreactive to propagate the chain. Convincing proof for this scheme is that deuterated allyl acetate, $CH_2{=}CHCD_2OAc$, polymerizes several times as fast to a much higher molecular weight product.

The first step out of this dilemma is that *relative* rates of reaction of different substrates towards the same radical can easily be measured by allowing them to react in competition. The first extensive body of data of this sort was first obtained in the 1940s by examining the composition of the copolymers obtained when two monomers were polymerized together, and the technique has been extended to many other systems. For example, the relative reactivity of different C—H bonds towards *t*-butoxy radicals may be measured by carrying out competitive chlorinations with *t*-butyl hypochlorite. Here the chain sequence is

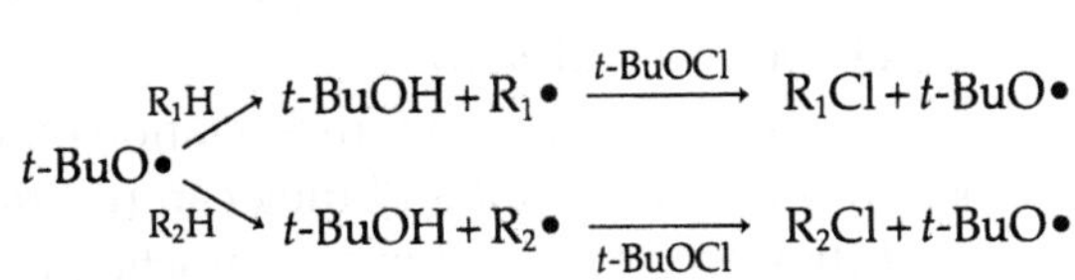

(1.40)

and the ratio of k_1/k_2 may be calculated from product analysis and correlated with substrate structure.

Determining the relative reactivities of different radicals towards the same substrate is much more difficult since it cannot be done by conventional steady-state kinetics and requires the direct measurement of individual rate constants. Many techniques have been devised to do this, usually requiring rather elaborate equipment. One of the oldest is the 'rotating sector method' in which the rate of a radical reaction photoinitiated by a flickering light is measured as the flickering rate is varied. A rather complicated mathematical analysis relates this to the average life-time of the radical chains from which rate constants can be deduced. Flow methods can also be used, particularly in the gas phase. Here radicals are introduced into a gas stream containing the desired substrate and their concentration determined spectroscopically at various points downstream. For liquid-phase reactions the most versatile technique is probably pulsed laser flash photolysis (or pulse radiolysis). Here radicals are produced by a short intense pulse of light (or high energy electrons) and either their decay or the appearance of products followed by fast spectroscopy over a time range of 10^{-9}–10^{-3} s. By these methods rate constants as fast as 10^{10} M^{-1} sec^{-1} can be determined and very extensive compilations are now available.

In the liquid phase, as rate constants become very large, $\geqslant 10^9$ $M^{-1}sec^{-1}$, the rate at which bimolecular reactions can actually occur becomes increasingly determined by the rate at which reactants can diffuse together. An important converse of this occurs when two reactive species are generated in close proximity so that their reaction with each other competes with their diffusion apart. Such reactions are called *cage reactions* and are important when radicals are produced in pairs by thermal or photochemical bond dissociation. Thus, in a simple peroxide dissociation to give two RO• radicals, a fraction recombine without ever diffusing out into the solution where they can be detected. This fraction increases with increasing solvent viscosity, so the apparent dissociation rate decreases. If the dissociation involves breaking more than one bond (via a concerted process or two steps in extremely rapid succession) as in the dissociation of an azo compound the starting material cannot be regenerated and increased viscosity has no effect on rate. However, the yield of available radicals, e.g. to start chain reactions, is reduced. Typically such yields lie between 10 and 80%.

A further consideration is that in thermal dissociations the spins of the two just-formed radicals are still paired and they can recombine without difficulty. However, if they are produced in a photochemical reaction by the dissociation of an excited triplet state they are now in an unpaired triplet state and must undergo spin inversion before they can combine. Similarly, if two radicals diffuse together from solution, three-quarters of the time they will be a triplet pair and the same barrier arises. Spin inversion and diffusion in and out of the solvent cage occur on similar time scales, and this leads to some very interesting phenomena, which cannot be discussed here but include chemically induced dynamic polarization (CIDNP) in NMR spectra and an effect of external magnetic fields on observed reaction rates.

The cage phenomenon is normally essentially over in 10^{-9} s but in a few cases the propagation steps of radical chain reactions are fast enough to compete. An example is in the radical chlorination of aliphatic hydrocarbons. Here reaction of a radical with Cl_2 yields an alkyl halide and Cl• in close proximity, and, particularly in dilute solution with no other substrate nearby, the two react to give a polychlorinated product before they can diffuse apart.

In some circumstances the fast recombination of radicals may be inhibited. Radicals trapped in solid matrices where they cannot diffuse together may persist for long times. In micellar solutions, water-insoluble radicals dissolved in the micelles may also have very long lives, a circumstance that is important in the success of emulsion polymerization reactions. Finally, it is plausible that the high specificity of enzyme-mediated reactions involving radical intermediates similarly involves radical and substrate being isolated and held together in proper orientation until reaction occurs.

STRUCTURE AND REACTIVITY

As pointed out in the introduction, the overarching consideration in determining the rates of the elementary steps in radical reactions are overall energetics, predictable from a knowledge of bond dissociation energies. To be observed (except in very high temperature processes) they cannot be strongly endothermic. However, the whole story is much more complicated than this. Even strongly exothermic elementary steps may have appreciable activation energies, and these may vary widely with structure. As an example, the reactions

$$X\bullet + H{-}CH_3 \rightarrow X{-}H + \bullet CH_3 \tag{1.41}$$

are almost thermoneutral for X• = Cl•, H• and CH_3•, but gas phase activation energies are respectively 3.9, 9, and 14.3 kcal mol^{-1}, corresponding to a range of about 10^7 in rate. It is hard to make generalizations about how the rates of isoenergetic radical reactions will vary as the atoms involved are changed (however, see the discussion of nucleophilic and electrophilic radicals below), but we can say a good deal about how the rate of a single reaction will change with changes in the structures of the radical and substrate. Three general principles emerge, which were first noted in copolymerization studies in the 1940s, but have now been greatly extended and refined.

First, overall energetics have a significant effect even in exothermic reactions. Reactions producing resonance stabilized radicals in which the odd electron is delocalized over the molecule are generally faster than when this is not the case. Conversely, delocalized resonance stabilized radicals react more sluggishly than those that are not delocalized. Other things being equal (see below) there is often a linear relation between RT ln k and ΔH in such systems although with only a modest slope, -0.3 to -0.5. Incidentally, a great deal is known about the structure of radicals and delocalization of their odd electrons from ESR spectra, again a topic not discussed here.

The second factor is steric hindrance, which is quite marked in radical additions to C=C double bonds. It is the major factor in determining the direction of addition, i.e. preferential addition to the least substituted carbon, accounts for the head-to-tail structure of vinyl polymers, and the the fact that non-terminal double bonds are less reactive than terminal ones. A single substituent at the point of addition generally reduces bond reactivity by a factor of 3–20 and depends somewhat on the size and shape of the attacking radical. Steric hindrance to radical displacements is harder to detect, but there are certainly cases where well shielded C—H bonds show decreased reactivity towards H abstraction. The most dramatic cases of steric hindrance are certainly those in which the normally diffusion controlled rate of radical coupling is suppressed in 'persistent' radicals, as discussed earlier.

While both a parallel between rate and energetics and some role of steric hindrance are to be expected, the third factor was quite unanticipated when first encountered, although it has since been amply validated as a general property of radical processes. This is that radicals with electron-withdrawing groups react preferentially with substrates with electron-supplying groups and vice versa so that some copolymers (e.g. that from styrene and maleic anhydride) had structures in which the two monomer units alternated along the chain. Stating the finding another way, there is a 'polar effect' in radical reactions. Radicals that can accept electrons and have relatively stable corresponding anions are *electrophilic* and attack points of high electron density. Radicals that can donate electrons and have stable cations are *nucleophilic* and prefer points of low electron density. The phenomenon was originally interpreted in terms of contributions of ionic structures lowering the energies of transition states in radical reactions, e.g. for the addition of a nucleophilic carbon radical to a carbonyl-conjugated double bond

$$\mathrm{R\bullet\cdots CH_2{=}CH\underset{\displaystyle R}{\underset{|}{C}}{=}O \leftrightarrow R^{+}\cdots CH_2\overset{\bullet}{\cdots}CH\cdots\underset{\displaystyle R}{\underset{|}{C}}\cdots\bar{O} \leftrightarrow R{-}CH{-}\overset{\bullet}{C}H\underset{\displaystyle R}{\underset{|}{C}}{=}O} \quad (1.42)$$

or, for the addition of an electrophilic halogen atom to an olefin

$$\mathrm{X\bullet\cdots CH_2{=}CHR \leftrightarrow X^{-}\cdots \overset{\bullet}{C}H_2\cdots\overset{+}{C}HR \leftrightarrow X{-}CH_2{-}\overset{\bullet}{C}HR} \quad (1.43)$$

In molecular orbital terms one would say that, with electrophilic radicals, overlap of a singly occupied molecular orbital (SOMO) of the radical with the highest occupied molecular orbital (HOMO) of the substrate lowers transition state energy. With nucleophilic radicals it is overlap between their SOMO with the lowest unoccupied molecular orbital (LUMO) of the substrate. Regardless of the explanation, the effect can be very large. Nucleophilic cyclohexyl radicals add to maleic anhydride 730 times more rapidly than to styrene, although the latter is almost certainly the more exothermic process and maleic anhydride is quite unreactive towards electrophilic radicals.

Similar effects arise in radical displacements. Thus for Cl• attack on C–H bonds one may write transition-state structures

$$\mathrm{Cl\bullet\cdots H{-}R \leftrightarrow Cl^{-}\cdots H\bullet\cdots {}^{+}R \leftrightarrow Cl{-}H\cdots\bullet R} \quad (1.44)$$

Chlorine atom reactions with alkanes are so rapid that they cannot be appreciably accelerated, but they are certainly retarded by electron-withdrawing groups on the substrate. Acetic acid or acetonitrile may be used as solvents for chlorination either with Cl_2 or *t*-butyl hypochlorite in which the chain carrier is the electrophilic *t*-butoxy radical and in the chlorination of long-chain acid derivatives substitution is diverted away from the acid group and the effect extends for a surprisingly long distance along the chain. Thus for heptanoyl chloride the distribution of isomeric products is reported to be $ClCOC_2(1\%)C_3(9\%)C_4(18\%)C_5(24\%)C_6(28\%)C_7(20\%)$. Other electron withdrawing groups have a similar effect.

Polar effects may account in part for the differing properties of radicals on different atoms remarked on earlier. Halogen atoms, thiyl radicals, RS•, alkoxy radicals RO•, and most of the other species that add to double bonds are electrophilic and react very rapidly with these electron-rich substrates. However, simple carbon radicals are nucleophilic and react more slowly. Peroxy radicals ROO• are a little puzzling in that they ought to be reasonably electrophilic but, as noted earlier, they react very slowly with either C—H or C=C with rate constants usually $<10^3\ M^{-1}\ sec^{-1}$. Their reactivities are, however, quite structure dependent. Electron withdrawing groups, which should make them more electrophilic, do make them more reactive. As extreme examples, acylperoxy radicals RC(O)OO• react roughly 5×10^4 times as rapidly as *t*-peroxy radicals R_3COO• with a given substrate, a fact that accounts for the extremely easy autoxidation of aldehydes. Other oxygen radicals, RO• and HO•, are much more reactive with rate constants of $>10^5$ and $>10^7$ $M^{-1}\ sec^{-1}$ respectively for either addition or H-abstraction.

SOLVENT EFFECTS

As noted earlier, the rates and paths of free radical reactions are generally quite solvent independent, a great simplification in predicting behavior in one system from data obtained in another. However, there are exceptions and this chapter is closed with some examples. Since they usually involve situations where the overall energetics of the reactions are altered, they give clues as to where others may be expected.

The radical chain chlorination of alkanes by molecular Cl_2 is very unselective with relative reactivities for primary, secondary, and tertiary C–H bonds being approximately 1:4:5.5 near room temperature in the gas phase and in several solvents. The simple explanation is that all reactions are exothermic and since reaction with primary C—H bonds occur at almost every collision, the other reactions cannot go much faster. In 1958 G. A. Russell reported that in benzene and other aromatic solvents the selectivity was greatly increased, and proposed that Cl• formed a π-complex with the aromatic which then became the reactive species. The selectivity was now quite temperature dependent, suggesting that the complexed Cl• was sufficiently stabilized so that the reactions were becoming endothermic and the differences in C—H bond dissociation energies were becoming important. Recent work involving fast laser spectroscopy has borne this out in detail, although there is still some debate as to the exact nature of the complex. At room temperature chlorine atoms react with 2,3-dimethylbutane with a rate constant of $3.3 \times 10^9\ M^{-1}\ sec^{-1}$. Free Cl• forms a spectroscopically visible complex with benzene with an equilibrium constant of about 200 corresponding to a stabilization of 3 kcal M^{-1} (since the complexing must involve some entropy loss, the change in ΔH is even larger). This complex reacts with 2,3-dimethylbutane with rate constant of only 4.8×10^7 and the tertiary: primary selectivity increases to about 120.

Another example is found in the competition between β-scission and substrate attack by *t*-butoxy radicals

$$t\text{-BuO}\bullet \begin{cases} \xrightarrow{k_d} t\text{-BuOH} + \text{R}\bullet \longrightarrow \text{etc.} \\ \xrightarrow{k_a} \text{CH}_3\text{COCH}_3 + \bullet\text{CH}_3 \longrightarrow \text{etc.} \end{cases} \tag{1.45}$$

At 40° in CCl_4 with cyclohexane as substrate, $k_d/k_a = 0.02$. In chlorinated olefins and aromatics the ratio is larger and reaches 0.34 in acetic acid. The proposed explanation that the transition state for β-scission can be solvated about the oxygen, of the radical, while that for the displacement reaction cannot, has been supported by fast laser spectroscopic measurements showing that β-scission of $C_6H_5CH_2CH_2O\bullet$ is, in fact, accelerated in hydrogen-bonding solvents.

Substrate solvation can also change the rates and energetics of radical reactions. The rate of reaction of peroxy radicals with phenols (involving attack on the O—H of the phenols and the crucial step in their behavior as antioxidants) is significantly slower in hydrogen-bonding solvents. As an extreme example, in the radical addition of HBr to olefins, equations (1.31)–(1.32), both steps are exothermic and fast with $\Delta H_{31} = -11 \text{ kcal M}^{-1}$ and $\Delta H_{32} = -7.5 \text{ kcal M}^{-1}$. However, in water containing HBr, Br• is almost entirely tied up via the equilibrium with relatively unreactive $Br_2\bullet^-$

$$\text{Br}\bullet + \text{Br}^- \leftrightarrow \text{Br}_2\bullet^- \tag{1.46}$$

which lies far to the right so its effective concentration and the rate of the addition step (equation (1.31)) must be greatly reduced. Further, the large heat of solution of HBr in water accompanying its ionization (20 kcal M^{-1}) would make the displacement (equation (1.32)) endothermic by $\cong 12.5 \text{ kcal M}^{-1}$. As far is known, the reaction has never been observed in water.

Finally, ionizing media facilitate the conversion of radicals and radical ions and may change the course of radical reactions. As noted earlier, 1,2-shifts of hydrogen are not generally observed in radical reactions. Intramolecular rearrangement of alkoxy radicals to α-hydroxyalkyl radicals is not seen in the gas phase or in non-polar solvents, although they are energetically favorable because of the difference in C—H and O—H bond strengths. They do occur readily in water. α-Hydroxy radicals are weak acids in water, and, because of the difference in bond strengths noted above and the fact that they give the same ketyl anion, alkoxy radicals must be considerably stronger. Presumably the rearrangement is brought about through ionization

$$HCR_2O\bullet \rightarrow H^+ + \bullet\dot{C}R_2{-}O^- \rightarrow \bullet CR_2{-}OH \tag{1.47}$$

analogous to the base-catalyzed conversion of an enol to a ketone.

CONCLUDING REMARKS

In this brief survey an attempt has been made to outline the major principles and characteristics of free radical reactions and to give a reasonable number of examples. Detailed references have not been provided, since the material comes from many sources gathered over a number of years. Readers desiring more information on specific topics can find it in a variety of places. Advanced physical organic texts give more detailed treatments. A recommended example is Lowry and Richardson's *Mechanism and Theory in Organic Chemistry* [1]. The last attempt to cover the entire field in a comprehensive manner is *Free Radicals* (two volumes) edited by J. K. Kochi [2]. Although it is over 20 years old it is a good source and has extensive data on most of the concepts discussed here.

More recent work will be found in monographs on more specific topics. As examples, Ingold and Roberts' *Free-Radical Substitution Reactions* [3] discusses valence expansion reactions in detail. Oxidation–reductions involving radicals and metal ions are reviewed by Sheldon and Kochi [4], and from a somewhat different point of view by Eberson [5]. A comprehensive review of the chemistry of organic peroxides appeared in 1983 [6], and an excellent exposition of the use of radical reactions in organic synthesis in 1986 [7]. Finally, a comprehensive treatment of the chemistry of oxygen radicals (including autoxidation) has appeared very recently [8].

As mentioned, the rate constants for a great many elementary radical reactions have now been measured. An extensive compilation for liquid-phase reactions has been published by Landolt Börnstein [9] and is being extended, and a useful, but older summary of gas-phase data has been edited by Kerr [10]. An up-to-date computer database for gas phase reactions is now available [11] and a similar one for liquid-phase reactions should be available shortly.

There have also been a large number of short reviews on more limited topics in review journals such as *Accounts of Chemical Research*, which provide good summaries and convenient access to the original literature.

REFERENCES

1. Lowry, T. H. and Richardson, K. S. (1981) *Mechanism and Theory in Organic Chemistry*, 2nd edn. Harper & Row, New York.
2. Kochi, J. K. (ed.) (1973) *Free Radicals*. Wiley-Interscience, New York.
3. Ingold, K. U. and Roberts, B. P. (1971) *Free-Radical Substitution Reactions*. Wiley-Interscience, New York.
4. Sheldon, R. A. and Kochi, J. K. (1981) *Metal Catalized Oxidations of Organic Compounds*. Academic Press, New York.
5. Eberson, L. (1987) *Electron Transfer Reactions on Organic Chemistry*. Springer-Verlag, New York.
6. Patai, S. (ed.) (1983) *The Chemistry of Peroxides*. J. Wiley & Sons, New York.
7. Giese, B. (1986) *Radicals in Organic Synthesis: Formation of Carbon–Carbon Bonds*. Pergamon Press, New York.
8. Foote, C. S., Valentine, J. S., Greenberg, A. and Liebman, J. F. (1995) *Active Oxygen in Chemistry*. Blackie A & P, Glasgow.
9. Landolt Börnstein (1983) *Numerical Data and Functional Relations in Science and Technology*, New Series, Vol. 13, (several subvolumes). H. Fischer (ed). Springer-Verlag, New York.
10. Kerr, J. A. (ed.) (1981) *Handbook of Bimolecular and Termolecular Gas Reactions*, (2 vol.). CRC Publishing Co., Boca Raton, Florida.
11. Mallard, W. G., Westley, F., Herron, J. T. and Frizzell, D. H. (1993) *NIST Chemical Kinetics Database – Ver 5.0*. NIST Standard Reference Data, Gaithersburg, MD.

2

Heats of Formation of Organic Free Radicals by Kinetic Methods

WING TSANG

INTRODUCTION

It has long been recognized that the key intermediates for the majority of the reactions of organic molecules in the gas phase, and frequently in the liquid phase as well, are free radicals. An important prerequisite for the description of the behavior of such systems is their thermodynamic properties. At first glance it may be surprising that thermodynamic properties should have any applicability to transient species that are present in trace quantities and disappear in short times. Indeed, if the kinetic properties of such systems are known, then thermodynamic properties are not really necessary and can in fact be derived from the kinetics. In reality, however, kinetic properties may be unavailable or difficult to measure. Thermodynamic properties serve as limits for kinetics and more generally as a basis for the estimation and evaluation of kinetic information [1]. More directly, through the equilibrium constant, rate constants for the reverse direction can be directly calculated from that in the forward direction. There are other physical situations where local thermodynamic equilibrium turns out to be a satisfactory approximation and kinetic information is not important.

Entropy and enthalpy are the two constituent parts of the thermodynamic properties of a molecule. For the radicals of concern here, the entropies can be estimated with a high degree of accuracy [2]. Thus in recent years most of the questions on the thermodynamic properties have involved uncertainties in the enthalpies of formation. Although computational chemistry is beginning to be used, experimental determinations and their interpretation remain the key issues. Indeed, a particular impetus for arriving at the best possible experimental values for the heats of formation is to provide a proper basis for comparison with the newly developing theory. Finally, the heats of formation of these radicals are used to derive group values and thus provide predictive capabilities for entire classes of molecules.

The exponential dependence of equilibrium or rate constants on the enthalpies of formation means that a 6 kJ/mol error in the activation energy or reaction enthalpy leads to an order of magnitude error at room temperature. The same difference in activation energy leads to a factor of 2.5 difference at 1000 K. For general trends of reactivity, high accuracy may not be really required. High accuracy data is necessary if one is concerned with rate or equilibrium constants. There is always the necessity of knowing as well as possible a few key numbers. The converse of this is that if equilibrium and kinetic measurements are used to determine enthalpies of formation then one can expect extremely accurate results. This will be the general theme of this review. Certainly, for appropriate systems it is well within the capabilities of modern kinetics to determine rate constants much better than the factors mentioned above.

The reactivity and hence the difficulty in isolating sufficient quantitites of free radicals means that many of the standard techniques for the determination of enthalpies of formation are not practical. For example, reliable heats of combustion can usually only be obtained from macroscopic samples of the highest purity. Direct determinations of the equilibrium constant for reactions involving free radicals can only occur under special circumstances. Measurements that approach these situations are now beginning to appear. All methods for determining the heats of formation of radicals have built into them assumptions which may or may not be justified. Failure to critically assess these assumptions has led to a great deal of controversy on these fundamental quantities. This is the justification for a review of a subject that has been covered by many earlier workers [2–4]. The gradual accumulation of high quality kinetic results means that for the radicals covered in this review, error limits are being steadily reduced. There is, unfortunately, in this area a history of apparent confirmation of erroneous data. Thus a critical examination of the experimental measurements is mandatory. The rationale for greater confidence in the newer values is that as experimental techniques and understanding improve, the interpretive element in the analysis of data has decreased. In systems where reaction times are short and concentrations of reactive intermediates small, contributions from unwanted side reactions are minimized. Furthermore, the data are now increasingly tied together by a plethora of experimental observations carried out under widely different conditions. Thus in some of the cases mentioned here pressures range from the sub-torr region to over an atmosphere and temperatures from 300 to 1200 K. Changing some of the values would now require the explanation and re-analysis of scores of experimental results. Indeed, if these values are in error it would imply serious fundamental faults in the current-understanding of chemical kinetics. Changes exceeding the estimated uncertainties in these new numbers will require much more work than in 1978 [5], when the first challenge to the then established numbers was proposed. Finally, since there are really no high accuracy direct methods for determining the enthalpy of formation of a short-lived transient specie, errors in that of the stable molecule from which it is formed can never be ignored.

The chapter begins with some necessary definitions and a brief discussion of the experimental methodology. Many of these details have been discussed exhaustively in earlier publications [2–4,6]. This is followed by a discussion of the present situation for a number of small and intermediate size radicals. The radicals that will be considered contain carbon, hydrogen and oxygen atoms. Specifically, methyl, ethyl, vinyl, propargyl, allyl, *n*-propyl, isopropyl, *s*-butyl, *t*-butyl, acetyl, hydroxymethyl, benzyl and phenoxy radicals are considered in detail. These include all the small and intermediate radicals for which there are replicable data from a variety of the experimental methodologies covered in this review. Aside from establishing the most reliable numbers, these results are indicative of the capabilities of the methods that are employed. Some of the problems encountered in deducing the enthalpies of formation by these methods will become clear. This is should permit the reader to assess the uncertainties of literature results on species not covered here. Some recommendations will be given in this regard.

DEFINITIONS AND PROCEDURES

The heat of formation of a compound at any temperature is defined as the enthalpy of reaction for the formation of that compound from its elements in their standard states at that temperature [7]. The focus of the work will be on the values at 298 K. The choice of the temperature was initially arbitrary. It serves as a common basis for comparison. Experimentally, it is in general not possible to make such a measurement directly, even for stable species. Hence the use of heats of combustion. This is in effect a transfer standard since the heats of reaction to form the oxides from elements in their standard states can and have been directly measured. Obviously, any reaction for which enthalpies of reaction can be accurately determined will also be satisfactory. Indeed, there is really no requirement that the reaction be driven to completion if heats and extent of reaction can be determined.

If equilibrium can be attained or calculated then the slope of the standard van't Hoff plot

$$\frac{d(\ln K_{equ})}{d(1/T)} = -\Delta H/R \tag{2.1}$$

where K_{equ} is the equilibrium constant, yields the heat of reaction. If the enthalpy of formation of all other species for a given reaction are known then that for a particular radical can be calculated. This is usually known as the 'second law method' for determining reaction enthalpy. A particular advantage of this approach is that the equilibrium constant need only be known to a multiplicative constant since the latter cannot contribute to the slope. Note that, except for correcting to the standard temperature, which is usually not a very severe requirement, very little need be known about the structural features or entropy of the molecule.

Alternatively, at any temperature

$$F = H^{\circ} - TS^{\circ} = -RT \ln K_{equ} \tag{2.2}$$

where F, H, and S are thermodynamic properties of reaction. Thus knowledge of the equilibrium constant and the entropies of all the species can lead to a value of the enthalpy of reaction. This is the 'third law method'. In general it is more reliable since it depends on only one measurement as opposed to the series of measurements that is necessary to determine a slope. The improvement in accuracy is a direct consequence of the heightened capability for the prediction of the properties of radicals, such as their structure and vibrational frequencies, which are directly used for the calculation of the entropy.

This chapter will consider reactions involving neutral species that create or destroy the free radicals of concern. In all cases kinetic measurements in the form of rate constants or rate expressions are the primary experimental data. Essentially, rate constants in both directions are used through detailed balance

$$K_{equ} = k_f(nRT)/k_r \tag{2.3}$$

where k_f is the rate constant in the forward direction, k_r is the rate constant in the reverse direction and n expresses the change in molarity of the reaction, to obtain the desired equilibrium constant.

There is a large volume of data on the enthalpies of formation of radicals as determined through the monitoring of chemical change as brought about by electrons, ions or radiation [3]. An important point to be remembered is that the enthalpy of formation is a thermodynamic quantity. Thus at any temperature, a Boltzmann distribution of molecules and products is necessary. At a sufficiently high pressure, nature automatically guarantees that such a distribution is established. At lower pressures the situation is much less certain. In the case of processes involving ions, electrons and photons, the nature of the distribution functions must be carefully considered and a methodology developed to convert results to standard heats of formation. It is important to realize that the sources of errors are different for the various types of experiments. This is true not only for the different methods under discussion here, but also within each type. Hence the importance of replication with a variety of starting products and experimental conditions.

In the case of attack by a neutral species, possible errors arise from uncertainties in mechanisms or the nature of the external interactions, while for the other cases they are due to uncertainties in the internal states of the molecules or products. It is interesting that the uncertainties responsible for errors in one technique are largely absent from those of the other. A very human error is to be fully aware of uncertainties in other determinations and to underestimate those of one's own measurements.

Two generic types of reactions have been used to derive heats of formation data from kinetic measurements. The first involves unimolecular decomposition (and the reverse AB ± A + B) of either a stable hydrocarbon or a radical. The other process involves abstraction processes (A + BX ± AX + B). It is only beginning to become possible to carry out experiments in two directions in the same apparatus or at exactly the same conditions. Practically all of the existing data are based on results where the equilibrium constant is determined from the results of two different kinetic experiments. As a result, there is always the need for some extrapolation. Frequently, the rate constants are measured only in one direction. Some assumption must be made regarding the rate constant in the reverse direction. Past practice has frequently used the assumption that the reverse of bond breaking or radical combination has no temperature dependence in concentration units. This is an arbitrary assumption, although it is now known that the error so introduced is small. With this methodology the second law treatment must be used. This can easily lead to 5–10 kJ/mol uncertainty in addition to any experimental artefacts that may lead to errors in the slope measurement. With the accumulation of kinetic data the assumption of zero activation energy can frequently be replaced by a measured or estimated value, thus lowering the uncertainty. When the rate expression for the reverse reaction is known, then both second and third law results can lead to results of the highest accuracy.

SOURCES OF INFORMATION

Thermal Decomposition and Radical Combination

The rate expressions for thermal decompositions of organic compounds provide a means of determining heats of formation of radicals, providing it can be related to the bond cleavage process. For most larger polyatomics it is generally still necessary to derive kinetic results from final stable product analysis. This is because in general it is still not possible to determine in real time the concentration of large polyatomic radicals at concentrations that are commensurate with 'clean' kinetics. Results must be interpreted in the framework of a postulated mechanism. A key factor for obtaining reliable results is the isolation of the products from the bond breaking reaction for study. The reactivity of the radicals and/or their decomposition products makes it very likely that they will induce decomposition by attacking the reactant. Thus unless such effects can either be eliminated or taken into account, results can be badly skewed. The possibility of surface reactions in static and flow reactors cannot be ignored. For the latter the proper definition of the reaction conditions is very important. The use of such information to determine enthalpies of formation has an unfortunate history. In essence, it proved to be experimentally difficult to obtain correct rate expressions. Hence many erroneous enthalpies of formation were reported. O'Neal and Benson has described some of the problems [2]. Modern understanding and developments in experimental methodology have led to greatly increased capability for the extraction of high accuracy information on enthalpies of formation of radicals from kinetic results. This is the main thrust of this chapter.

Single pulse shock tube studies [6] offer an ideal means of circumventing many of the problems in deriving accurate bond cleavage rate expressions from pyrolytic studies. Figure 2.1 is a schematic of such an apparatus and the wave diagram that describes the processes of interest. Of prime importance is the complete absence of any surface contributions, since heating is by the shock wave, the walls are cold and in any case the short heating time (less than 1 ms) precludes reactive species from diffusing into or from the wall. The experimental procedure

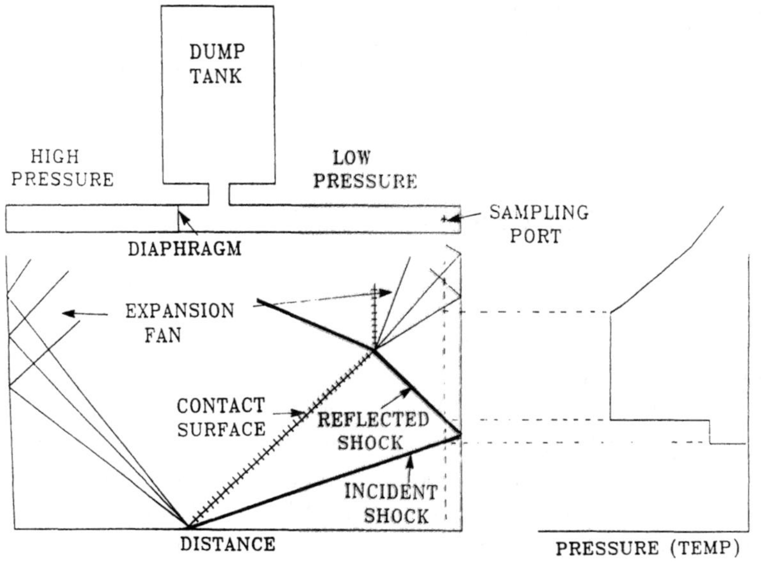

FIGURE 2.1 Schematic of single pulse shock tube and consequences of wave processes.

involves shock heating dilute quantities (order of 100 p.p.m.) of the compound to be pyrolyzed in the presence of large excesses (1% or more) of a chemical inhibitor such as mesitylene. The inhibitor serves to convert reactive radicals such as H-atoms and CH_3 to much less reactive benzylic radicals. In the short time (<1 ms) available for reaction, such radicals can only react with other radicals. Larger radicals without resonance stabilization are decomposed rapidly and quantitatively into stable unsaturated compounds and H-atoms or methyl radicals. All possible chain processes are thereby terminated and the initial reactions are isolated for study. The quantitative conversion of the larger polyatomic radicals into unsaturated compounds means that the yields of the latter can be directly associated with bond cleavage. A particular advantage of making a total product analysis is that it represents a crucial test of the mechanism.

The isolation of individual reactions for study also means that it is possible to study several unimolecular decompositions simultaneously. This is the basis of the internal standard method. It involves decomposing together two molecules, one the standard and the other whose rate constants and expression are to be determined. The rate expression for the decomposition of the standard is known. Thus from the extent of decomposition of the standard, a reaction temperature can be calculated.

This removes the uncertainties in the calculated reaction temperature (from the shock velocity and the gas dynamic properties of the system) that are inherent in shock tube studies and thus simultaneously reduces the scatter as well as systematic errors.

A large volume of rate expressions for bond breaking have been determined from such single pulse shock tube studies. When combined with lower temperature data on combination of radicals, equilibrium constants for bond cleavage can be determined and from this the heats of formation of the radicals that are formed. The interesting feature of these combination reactions, which presumably proceed without a barrier, is the existence of increasing negative temperature dependence in the rate constants with methylation. They may be as large as $(1/T)^{1.5}$ for the combination of 2-*t*-butyl radicals to $(1/T)^{0}$ for the combination of 2-ethyl radicals [6]. The study of the combination of large polyatomic radicals is not easy. The best results are derived from direct spectroscopic determinations. The bimolecular nature of these processes means that the measured rate constant is dependent on the calibration of the radical concentration. Nevertheless, the results from a variety of methods do seem to form a consistent pattern.

An important problem in resolving the two sets of data is the wide divergence in the temperature ranges covered. The shock tube results are in the range of 1000–1200 K, while the combination data covers the range of 300–600 K. Thus considerable extrapolation is needed to bring all the results into the same temperature range. Nevertheless it was from these results that the scientists obtained the first inkling that the generally accepted (for close to 30 years) heats of formation of the three prototypical alkyl radicals, ethyl, isopropyl and *t*-butyl were in error [5].

Another commonly used method is the very low pressure pyrolysis (VLPP) technique. When operated near ambient temperatures it is termed very low pressure reactor (VLPR) technique. A schematic of such an instrument can be found in Figure 2.2. The VLPR technique can be used to determine rate constants for radical attack on stable compounds and thus offers information on those types of equilibria. For pyrolysis, the VLPP technique represents a thermal activation [4] device where the requisite energy for dissociation is obtained by collisions with the wall. As a means of obtaining the enthalpies of formation it offers the possibility, in some cases, of mechanistically cleaner processes. Contributions from surface processes must always be considered. An additional problem is that the results derived from such studies are obtained on the basis (a) of an assumed A-factor and (b) that the wall is a strong collider. This latter is probably a valid assumption but the former means that there is a range of A-factors and activation energies that are equally compatible

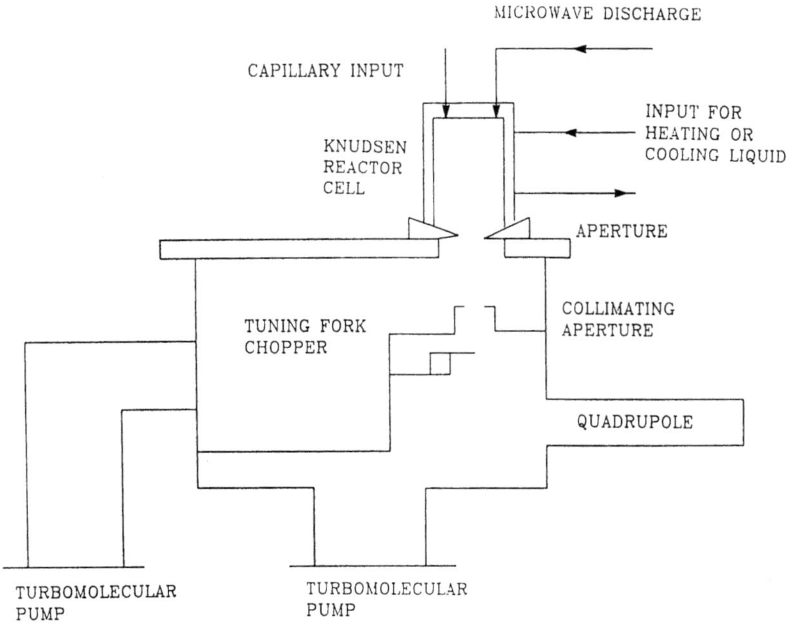

FIGURE 2.2 Schematic of very low pressure reactor (VLPR) apparatus used to study radical reactions. For pyrolysis, the reactor is resistively heated.

with the experimental observations. Thus an added uncertainty is introduced into the results on the reaction enthalpies. Nevertheless, it does appear that the results are compatible with the rate expressions derived from the experiments at pressures such that limiting high pressure values are obtained.

A fundamental problem with the determination of heats of formation by thermal decomposition is that reactions must be carried out at higher temperatures. The enthalpies of reaction are therefore less sensitive to rate or equilibrium constants. It is always necessary to correct the data to 298 K. This represents another possible uncertainty. The advantage is that with a suitable choice of parent molecule one can study virtually any system. However, two radicals are generated and the value of one is dependent on that of the other.

Radical Decomposition

At temperatures considerably lower than that achieved in the usual thermal decomposition experiments described above, organic radicals can combine or decompose. In practice, the radicals can be created either photochemically or through the decomposition of a suitable precursor. Then when one measures the ratio of decomposition (in terms of the stable unsaturate that is formed) and combination products of the radical, this is related to the rate constants for decomposition and combination of the radical via the relation

$$k_d/k_r = \text{decomposition product}/(\text{combination product})^{1/2} \tag{2.4}$$

If the rate expression for combination is known, then the decomposition rate expression can be derived. There is an extensive literature on such studies carried out many years ago. The results were all reported in terms of the ratio of rate constants given above. The rate expressions showed large scatter (as large as 3 orders of magnitude in the A-factor and 20 kJ/mol in activation energies). This lack of replicability demonstrates that the actual process is more complex than the simple mechanism outlined above. In addition, for many of the smaller radicals the reactions are in the unimolecular pressure dependent region when what is desired is the high pressure rate expressions. However, when one compares the actual experimental rate constants the variations are no more than a factor of 2. This is of suitable accuracy for application of the third law technique. If entropies can be accurately calculated this is equivalent to an uncertainty of about 2 kJ/mol.

The necessary complement to such measurements are the radical addition processes. For the present purposes interest need only be focused on addition of methyl and hydrogen atoms to olefins. The former represent a class of reactions that have been thoroughly studied nearly 40 years ago in terms of the ratio of addition to methyl combination [9]. The latter have been directly studied more recently by following the decay of the resonance fluorescence of hydrogen atoms in the presence of olefins [10]. These represent data of high accuracy. The range covered in the decomposition process is 500–800 K. For the addition process, the range is 300–450 K. Thus they almost overlap and extrapolation of the lower temperature data to the higher temperature region should not lead to serious uncertainties.

Brouard *et al.* [11] have extended resonance fluorescence detection of hydrogen atoms in the presence of ethylene to 825 K. At this temperature the system is actually equilibrating. In this sense the derived values represent true equilibrium constant measurements. Clearly this should be the direction of future work. However, these are difficult experiments. For more complex molecules the presence of alternative channels will become more important and thus mechanistic problems will once more come to the fore.

Abstraction Processes

For many years the standard method for determining radical heats of formation was by the iodination technique [2]. This involve determination of the rate constants for either the reaction

$$I + RI \rightarrow I_2 + R$$

or

$$I + RH \rightarrow R + HI$$

It is based on the overall process

$$RI + HI \leftrightarrow RH + I_2$$

which proceeds by a free radical chain mechanism. Benson has analyzed the process in great detail and shown how the rate constants and expressions for the reactions involving atomic iodine (the endothermic direction) can be extracted from the temporal behavior of its formation. Arguments are then presented for the assumption that the reverse of the two processes have activation energies of 0 ± 4 and 4 ± 4 kJ/mol. On this basis a second law treatment then leads to the enthalpy of reaction for the processes. Next, from the known enthalpies of formation of the other species in the reaction those of the appropriate radicals are calculated. These results were apparently supported by similar studies involving bromine atoms. Consequently, for many years these were the results that set the standards for the values of the enthalpies of formation of most organic radicals. The last summary of these values can be found in the review by McMillan and Golden in 1982 [4].

Advances in the determination in real time of atoms and radicals have led to a plethora of new and undoubtedly more reliable measurements on the rate constants for abstraction processes involving bromine, chlorine and iodine atoms [3]. Particularly noteworthy is the work of Gutman [12] who developed a photoionization mass spectroscopic method (PIMS) for the direct detection of organic radicals. A schematic of this apparatus can be found in Figure 2.3. It is thus possible to determine with great accuracy the rate of reaction of organic radicals with hydrogen halides (the more exothermic direction). These are direct measurements and the conditions can be set so that mechanistic artefacts are minimized. In the case of the alkyl radicals, the reproducibility of the results in terms of the heats of formation derived from all those except possibly for *t*-butyl radical proved to be excellent. In some of the cases the same results have been obtained from second and third law treatments. The convergence of all of

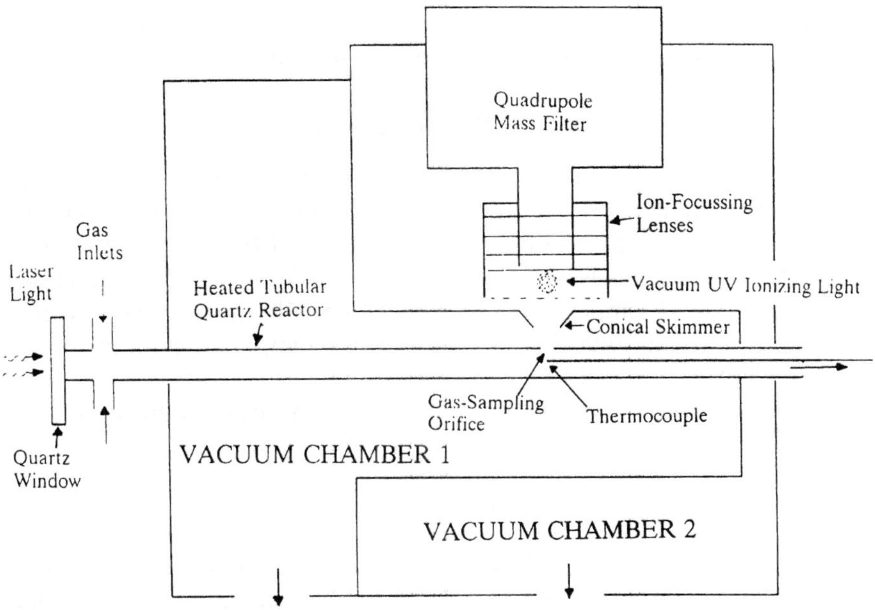

FIGURE 2.3 Apparatus for determining radicals via high temperature photoionization mass spectrometry.

these results makes it likely that any further changes will be within the estimated error limits. A distinct advantage of using these abstraction processes is that the studies are frequently carried out at or near room temperature, where enthalpies of reaction determination are least sensitive to possible errors in measurements and entropy estimates.

Seakins and Pilling [13] and Parmar and Benson [14] have determined the approach to equilibrium for the reactions Br + $iC_4H_{10} \leftrightarrow$ HBr + tC_4H_9 and Cl + $C_2D_6 \leftrightarrow$ DCl + C_2D_5, respectively. These represent the closest approach to a direct measurement of the equilibrium constant for studies involving abstraction reactions. The results they obtained are in good agreement with that determined from a combination of rate measurements far from equilibrium and thus represent very satisfactory confirmation of the results. As noted earlier, they are not simple experiments and one must be very careful that mechanistic artefacts are not making a contribution.

A very interesting consequence of these studies is the conclusion that the radical attack on many of the hydrogen halides [14], an exothermic reaction, has a negative temperature dependence. This is similar to the situation for radical combination. However, in the latter case one can assign the effect in a very natural manner to the formation of an intermediate complex. It is difficult to make a similar assignment for what appears on the surface to be an abstraction reaction.

Radical Buffer Technique

In its most direct form this is not a kinetic method. As used by Castelhano and Griller [15] it is the only procedure where a true equilibrium constant is measured. In an alternative approach, as used by Hiatt and Benson [16], the method leads to a determination of the rate constants for decomposition at temperatures that are essentially inaccessible to direct studies and yields data that cover the same temperature range as the radical combination results.

In both cases advantage is taken of the fact that the equilibrium for the reaction

$$R + SI \pm RI + S \qquad K_{RS}$$

where R and S are alkyl radicals and K_{RS} is the equilibrium constant, is rapidly attained. Castelhano and Griller carried out studies in isooctane solution and made direct measurements of the concentration of the two radicals via electron paramagnetic resonance spectroscopy. They demonstrated that for alkyl radicals equilibrium is attained and that the competing radical combination processes are unimportant. From the relative concentrations of the radicals as determined by electron paramagnetic resonance an equilibrium constant can be determined. With methyl as one of the radicals and since the thermodynamic properties of the iodides are also known, and enthalpy of formation of the other radical can be calculated using the third law procedure. An important uncertainty is often the effect of the solvent on the heat of formation. In addition the heat of formation of methyl used was 143.9 kJ/mol, instead of the more generally accepted value of 146.9 kJ/mol and some of the entropy values used were also out of date. This has been corrected for in the tables by giving the numbers in parentheses. When so treated the results are in remarkable agreement with gas phase results and this is indicative of the unimportance of solvent effects in non-polar media on the thermodynamic properties of radicals.

The work of Hiatt and Benson is of the gas phase. The reactions

$$2R \rightarrow R_2 \qquad k_r$$

and

$$2S \rightarrow S_2 \qquad k_s$$

where k_r and k_s are the individual rate constants to be considered. These rate constants are related to the observables and the equilibrium constant by the relation

$$k_{RS}\,(k_s/k_r)^{1/2} = (S_2/R_2)^{1/2}(RI/SI)$$

where the term on the right hand side of the equation is measured and K_{RS} is calculable from the basis of the thermodynamic properties of the molecules and radicals in question. Thus relative rate constants for association can be determined. This assumes that the thermodynamic properties of the radicals are known. However, it is possible to convert this to relative rate constants for the dissociation process and remove the dependence on the thermodynamic properties of the radicals in question. Consider the reverse of the two reactions described above and designating them by k_{-r} and k_{-s}, then

$$k_r/k_{-r} = K_{R2}/K_R^2$$

$$k_s/k_{-s} = K_{S2}/K_S^2$$

where K_{R2}, K_R, K_{S2} and K_S are the equilibrium constants of formation. Since

$$K_{RS} = K_{RI} \times K_S/K_{SI} \times K_R$$

where K_{RI} and K_{SI} are the equilibrium constants of formation for RI and SI, then upon substitution one obtains the relation [5]

$$[K_{RI}^2 K_{S2}/K_{SI}^2 K_{R2}]^{1/2} \times (k_{-s}/k_{-r})^{1/2} = (S_2/R_2)^{1/2}(RI/SI).$$

With this equation only with the thermodynamic properties of stable species, RI, SI, R_2 and S_2 are used to obtain a ratio of rate constants for the decomposition processes. No assumptions need be made on the thermodynamic properties of the radicals. Therefore uncertainties in this regard are no longer a problem. The studies of Hiatt and Benson can be considered more directly related to the decomposition that the combination rate constants in the sense that values of the latter depend on the thermodynamic properties of the alkyl radicals.

The 'standard' reaction used by Hiatt and Benson [17] is the methyl combination process. In the present case use must be made of the rate expression for the ethane decomposition process. This is no longer in question. The result is then rate expressions for ethyl, isopropyl and *t*-butyl radicals at temperatures near 400 K with rate constants in the $10^{-25}\,s^{-1}$ range. It overlaps directly with measured recombination rate constants. Extrapolations are no longer needed and a third law treatment should lead to extremely accurate enthalpies of formation. Probably the chief source of uncertainties are now the heats of formation of the closed shell species.

SPECIFIC RADICALS

In the following experimental results are summarized on the enthalpies of formation of a number of radicals for which there have been kinetic determinations via a number of different approaches. They are indicative of the plethora of existing results and the type of agreement that can be expected. The results are presented in the form of tables beginning in most cases with the recommendation of McMillen and Golden in 1982 [4]. For earlier data, the reader is referred to their review. Although it represented a correct view of the state of the art at that time, it will be clear that a far too optimistic view of the accuracy of existing determination of the enthalpies of formation was adopted. The uncertainty limits given here do not represent the results of a statistical analysis but represent our best estimates, based on the intercomparison of data sets and methods, of uncertainties in the reaction mechanisms.

Methyl

This is the simplest alkyl radical. It is thermally stable and is ubiquitous in reacting organic systems. An enormous variety of methods have been used to determine its heat of formation. At the present time there do not appear to be any serious controversies. A summary of the latest results from kinetic measurements can be found in Table 2.1. An average of the latest results lead to an enthalpy of formation of 147 ± 1 kJ/mol and is the recommended number. Within the error limits it is in accord with the recommendation of McMillen and Golden. It is also in accord with the values deduced by a variety of non-kinetic methods and in theory [3]. Thus the value of the enthalpy of formation of methyl can and had been used in a variety of kinetic experiments where enthalpies of formation of other organic radicals are determined.

Vinyl

There is a scarcity of reliable kinetic data that can be brought to bear on the heat of formation of this radical. Until very recently the existing kinetic data were in serious disagreement with results derived from other techniques such as photoionization mass spectrometry, gas phase acidity scales and theory [3]. The results from the last three methods

TABLE 2.1 Summary of data on methyl heat of formation

Year	Authors	Methods	Value (kJ/mol)
1982	McMillen and Golden [4]	Chlorination kinetics. Based on work of Dobis and Benson [18]	146.8 ± 0.8
1988	Russell *et al.*[19]	$CH_3 + HCl \leftrightarrow CH + 4Cl$ (296–495) mass spectrometry detection combined with literature values for reverse chlorine abstraction rate constant.	
		Second Law	145.2 ± 2.5
		Third Law	145.6 ± 1.2
1988	Russell *et al.*[20]	$CH_3 + HBr \leftrightarrow CH_4 + Br$ (295–532) ethyl via mass spectrometry and literature values leading to reverse Second Law	148.3 ± 3
1990	Seetula *et al.*[21]	$CH_3 + HI \leftrightarrow CH_4 + I$ (292–648) mass spectrometry detection of methyl combined with measurement of Goy and Pritchard [22] for reverse. Third Law	146 ± 1
1991	Nicovich *et al.*[23]	$CH_3 + HBr \leftrightarrow CH_4 + Br$ Resonance fluorescence detection of Br. Literature value from Comber and Whittle for reverse (257–430) [24] Third Law	147.7 ± 2.5
Recommended value			**147 ± 1**

were in substantial agreement with or approximately 298 kJ/mol, while the kinetic studies lead to a value of 277 kJ/mol. The kinetic result is derived from the equilibrium constant for the reaction $Cl + C_2H_4 \leftrightarrow C_2H_3 + HCl$. It is, however, entirely dependent on the determination by Parmar and Benson [25] for the rate constant of the reaction $Cl + C_2H_4 \rightarrow C_2H_3 + HCl$. Most recently, Kaiser and Wallington [26] have studied the rate constant for this reaction at sufficiently low pressures so that the rate constants for the addition reaction are well into the fall-off region and substantially smaller than those reported by Parmar and Benson [25]. They have also identified a product at 383 K, vinyl chloride, which in their system can be associated with the chlorine abstraction reaction and used to derive a maximum rate constant for this process. When their rate constant is combined with that of Russell *et al.* [27] for the reverse reaction, on the basis of a third law approach, one obtains a minimum value of 299 kJ/mol for the heat of formation. This is now in excellent agreement with the theoretical number. The earlier measurement of Parmar and Benson of the metathesis process in a very low pressure reactor must be in error by close to 4 orders of magnitude. Clearly, the error must arise from the interpretation of the mechanism. The general situation is summarized in Table 2.2.

TABLE 2.2 Summary of data on vinyl heat of formation

Year	Authors	Methods	Value (kJ/mol)
1982	McMillen and Golden [4]	Iodination, (Second Law). Theory and ion equilibria	295 ± 8
1988	Parmar and Benson [25]	$C_2H_4 + Cl \leftrightarrow C_2H_3 + HCl$ VLPR experiment. Kinetics of Cl attack on ethylene in the absence and presence of HCl. Third Law	277 ± 2
1989	Russell *et al.*[21]	$C_2H_3 + HCl \leftrightarrow C_2H_4 + Cl$ Mass spectrometry detection of vinyl. Reverse rate from that of Parmar and Benson.	
		Second Law	281 ± 2.5
		Third Law	280 ± 1.2
1995	Kaiser and Wallington [26]	$C_2H_4 + Cl \rightarrow C_2H_3 + HCl$ Product analysis. Combined with results of Russell *et al.* and Third law analysis	>299 ± 5
Recommended Value			**299 ± 5**

Cui *et al.* [28] have determined the vinyl C—H bond strength in propene through the thermal decomposition of 1-phenyl-2-butene. They find a bond strength of 459 ± 10 kJ/mol. This is in substantial agreement with the heat of formation of vinyl given in Table 2.2.

Ethyl

Table 2.3 summarizes the recent data on the enthalpy of formation of ethyl as determined from kinetic methods. The general trends given in this table are repeated in many of the subsequent tables. The older numbers from iodination studies are inevitably smaller by about 10–20 kJ/mol. In 1978 [5] it was first pointed out that the existing data on the thermal decomposition of *n*-butane was incompatible with this value. Since the mid 1980s there has been a virtual flood of kinetic results that are in complete agreement with the suggested higher value. There can be little doubt that the enthalpy of formation of the ethyl radical is 119 ± 2 kJ/mol. Note that this value is based on completely different experiments dealing with the equilibrium constant for seven independent reactions that cover a temperature range from near room temperature to over 1000 K. Indeed, these results attest to the central role of enthalpies of formation to chemical kinetics. Finally, it should be noted that this high value (in comparison to the earlier recommendations) is now in agreement with data from a variety of non-kinetic methods.

Propargyl

Propargyl radical has been the focus of much recent attention as the main source of benzene in high temperature hydrocarbon systems. It is particularly interesting in the context of its resonance energy in comparison to allyl. The data summarized in Table 2.4 contains results on the decomposition of different compounds and in completely different experimental studies. The agreement in the numbers is very good and indicates that the estimated uncertainty limits are realistic. The recommendation is 339 ± 4 kJ/mol. The first result in Table 2.4 has been given less weight due to the fact that the final product from the decomposition, propene, can be formed from molecular as well as a bond dissociation process. This

TABLE 2.3 Summary of data on ethyl heat of formation

Year	Authors	Methods	Value (kJ/mol)
1982	McMillen and Golden [4]	Iodination second law	108.4 ± 4
1978	Tsang [6]	$C_4H_{10} \leftrightarrow 2C_2H_5$ (354–1200) Reinterpretation of radical buffer study of Hiatt and Benson. Results from shock tube study Third Law. Reverse rate from Parkes and Quinn [29]	119.2 ± 6
1982	Castelhano and Griller [15]	$CH_3 + C_2H_5I \leftrightarrow CH_3I + C_2H_5$ Radical buffer in isooctane solution. Electron paramagnetic resonance detection of radical. Third Law	117.2 (120.1)
1984	Cao and Back [30]	$C_2H_5 + H_2 \leftrightarrow C_2H_6 + H$ (700 K). Analysis of existing data. Third Law	117.2 ± 4
1984	Pacey and Wimalasena [31]	$2C_2H_5 \leftrightarrow C_4H_{10}$ (902) Ethyl combination in ethane decomposition and literature value of reverse. Third Law	118.8 ± 2
1986	Brouard *et al.* [11]	$C_2H_5 \leftrightarrow C_2H_4 + H$ (775–825) H-atom via resonance fluorescence in equilibrating system. Third law	118.9 ± 2
1988	Russel *et al.* [20]	$C_2H_5 + HBr \leftrightarrow C_2H_6 + Br$ Ethyl via PIMS and literature value of reverse (295–532). Second and Third Law. Based on rate constants for Br attack from Fettis *et al.* [32]	120.1 ± 3
1989	Parmar and Benson [14]	$C_2D_5 + DCl \leftrightarrow C_2D_6 + Cl$, VLPR Equilibrating system. Third Law	118.5 ± 6

continued

TABLE 2.3 *Continued*

Year	Authors	Methods	Value (kJ/mol)
1990	Seetula *et al.* [21]	$C_2H_5 + HI \leftrightarrow C_2H_6 + I$ (295–648) Ethyl via PIMS and rate constant of Knox and Musgrove [33] for the reverse. Third Law	117.2 ± 2
1991	Nicovich *et al.* [23]	$C_2H_5 + HBr \leftrightarrow C_2H_6 + Br$ Resonance fluorescence detection of Br literature value for reverse (257–430). Third Law	121.8 ± 2.5
1992	Seakins *et al.* [34]	$C_2H_5 + HBr \leftrightarrow C_2H_6 + Br$ Kinetics in both directions. PIMS detection of ethyl and resonance fluorescence detection of Br. Second and Third Law	121 ± 1.5
Recommended value			**119 ± 2**

introduces an element of ambiguity into the interpretation of the results. These data lead to a resonance energy that is about 4 kJmol/mol smaller than that for allyl radical. Although this is within the assigned

TABLE 2.4 Summary of data on propargyl radical

Year	Authors	Methods	Value (kJ/mol)
1970	Tsang [35]	$C_3H_3iC_3H_7 \leftrightarrow C_3H_3 + iC_3H_7$ Single pulse shock tube study near 1100 K. Second Law	332 ± 6
1978	Tsang [36]	$1\text{-}C_6H_{10} \leftrightarrow C_3H_3 + C_3H_7$ Single pulse shock tube study near 1100 K. Second Law	338 ± 6
1979	King and Nguyen [37]	$C_6H_5CH_2CH_2CCH_3 \leftrightarrow C_6H_5CH_2 + C_3H_3$ VLPP results near 1000 K. Second Law	341 ± 10
1978	King [38]	$1\text{-}C_4H_8 \leftrightarrow C_3H_5 + CH_3$ VLPP results near 1100 K. Second Law	341 ± 4
Recommended value			**339 ± 4**

uncertainty limits it should be noted that comparison of results from the decomposition of 1-alkenes and 1-alkynes are consistent with such a number.

Allyl

Table 2.5 contains a summary of kinetic results pertaining to the enthalpy of formation of allyl. All the kinetic results are from pyrolysis experiments at higher temperatures. Aside from bond breaking allyl decomposition equilibria is also used. Uncertainties can be expected to be larger for the reasons given earlier. Nevertheless all data can be fitted within the estimated 3 kJ/mol uncertainty limit. Note that the data for 1-pentene and allylbromide decomposition are all based on a second law treatment and the assumption of zero activation energy for the combination

TABLE 2.5 Summary of data on allyl heat of formation

Year	Authors	Methods	Value (kJ/mol)
1982	McMillen and Golden [4]	Iodination. Second law	163.6 ± 6
1978	Tsang [39]	$nC_5H_{10} \leftrightarrow C_3H_5 + C_2H_5$ Single pulse shock tube study. Second Law	171.1 ± 6
1985	Tsang [40]	$C_3H_5Br \leftrightarrow C_3H_5 + Br$ Single pulse shock tube study. Second Law	173.6 ± 4
1991	Roth *et al.* [40]	1,5 $C_6H_{10} \leftrightarrow 2C_3H_5$ Shock tube study using detection of allyl. Second and Third Law [41]	167 ± 2
1992	Tsang and Walker [41]	$C_3H_5 \leftrightarrow C_3H_4 + H$ Single pulse shock tube study. Third Law treatment. Combination with low temperature study of Wagner and Zellner [43]	171.1 ± 4
Recommended value			**171 ± 3**

process. It may well be that the small discrepancies noted here are due to a small negative activation energy. The 1-pentene results depend on the enthalpy of formation value used for the ethyl radical. The value recommended here, 171 ± 3 kJ/mol, is the average of the values in Table 2.5. It implies an allyl resonance energy of 54 kJmol/mol.

n-Propyl

The high value for the enthalpy of formation of ethyl radical calls into question the earlier recommendation on the enthalpy of formation of *n*-propyl radical, since it is well established that reactivity of primary radicals are similar. A summary of work beginning with the review of McMillen and Golden [4] can be found in Table 2.6 and fully confirms the suspicion that their recommendation must be in error. Four equilibria were used in arriving at the recommended enthalpy of formation and the temperatures ranged from near room temperature to over 1000 K. It is difficult to carry out abstraction studies to determine this heat of formation due to the presence of secondary hydrogens in propane. Thus, really accurate results from abstraction results are not available. The recommended enthalpy of formation of 100 ± 2 kJ/mol is clearly compatible with similar bond strength for primary C—H bonds. It certainly

TABLE 2.6 Summary of data on *n*-propyl heat of formation

Year	Authors	Methods	Value (kJ/mol)
1982	McMillen and Golden [4]	Iodination. Second Law	87.9 ± 4
1982	Castelhano and Griller [15]	$C_2H_5 + nC_3H_7I \leftrightarrow C_2H_5I + nC_3H_7$ $nC_3H_7 + iC_3H_7I \leftrightarrow nC_3H_7 + iC_3H_7$ Third Law. In isooctane solution. Electron paramagnetic detection of radicals	95.4 (98.3)
1985	Tsang [44]	$1\text{-}C_6H_{12} \leftrightarrow C_3H_5 + nC_3H_7$ Single pulse shock tube results at 1100 K and combination rate from geometric mean rule. Third Law $nC_3H_7 \leftrightarrow C_2H_4 + CH_3$ Best fit of all existing data in both directions	100.4 ± 6 100 ± 2
Recommended value			**100 ± 2**

justifies the use of the same value of 422 kJ/mol as the bond strengths for all such bonds.

Isopropyl

Table 2.7 contains a summary of available measurements on isopropyl radical. The situation is very much the same as in the ethyl radical. The older iodination results are clearly too low. Determinations from five independent equilibria ranging in temperatures from 300 to over 1000 K lead to very similar results. The recommendation of 90 ± 2 kJ/mol is more strongly weighed towards the abstraction equilibria.

s-Butyl

Enthalpy of formation data for *s*-butyl radical are summarized in Table 2.8. As in the case of isopropyl radical, results from multiple equilibria determinations ranging from room temperature to over 1000 K lead to similar values that are appreciably higher than the recommendations of McMillen and Golden [4]. The recommended enthalpy of formation, 69 ± 2 kJ/mol, leads to a C—H bond strength that is within 1 kJ/mol of that for the secondary bond in propane. This is within the stated error limits. For larger secondary radicals, one can confidently estimate enthalpies on the basis of group contributions.

t-Butyl

The heat of formation of this radical has been the center of considerable controversy. Evidence for the continuing interest can be found in the extensive number of determinations listed in Table 2.9. Gutman [12] has summarized the present situation in great detail. The remaining discrepancy is in the rate constant for *t*-butyl attack on HBr and HI. The measurement of Muller-Markgraf *et al.* [50] led to rate constants that are factors of as much as 100 smaller than those from Gutman's group. In addition, the temperature dependence is positive and in accord with a small activation energy. The consequence is consistency with the 38 kJ/mol enthalpy of formation recommended by McMillen and Golden. All other results are consistent with much higher values of the enthalpy of formation and Gutman has suggested that the results of Muller-Markgraf are consistent with the failure to take into account heterogeneous loss of *t*-butyl radical in the experiments of Muller-Markgraf *et al.* Particularly impressive is that this wall loss was consistent with the observed wall loss from Gutman's experiments. Furthermore, the high value is consistent with all the other equilibria that have been studied. 48 ± 3 kJ/mol is therefore suggested for the enthalpy

TABLE 2.7 Summary of Data on Isopropyl Radical

Year	Authors	Methods	Value (kJ/mol)
1982	McMillen and Golden [4]	Iodination. Second Law	76.2 ± 4
1978	Tsang [5]	$(iC_3H_7)_2 \leftrightarrow 2iC_3H_7$ Single pulse shock tube study plus reinterpretation of radical buffer study of Hiatt and Benson for decomposition rate constant. Combined with combination measurements of Parkes and Quinn. Third Law (415–1200)	90 ± 3
1982	Castelhano and Griller [15]	$C_2H_5 + iC_3H_7I \rightarrow C_2H_5I + iC_3H_7$ Radical buffer study in isooctane solution. Electron paramagnetic resonance detection of radical. Third Law	80.3 (85.4)
1985	Tsang [44]	$iC_3H_7 \leftrightarrow iC_3H_6 + H$ (523–813) Analysis of all existing data on rate constants in both directions. Third Law	89.5 ± 2
1988	Russell *et al.* [20]	$iC_3H_7 + HBr \rightarrow C_3H_8 + Br$ Isopropyl via PIMS and literature value from Fettis and Knox [46] of the reverse (295–532) Second and Third Law	88 ± 3
1990	Seetula *et al.* [21]	$iC_3H_7 + HI \leftrightarrow C_3H_8 + I$ Isopropyl via PIMS and literature value of reverse (295–648). Third Law	91.2 ± 2.5
1992	Seakins *et al.* [34]	$iC_3H_7 + HBr \leftrightarrow C_3H_8 + Br$ Kinetics in both directions. PIMS detection of ethyl and resonance fluorescence detection of Br. Second and Third Law	90 ± 1.7
Recommended value			**88 ± 2**

TABLE 2.8 Summary of data on *s*-butyl heat of formation

Year	Authors	Methods	Value (kJ/mol)
1982	McMillen and Golden [4]	Iodination. Second law	54.4 ± 8
1985	Castelhano and Griller [15]	$C_2H_5 + sC_4H_9I \leftrightarrow C_2H_6 + sC_4H_9$ $iC_3H_7 + sC_4H_9I \leftrightarrow C_3H_8 + sC_4H_9$ Radical buffer in isooctane solution. Electron paramagnetic resonance detection of radical. Third Law	58.2 (65)
1985	Tsang [5]	$(sC_4H_9)_2 \leftrightarrow 2sC_4H_9$ (1100) Single pulse shock tube on thermal decomposition combined with extrapolated combination rate constant for isopropyl	71.1 ± 3.5
		$sC_4H_9 \leftrightarrow CH_3 + C_3H_6$ Best fit of existing data in both directions 500. Third Law	70.3 ± 3.5
1989	Seetula and Gutman [44]	$sC_4H_9 + HBr \leftrightarrow sC_4H_9H + Br$ PIMS detection of sC_4H_9 Reverse rate constant from data of Fettis and Knox [45]. Third Law	66.7 ± 3
1990	Seetula *et al.* [21]	$sC_4H_9 + HI \leftrightarrow C_4H_{10} + I$ PIMS detection of sC_4H_9. Reverse rate constant from data of Chekov *et al* [47]. Third Law	71 ± 1.6
1992	Seakins *et al.* [34]	$sC_4H_9 + HBr \leftrightarrow nC_4H_{10} + Br$ Kinetics in both directions. Mass spectrometry detection of *s*-butyl and resonance fluoresence detection of Br. Second and Third Law	67.5 ± 2.2
Recommended value			**69 ± 2**

of formation of *t*-butyl. There is still some uncertainty regarding the entropy of *t*-butyl. The present numbers are based on the analysis of Knyazev *et al.* [63] and leads to an entropy of 316 J/mol K. This assumes a 6 kJ/mol barrier to internal rotation. A lower barrier will lead to larger

TABLE 2.9 Summary of data on *t*-butyl radical

Year	Authors	Methods	Value (kJ/mol)
1982	McMillen and Golden (4)	Iodination Second law	38.9 ± 4
1978	Tsang [5]	$(tC_4H_9)_2 \leftrightarrow 2tC_4H_9$ Reinterpretation of radical buffer study of Hiatt and Benson to yield decomposition rate. Combination rate of Parkes and Quinn (373). Third Law	48.1 ± 6
1982	Castelhano and Griller [15]	$iC_3H_7 + tC_4H_9I \leftrightarrow C_3H_8 + tC_4H_9$ $sC_4H_9 + tC_4H_9I \leftrightarrow C_4H_{10} + tC_4H_9$ Radical buffer study in isooctane solution. Electron paramagnetic resonance detection of radicals. Third Law	39.3 (43.5)
1985	Tsang [44]	$(tC_4H_9)_2 \leftrightarrow 2tC_4H_9$ (750–1100) Analysis of data on high temperature decomposition and low temperature combination rate. Third Law	45.6 ± 3
		$C(CH_3)4 \leftrightarrow CH_3 + tC_4H_9$ (700–1100) Analysis of data on high temperature decomposition rate and low temperature combination rates and geometric mean rule. Third Law	50.6 ± 4
		$t\text{-}C_4H_9 \leftrightarrow iC_4H_8 + H$ (600–800) Analysis of decomposition rate and low temperature addition rate. Third Law	47.7 ± 3
1987	Benson *et al.* [48]	$Br + iC_4H_{10} \leftrightarrow tC_4H_9 + HBr$ VLPR studies. Third Law	45.2
1989	Russell *et al.* [20]	$tC_4H_9 + HBr \leftrightarrow iC_4H_{10} + Br$ *t*-Butyl for abstraction from HBr via PIMS (295–532). Br for abstraction from iC_4H_{10} *via* PIMS (533–710) Resonance fluorescence detection of Br (298–478) Second and Third Law	48.5 ± 2 (Second Law) 47.7 ± 3 (Third Law)

continued

TABLE 2.9 *Continued*

Year	Authors	Methods	Value (kJ/mol)
1989	Muller-Markgraf *et al.* [49]	*t*-C_4H_9 + HBr ↔ *i*C_4H_{10} + Br (285–384) VLPR study of HBr abstraction and rate constant of reverse from Russell *et al.* and Benson *et al.* Third Law	38.5 ± 2
1991	Seetula *et al.* [54]	*t*-C_4H_9 + HI ↔ *i*C_4H_{10} + I *t*-butyl via PIMS and literature value of reverse. Third Law	49.4 ± 2
1991	Nicovich *et al.* [23]	*t*C_4H_9 + HBr ↔ *i*C_4H_{10} + Br Resonance fluorescence detection of bromine. Literature value for reverse (257–430) Second and Third Law	50.6 ± 3
1991	Seakins and Pilling [13]	*t*C_4H_9 + HBr ↔ *i*C_4H_{10} + Br Resonance fluorescence detection of bromine, rate constant for HBr abstraction at 300 K. Equilibration experiments at 573 K, 641 K. Third Law	47.3 ± 3
1992	Seakins *et al.* [34]	*t*C_4H_9 + HBr ↔ *i*C_4H_{10} + Br Kinetics in both directions. PIMS detection of *t*-butyl and resonance fluoresence detection of Br. Second and Third Law	51.3 ± 1.3
Recommended value			**48 ± 3**

entropies and higher values of the enthalpy of formation. We have tried to compensate for this by assigning a slightly larger uncertainty to the recommended value.

Acetyl

There is no question that the recommended numbers of McMillen and Golden are too low. A summary of past measurements can be found in Table 2.10. A higher value of -12 ± 3 kJ/mol will encompass all the

TABLE 2.10 Summary of data on acetyl radical

Year	Authors	Methods	Value (kJ/mol)
1982	McMillen and Golden [4]	Iodination. Second Law	−24.3 ± 2
1974	Watkins and Ward [50]	$CH_3CO \leftrightarrow CH_3 + CO$ Acetyl formation and decomposition in varying amounts of CO. Second Law and Third Law	−14 ± 3 −11 ± 3
·1984	Tsang [51]	$CH_3COsC_4H_9 \leftrightarrow CH_3CO + sC_4H_9$. Single pulse shock tube study at 1100. Second Law	−13.8 ± 6
1992	Niiranen *et al.* [52]	$CH_3CO + HBr \leftrightarrow CH_3CHO + Br$ CH_3CO detection by PIMS. Reverse reaction from work of Nicovich *et al.* on Br, via resonance fluorescence, attack on acetaldehyde. Second Law and Third Law	−10.3 ± 1.6 −9.9 ± 1.1
Recommended Value			**− 12 ± 3**

more recent measurements. As in the other cases they involve determinations from room temperature to over 1000 K and three completely independent equilibria.

Hydroxymethyl

The situation here is the same as that for the acetyl radical. There is very impressive agreement between the two independent determinations. Both studies are carried out at the low temperatures where accuracy can be expected to be high. The recommended value is therefore -9 ± 4 kJ/mol or 15 kJ/mol larger than the original recommendation of McMillan and Golden (see Table 2.11).

Benzyl

This is the first of the resonance stabilized aromatic species. All the results are from higher temperature studies. Thus the accuracy is less than that from lower temperature studies. An increasingly important uncertainty is the entropy of the radical. In fact, the results of Hippler and Troe [59] are

TABLE 2.11 Summary of data on hydroxymethyl radical

Year	Authors	Methods	Value (kJ/mol)
1982	McMillen and Golden [4]	Iodination. Second Law	− 24.3 ± 2
1992	Seetula and Gutman [54]	$CH_2OH + HBr \leftrightarrow CH_3OH + Br$ $CH_2OH + HI \leftrightarrow CH_3OH + I$ PIMS detection CH_2OH for halogen production. Reverse rate constants from the data of Buckley and Whittle [55] and Cruickshank and Benson [56] for Br and I respectively. Third Law	−8.9 ± 1.8
1993	Dobe *et al.* [57]	$CH_2OH + HCl \leftrightarrow CH_3OH + Cl$ Laser magnetic resonance detection of HO_2 from $CH_2OH + O_2$ (500–812) for chlorine forming reaction. Electron paramagnetic detection of Cl for reverse (300). Third Law	−9 ± 6
Recommended value			**−9 ± 4**

in agreement with that of Tsang and Walker [42] had they used an entropy as deduced following the prescription of O'Neal and Benson [2]. The recommended enthalpy of formation is therefore 207 ± 5 kJ/mol. Note that the accuracy of a low temperature determination of the enthalpy would be much less influenced by contributions from uncertainties in the entropy. When this is coupled with the high temperature data, then information about the entropy can also be inferred. In any case, with the present result the resonance energy of the benzyl radical is 47 kJ/mol or slightly smaller than that for allyl (see Table 2.12).

Phenoxy

All the enthalpies of formation for this radical have been determined from thermal decomposition studies at high temperatures with correspondingly lower accuracies. The entropy also has uncertainties. However in the second law treatment the results are less dependent on

TABLE 2.12 Summary of data on benzyl radical

Year	Authors	Methods	Value (kJ/mol)
1982	McMillen and Golden [4]	Iodination. Second Law	200 ± 6
1990	Hippler and Troe [58]	$C_6H_5CH_3 \leftrightarrow C_6H_5CH_2 + H$ 1200–1800 K $C_6H_5CH_2I \leftrightarrow C_6H_5CH_2 + I$ 750–950 $(C_6H_5CH_2)_2 \leftrightarrow 2C_6H_5CH_2$ 900–1500 Benzyl radicals directly measured Third Law determination	210.5 ± 4
1990	Walker and Tsang [59]	$C_6H_5CH_2nC_4H_9 \leftrightarrow C_6H_5CH_2 + nC_4H_9$. Single pulse shock tube study. Second Law	203 ± 6
Recommended value			**207 ± 4**

this variable. The VLPP results depend on an assumed A-factor. An enthalpy of formation is recommended of 54 ± 6 kJ/mol (see Table 2.13).

GENERAL COMMENTS

The results on the thirteen radicals considered here should give the reader some idea of capabilities of kinetic techniques for the determination of the enthalpies of formation of organic radicals. It is clear that the more recent data yield highly accurate and reproducible results. It is particularly gratifying that the different methods can replicate results with such accuracy. Nevertheless the experiments are not easy and considerable care must be taken in carrying them out and in the interpretation of the results. The situation with the vinyl radical is illustrative of possible problems.

These results also suggest the proper approach regarding adjustments of existing kinetic data to obtain optimum values of the enthalpies of formation. The main problem with the old iodination results is the assumption of the lack of temperature dependence in the reverse exothermic direction. As a result, an addition of between 8 to 12 kJ/mol to much of the published result will probably give more accurate values. This is illustrated in the cases discussed earlier. This implies, however,

TABLE 2.13 Summary of data on phenoxy radical

Year	Authors	Methods	Value (kJ/mol)
1982	McMillen and Golden [4]	VLPP Experiment. Second Law	48 ± 8
1986	Lin and Lin [60]	$C_6H_5OCH_3 \leftrightarrow C_6H_5O + CH_3$ Shock tube study near 1200 K. Enthalpy compatible with $1.2 \times 10^{16} s^{-1}$ A-factor. Second Law	57.7 ± 8
1989	Suryan *et al.* [61]	$C_6H_5OCH_3 \leftrightarrow C_6H_5O + CH_3$ VLPP experiment near 1100 K. Second Law	47.7 ± 8
1990	Tsang and Walker [66]	$C_6H_5OnC_4H_9 \leftrightarrow C_6H_5O + nC_4H_9$ Single pulse shock tube study near 1100 K. Second Law	55.3 ± 6
Recommended Value			**54 ± 6**

that these radical abstraction reactions have negative activation energies. The exact meaning of such phenomena is unclear. Nevertheless, the experimental evidence for such an effect is very strong.

It is now clear that the comparative rate single pulse shock tube technique leads to extremely accurate rate expressions. When this is coupled with the general observations regarding the temperature dependence of radical combination, essentially null for primary radicals to $T^{0.5}$ and $T^{1.5}$ for two isopropyl and *t*-butyl radicals respectively, very accurate enthalpies of formation can be determined. A very noteworthy feature of the single pulse shock tube work is the relative constancy of the A-factors for a particular type of reaction. Table 2.14 contains results for a series of substituted ethanes. The scatter in the A-factor per bond is no more than a factor of 1.4. The activation energy, however, varies by 55 kJ/mol and is equivalent to a change in rate constant of a factor of 400. The implication is that the differences in the rate constants are probably entirely due to changes in the activation energy. Certainly the results in Table 2.14 suggest that this will introduce an error no greater than 2 kJ/mol. It appears that in gas phase reactions for bond breaking there are no compensation effects.

This has some very important consequences. Table 2.15 contains data

TABLE 2.14 Experimentally measured rate expressions for alkane decomposition [12]

$k(iC_3H_7\text{–}tC_4H_9 \rightarrow iC_3H_7 + tC_4H_9)$	$= 2.5 \times 10^{16} \exp(-36\,800/T)\ s^{-1}$
$k(CH_3\text{–}tC_4H_9 \rightarrow CH_3 + tC_4H_9)$	$= 10^{17} \exp(-41\,100/T)\ s^{-1}$
$k(C_2H_5\text{–}tC_4H_9 \rightarrow C_2H_5 + tC_4H_9)$	$= 6 \times 10^{16} \exp(-38\,800/T)\ s^{-1}$
$k(iC_3H_7\text{–}tC_5H_1^1 \rightarrow iC_3H_7 + tC_5H_1^1)$	$= 2.3 \times 10^{16} \exp(-35\,900/T)\ s^{-1}$
$k(sC_4H_9\text{–}sC_4H_9 \rightarrow sC_4H_9 + tC_4H_9)$	$= 3 \times 10^{16} \exp(-36\,400/T)\ s^{-1}$
$k(sC_4H_9\text{–}sC_4H_9 \rightarrow sC_4H_9 + sC_4H_9)$	$= 3.5 \times 10^{16} \exp(-37\,900/T)\ s^{-1}$
$k(cC_6H_1^1\text{–}tC_4H_9 \rightarrow cC_6H_1^1 + tC_4H_9)$	$= 3 \times 10^{16} \exp(-37\,400/T)\ s^{-1}$
$k(tC_4H_9\text{–}tC_4H_9 \rightarrow tC_4H_9 + tC_4H_9)$	$= 3 \times 10^{16} \exp(-34\,500/T)\ s^{-1}$

on the decomposition of a number of alkynes. Other than the rate expressions, also included are the rate constants at the temperature where the determinations were carried out. Although the scatter is larger than the data in Table 2.14, once again the A-factors cover a narrow range. The larger differences in the A-factor, particularly in the case where isopropyl radical is produced, is undoubtedly due to the possibility of a molecular channel that yields the same products. The difference in value from that of the alkanes can in fact be readily rationalized in terms of the stiffer structure of the resonance stabilized propargyl radical. Of particular interest are the rate constants. If it assumed that this is entirely due to changes in the activation energy, then we are in a position to determine small differences in bond energies very accurately. Thus from the data on hexyne-1, heptyne-2 and 5-methylhexyne-1 decomposition one can conclude that the C—C bond strengths for the bond being broken do not differ by more than 2 kJ/mol. From this it is inferred that the primary and secondary C—H bond strength in isobutane and propane and *n*-butane and isopropane do not differ by more than this quantity. This is as expected. More interestingly, it appears that the resonance energy in

TABLE 2.15 Experimentally measured rate expressions and rate constants for the bond breaking reaction of some alkynes. [31]

Reaction	Rate Expression s^{-1}	Rate Constant s^{-1} (1100 K)
$k(HCCCH_2\text{–}nC_3H_7 \rightarrow HCCCH_2 + nC_3H_7)$	$10^{15.9} \exp(-36\,300/T)$	37
$k(HCCCH_2\text{–}iC_4H_9 \rightarrow HCCCH_2 + iC_4H_9)$	$10^{16.1} \exp(-36\,700/T)$	41
$k(CH_3CCCH_2\text{–}nC_3H_7 \rightarrow CH_3CCCH_2 + nC_3H_7)$	$10^{16.2} \exp(-36\,800/T)$	47
$k(HCCCH_2\text{–}sC_4H_9 \rightarrow HCCCH_2 + sC_4H_9)$	$10^{15.9} \exp(-35\,000/T)$	121
$k(HCCCH_2\text{–}iC_3H_7 \rightarrow HCCCH_2 + iC_3H_7)$	$10^{15.7} \exp(-34\,800/T)$	91

TABLE 2.16 Bond Energies R_1–R_2 for some organic compounds (kJ/mol)

R_1/R_2	218 H	147 ± 1 CH_3	119 ± 2 C_2H_5	88 ± 2 iC_3H_7	48 ± 3 tC_4H_9	39 OH	83 F	121 Cl	112 Br	107 I	188 NH_2
CH_3 (147 ± 1)	440	378	370	370	360	388	464	354	296	240	358
C_2H_5 (119 ± 2)	422	370	364	362	352	393	463	352	295	234	353
nC_3H_7 (100 ± 2)	422	373	365	362	353	397	464	351	299	237	360
iC_3H_7 (90 ± 2)	412	372	364	361	343	403	462	358	299	239	362
sC_4H_9 (69 ± 2)	413	370	362	356	336	401	—	351	301	238	361
iC_4H_9 (70 ± 2) [64]	422	371	363	360	341	393	—	350	294	239	356
tC_4H_9 (48 ± 3)	400	360	352	341	321	400	—	352	293	226	356
tC_5H_{11} (28 ± 3) [66]	402	361	354	334	314	398	—	352	—	228	343
C_2H_3 (299 ± 5)	464	425	418	414	407	—	520	385	332	276	—
C_3H_5 (171 ± 3)	368	318	311	309	302	342	408	292	233	182	313
C_2H (556 ± 8) [65]	548	518	510	509	499	—	531	—	—	—	—
C_3H_3 (339 ± 4)	372	321	314	310	—	337	—	297	—	—	—
C_6H_5 (339 ± 8) [67]	474	436	428	423	408	475	538	405	346	281	440
$C_6H_5CH_2$ (207 ± 4)	375	324	318	317	309	347	—	—	252	151	—
CH_3O (17 ± 4) [65]	437	346	350	355	—	185	—	194	—	—	228
CH_2OH (−9 ± 4)	411	373	367	363	—	—	—	—	—	—	—
HCO (42 ± 4) [65]	376	355	349	341	328	456	497	—	—	—	—
CH_3CO (−12 ± 3)	372	352	345	338	325	463	514	353	290	221	415
C_6H_5O (54 ± 6)	368	268	274	280	—	—	—	—	—	—	—

progargyl and methylpropargyl radicals are within 2 kJ/mol of each other.

These observations furnish a basis for adjusting the vast existing literature [61] on thermal decomposition of many organics under totally inhibited conditions so as to arrive at much better enthalpies of formation. It involves using just the rate constant at the midpoint of the range studied and scaling the rate expression to produce the characteristic A-factors obtained from comparable comparative rate single pulse shock tube experiments. It is our experience that in general one obtains results that are within 6 kJ/mol of the correct values. The same type of scaling should also bring much of the VLPP results into line. In addition, for small differences in rate constants, as before it should be possible to attribute these largely to activation energy effects.

From the above, it is clear that the advances in the techniques and understanding of chemical kinetics during the last two decades have been such that it is now possible to define much more clearly the thermodynamic properties of the important radical intermediates. One expects that in the next few years there will be continual increases in accuracy and increasing capabilities of validating results through various thermochemical cycles in the same manner as with stable compounds. Even today, it is possible to define values with appropriate uncertainty limits. This will then permit tests and calibration of results from other methods as well as the use of these numbers to advance the state of the art of predictive chemical kinetics itself.

Finally, Table 2.16 summarizes bond energies for a variety of organic compounds using the heats of formation derived by the kinetic methods described in the paper. A number of other radicals are included in order to make the table more complete. The reader is referred to similar tables in the literature [1] for comparison. In general, for the higher alkyl radicals there is an increase in the 10–20 kJ/mol range per species. This is doubled when a C—C bond is broken. Such differences are enormous from a kinetic point of view. Hence it is not surprising that the first differences were noted from simple bond breaking studies using the single pulse shock tube.

REFERENCES

1. Benson, S. W. (1976) *Thermochemical Kinetics*. John Wiley and Sons, New York.
2. O'Neal, H. E. and Benson, S. W. (1993) In *Free Radicals*, Vol 2, J. K. Kochi (ed.) John Wiley and Sons, New York, p. 272.

3. Berkowitz, J., Ellison, G. B. and Gutman, D. J. (1994) *Phys. Chem.*, **98**, 2744.
4. McMillen, D. F. and Golden, D. M. (1982) *Annu. Rev. Phys. Chem.*, **33**, 493.
5. Tsang, W. (1978) *Int. J. Chem. Kin.*, **10**, 821.
6. Tsang, W. (1981) Comparative Rate Single Pulse Shock Tube in the Thermal Stability of Polyatomic Molecules. In *Shock Tubes in Chemistry*, A. Lifshitz (ed.), Marcel Dekker, New York, p. 59.
7. Stull, D. R., Westrum, E. F. jr., and Sinke, G. C. (1969) *The Chemical Thermodynamics of Organic Compounds.* John Wiley and Sons, New York.
8. Kerr, J. A. and Lloyd, A. C. (1968) *Quart. Rev.*, **22**, 549.
9. Kerr, J. A. and Parsonage, M. J. (1972) *Evaluated Kinetic Data on Gas Phase Addition Reactions.* Butterworths, London.
10. Harris, G. W. and Pitts, J. N. (1982) *J. Chem. Phys.*, **77**, 3995.
11. Brouard, M., Lightfoot, P. D. and Pilling, J. (1986) *Phys. Chem.*, **90**, 445.
12. Gutman, D. (1990) *Acc. Chem. Res.*, **23**, 375.
13. Seakins, P. W. and Pilling, M. J. (1991) *J. Phys. Chem.*, **95**, 9874.
14. Parmar, S. S. and Benson, S. W. (1989) *J. Amer. Chem. Soc.*, **111**, 57.
15. Castelhano and Griller (1982) *J. Amer. Chem. Soc.*, **104**, 3655.
16. Hiatt, R. and Benson, S. W. (1972) *J. Amer. Chem. Soc.*, **94**, 25, 6886.
17. Hiatt, R. and Benson, S. W. (1972) *Int. J. Chem. Kin.*, **4**, 151, 479.
18. Dobis, O. and Benson, S. W. (1987) *Int. J. Chem. Kin.*, **19**, 691.
19. Russell, J. J., Seetula, J. A., Senkan, S. M. and Gutman, D. (1988) *Int. J. Chem. Kin.*, **20**, 759.
20. Russell, J. J., Seetula, J. A. and Gutman, D. (1988) *J. Amer. Chem. Soc.*, **110**, 3092.
21. Seetula, J. A., Russell, J. J. and Gutman, D. (1990) *J. Amer. Chem. Soc.*, **112**, 1347.
22. Goy, C. A. and Pritchard, H. O. (1965) *J. Phys. Chem.*, **69**, 3040.
23. Nicovich, J. N., van Dijk, C. A., Kreutter, K. D. and Wine, P. H. (1991) *J. Phys. Chem.*, **95**, 9890.
24. Comber, J. W. and Whittle, E. (1966) *Trans. Faraday Soc.*, **62**, 1553.
25. Parmar, S. S. and Benson, S. W. (1988) *J. Phys. Chem.*, **92**, 2652.
26. Kaiser, E. W. and Wallington, T. J. (1995) Kinetics of the Reactions of Cl with $C_2H_4(k_1)$ and $C_2H_2(k_2)$: An upper limit to the Vinyl Radical Yield, *J. Phys. Chem.*, in press.
27. Russell, J. J., Senkan, S. M., Seetula, J. A. and Gutman, D. (1989) *J. Phys. Chem.*, **93**, 5184.
28. Cui, J. P., He, Y. Z. and Tsang, W. (1988) *Energy and Fuel*, **2**, 1086.
29. Parkes, D. and Quinn, C. P. (1953) *J. Chem. Soc., Faraday Trans.*, **1**, 72.
30. Cao, J. R. and Back (1984) *Int. J. Chem. Kin.*, **16**, 961.
31. Pacey, P. D. and Wimalaseba, J. H. (1984) *J. Phys. Chem.*, **88**, 5657.

32. FETTIS, G. C., KNOX, J. H. and TROTMAN-DICKENSON, A. F. (1960) *J. Chem. Soc.*, 4177.
33. KNOX, J. H. and MUSGRAVE, R. G. (1967) *Trans. Faraday Soc.*, **63**, 2201.
34. SEAKINS, P. W., PILLING, M. J., NIRANEN, J. T., GUTMAN, D. and KRASNOPEROV, L. N. (1992) *J. Phys. Chem.*, **96**, 9847.
35. TSANG, W. (1970) *Int. J. Chem. Kin.*, **2**, 23.
36. TSANG, W. (1978) *Int. J. Chem. Kin.*, **10**, 687.
37. KING, K. D. and NGUYEN, T. T. (1979) *J. Phys. Chem.*, **83**, 1940.
38. KING, K. D. (1978) *Int. J. Chem. Kin.*, **10**, 545.
39. TSANG, W. (1978) *Int. J. Chem. Kin.*, **10**, 599.
40. ROTH, W. R., BAUER, F., BEITAT, A., EBBRECHT, T. and WUSTFELD, M. (1991) *Chem. Ber.*, **124**, 1453.
41. TULLOCH, J. M., MACPHERSON, M. T., MORGAN C. A. and PILLING, M. J. (1982) *J. Phys. Chem.*, **86**, 3812.
42. TSANG, W. and WALKER, J. A. (1992) *J. Phys. Chem.*, **96**, 8378.
43. WAGNER, H. G. and ZELLNER, R. (1992) *Ber. Bunsenges Physik Chem*, **76**, 667.
44. TSANG, W., J. (1985) *Amer. Chem. Soc.*, **107**, 2872.
45. FETTIS, G. C. and KNOX, J. H. (1964) In *Progress in Reaction Kinetics*, Porter, G. (ed.), Pergamon, New York, Chapter 1.
46. SEETULA, J. A. and GUTMAN, D. (1990) *J. Phys. Chem.*, **94**, 7529.
47. CHEKOV, E. E., TSAILINGOL'S, A. L. and IOFFE, I. I. (1968) *Chem. Abstr.*, **68**, 48759k.
48. BENSON, S. W., KONDÓ, O. and MARSHALL, R. M. (1987) *Int. J. Chem. Kin.*, **19**, 829.
49. MULLER-MARKGRAF, W., ROSSI, M. J. and GOLDEN, D. M. (1989) *J. Amer. Chem. Soc.*, **111**, 956.
50. WATKINS, K. W. and WARD, W. W. (1974) *Int. J. Chem. Kin.*, **6**, 855.
51. TSANG, W. (1984) *Int. J. Chem. Kin.*, **16**, 1543.
52. NIIRANEN, J. T., GUTMAN, D. and KRASNOPEROV, L. (1992) *J. Phys. Chem.*, **96**, 5881.
53. NICOVICH, J. M., SHACKELFORD, C. J. and WINE, P. J. (1990) *J. Photochem. Photobio. A*, **51**, 141.
54. SEETULA, J. A. and GUTMAN, D. J. (1992) *Phys. Chem.*, **96**, 5401.
55. BUCKLEY, E. and WHITTLE, E. (1966) *Trans. Faraday Soc.*, **58**, 536.
56. CRUICKSHANK, F. R. and BENSON, S. W. (1969) *J. Phys. Chem.*, **73**, 733.
57. DOBE, S., OTTING, M., TEMPS, F., WAGNER, H. G. and ZIEMER, H. (1993) *Ber. Bunsenges. Phys. Chem.*, **97**, 887.
58. HIPPLER, H. and TROE, J. (1990) *J. Phys. Chem.*, **94**, 3803.
59. WALKER, J. A. and TSANG, W. (1990) *J. Phys. Chem.*, **94**, 3324.
60. LIN, C. Y. and LIN, M. C. (1986) *J. Phys. Chem.*, **90**, 425.

61. SURYAN, M. M., KAFAFI, S. A. and STEIN, S. E. (1989) *J. Amer. Chem. Soc.*, **111,** 1423.

62. KNYAZEV, V. D., DUBINSKY, I. A., SLAGLE, I. R. and GUTMAN, D. (1995) The Unimolecular Decomposition of *tert*-butyl Radicals, *J. Phys. Chem.*, in press.

63. SZWARC, M. (1950) *Chem. Rev.*, **47,** 75.

64. TSANG, W. (1990) *J. Phys. Chem. Ref. Data*, **19,** 1.

65. DEMORE, W. B., SANDER, S. P., GOLDEN, D. M., HAMPSON, R. F., KURYLO, M. J., HOWARD, C. J., RAVISHANKARA, A. R., KOLB, C. J. and MOLINA, M. J. *Chemical Kinetics and Photochemical Data for Use in Stratospheric Modeling, Evaluation No. 10*. JPL Publication 92–20, Jet Propulsion Laboratory, California Institute of Technology, Pasadena, California.

66. TSANG, W. and WALKER, J. A., unpublished results.

67. ROBAUGH, D. and TSANG, W. (1986) *J. Phys. Chem.*, **90,** 5363.

3

Thermochemical Data for Free Radicals from Studies of Ions

JOHN C. TRAEGER and BARBARA M. KOMPE

INTRODUCTION

Radicals occur widely in chemistry, playing a vital role in many important liquid, solid-state and gas-phase reactions [1]. A radical is a chemical species containing one or more unpaired electrons and may be either charged (radical cation or radical anion) or uncharged (free radical). A biradical, or diradical, is a species containing two separate free electrons. Because radicals are highly reactive they are generally difficult to isolate and detect, and when they can be isolated they are usually present in very low concentrations, making any direct experimental observation quite demanding.

Free radicals are neutral species and so traditional techniques based on detection via the presence of a charged particle are not directly applicable. They can, however, be detected by the presence of the unpaired electron, using a method such as ESR spectroscopy, and there have been a large number of studies concerning the detection of liquid-phase radicals using this particular technique [2]. In addition, various gas-phase free radical spectroscopic studies have also been made [3].

The major source of energetics information for radicals has tended to come from kinetic studies in which reaction rates are monitored and the corresponding Arrhenius parameters evaluated [4]. This is often done by indirectly following the radical concentrations via associated stable reactants and/or products. From a knowledge of the thermodynamic parameters for the other reaction species involved, thermochemical information, such as the enthalpy of formation for a radical, may then be obtained from the kinetically derived enthalpy of reaction. It is also possible to obtain such energetic data indirectly via the detection of ions related to the radical of interest, or as the neutral lost in a unimolecular fragmentation process [5,6]. This chapter examines how gas-phase ion studies may be used to obtain thermochemical information for free radicals.

TYPES OF MEASUREMENTS INVOLVING IONS

There are a number of different thermochemical experiments involving ions in which free radicals participate [7]. Probably the most common is the measurement of ionization and appearance energies, which are related to the radical by a thermochemical cycle for the particular dissociation or reaction. Translational (kinetic) energy releases can provide information about radicals in experiments in which neutralization or charge inversion of cations occurs. It is also possible that radicals may be involved in ion–molecule reactions. In some kinetic experiments the concentrations of free radicals can be conveniently followed by ionizing and detecting the radical as its positive ion.

Ionization Energies and Electron Affinities

When an ion is produced by the input of energy, or by reaction between two species, it can be analyzed by virtue of the fact that it carries a charge. Typically this is carried out with a mass spectrometer in which the resultant ion beam is passed through a mass analyzer and detected according to its mass-to-charge ratio. Although the ion can be either positively or negatively charged, most studies have been made of the former.

A free radical may be ionized by either electron or photon impact, or conversely, a negatively charged ion may form a free radical by removal of an electron. The minimum energy required to remove an electron from a radical and form a cation is referred to as the ionization energy, IE (also called ionization potential, IP, or I), whereas the minimum energy required to remove an electron from an anion and form a free radical is referred to as the electron affinity, EA (sometimes designated by A or E), which is essentially the ionization energy for the corresponding negative ion.

The ionization energy is usually obtained from the threshold of an ionization efficiency curve, which is basically a plot of the mass selected ion abundance as a function of the ionizing beam energy [8]. If this energy corresponds to the difference between the ground rovibronic states for both the neutral and the ion, then the ionization energy is called adiabatic. The energy change associated with the most probable ionizing transition is the vertical ionization energy, which corresponds to formation of an ion with a geometry essentially the same as that for the neutral precursor. The different nature of these two types of ionization energy is illustrated in Figure 3.1.

Ionization of a molecule or radical is Franck–Condon controlled. According to the Franck–Condon principle, the time required for an electronic transition during ionization is much less than the time required for a vibration, so that atoms in the ion will probably have the same internuclear geometry as the neutral precursor molecule. If the respective potential energy wells have minima at similar internuclear separations, an adiabatic ($0 \rightarrow 0$) transition will occur with a high probability. However, if there is a large change in structure along the direction of the normal mode of the excited vibration, it is most likely that the potential energy well minima will occur at different internuclear separations and the most intense (probable) transition will be to a higher energy vibrational and rotational level of the particular electronic state of the ion [8],

FIGURE 3.1 Potential energy curves for a molecule M and its corresponding positive ion $M^{+\bullet}$.

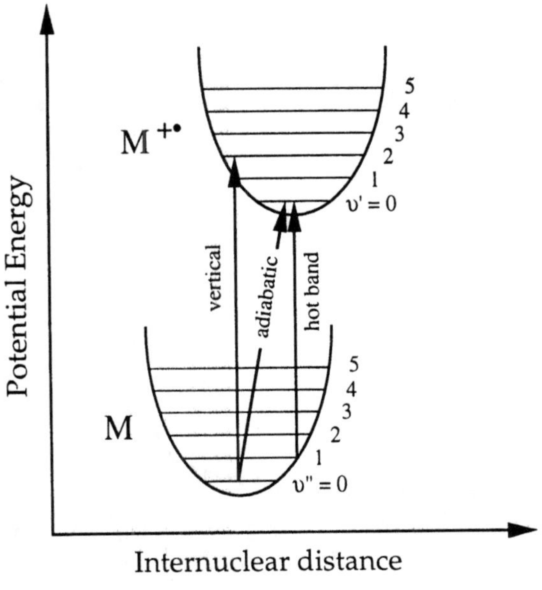

as shown in Figure 3.1. Thus, the vertical ionization energy will only be the same as the adiabatic ionization energy for a negligible geometry change following ionization. In other cases it will be greater than the adiabatic value (see Figure 3.1), with the actual difference depending on the particular change in geometry. From a thermochemical viewpoint, the adiabatic ionization energy is the more useful parameter, although it may be difficult to detect if few neutrals are undergoing the adiabatic transition.

The precursor molecules will usually be present at some particular finite temperature. Because of this, their thermal energy will produce a distribution of vibrational and rotational states, even if they are in the ground electronic state. Ionization of neutrals from these excited vibronic states requires less energy than the adiabatic transition, resulting in ions being produced below their true threshold energy (Figure 3.1). It is responsible for the 'hot band' structure observed in the pre-threshold region of ionization efficiency curves [9]. For a molecule at room temperature this will typically extend over a range of approximately 0.1 to 0.2 eV (Figure 3.2), although it will increase at the elevated temperatures often used to prepare a free radical for study. The use of supersonic molecular beams for sampling significantly reduces the population of

FIGURE 3.2 PIE curve for $C_4H_9I^{+\bullet}$ from *t*-butyl iodide. Note the pre-threshold 'hot' band structure that extends ≅0.1 − 0.2 eV below the adiabatic IE.

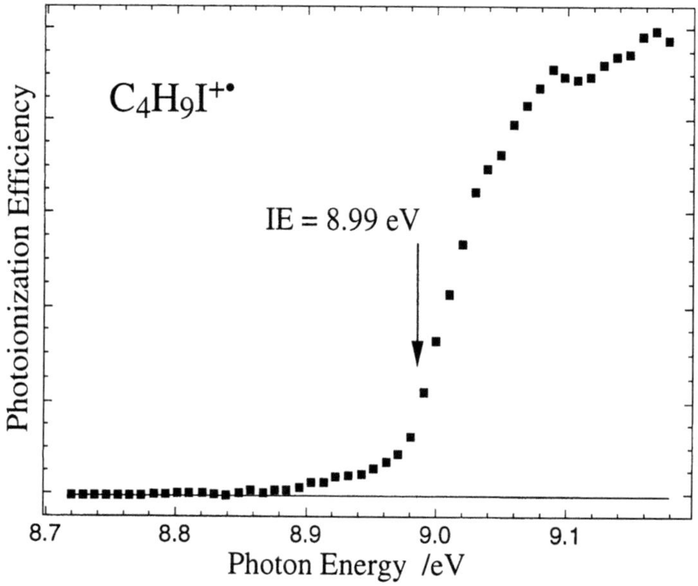

excited vibronic states, and therefore the interference from such hot bands [10,11].

When a variable energy beam of electrons or photons is used for ionization, it is possible for an autoionization process to be observed in addition to the direct ionization process. This is a two-step reaction in which the neutral molecule or radical absorbs energy greater than the minimum required to form an ion. Instead of ionizing immediately, an excited neutral state, known as a Rydberg state, is formed which can subsequently eject an electron to form a positive ion. Autoionization is observed as fine structure in the ionization efficiency curve and may complicate the assignment of the adiabatic transition [8].

The easiest way to directly ionize a free radical is by electron impact and the earliest mass spectrometric studies were made in this way [12,13].

$$R^{\bullet} + e^{-} \rightarrow R^{+} + 2e^{-} \tag{3.1}$$

To extract accurate thermochemical information from the experimental data, it is essential to control precisely the energy of the ionizing electron beam. There are two main difficulties associated with such electron impact studies. First, it is necessary to constantly calibrate the absolute electron energy with an internal reference gas. More importantly, however, it is not possible to easily produce a beam of electrons with a narrow spread in energies. This is largely due to the high filament temperature required for thermionic emission, which results in a quasi-Maxwell–Boltzmann distribution of kinetic energies. The typical energy spread for a conventional electron beam is >1 eV, which smears out the actual ionization onset and increases the uncertainty associated with its determination.

Various experimental and analytical methods have been used to overcome the energy spread problem inherent in electron impact measurements [8]. Probably the most successful technique that has been applied to free radical studies has been the use of an electron monochromator in which the electron beam energy is narrowly defined by passage through an electron velocity selector. Several types of electrostatic analyzer have been interfaced with a mass spectrometer, including the 127° electric sector [14] and the double focusing hemispherical condenser [15]. Although full-width at half-maximum (FWHM) resolutions of 0.005 eV have been achieved [16], typical operating values are of the order of 0.07–0.10 eV. Extensive monoenergetic electron impact studies of free radical ionization energies have been made by Lossing and coworkers [17–22].

Another means of obtaining free radical ionization energies from ionization efficiency curves is by photon impact, [23,24] with numerous studies having been made over the last thirty years [25–32].

$$R^{\bullet} + h\nu \rightarrow R^{+} + e^{-} \tag{3.2}$$

The free radicals may be prepared by pyrolysis of appropriate precursors, such as nitrite compounds [25].

$$\begin{aligned} RCH_2ONO &\rightarrow RCH_2O^{\bullet} + {}^{\bullet}NO \\ RCH_2O^{\bullet} &\rightarrow R^{\bullet} + CH_2O \end{aligned} \tag{3.3}$$

However, the elevated temperatures required increase the hot band structure in the photoionization efficiency curve and may complicate any energetic analysis. It is also possible to produce free radicals via abstraction reactions involving atoms, such as H or F, produced in a microwave discharge [28,32].

Photon beams of variable wavelength (energy) can be produced using a suitable light source coupled to a monochromator [33]. The most common light sources are the hydrogen discharge, which produces a many-lined (pseudocontinuum) emission spectrum in the 7.7–14.5 eV range, and the helium Hopfield continuum, which provides useful energies in the range of approximately 12.5–20.6 eV [34]. Care must be taken when using a discrete-line light source to obtain a photoionization efficiency curve as it is possible for false structure to be generated [35]. A typical photon flux is $\cong 10^{9}\,s^{-1}$, which is quite low when compared to an electron velocity selector ($\cong 10^{11}\,s^{-1}$) or a conventional non-monoenergetic electron source ($\cong 10^{13}\,s^{-1}$). Synchrotron radiation, in which light is emitted from rapidly moving electrons accelerated in a circular path, has also been used for photoionization experiments [36,37]. Although it has the advantage of covering a broad energy range with high intensity ($\cong 10^{13}\,s^{-1}$), specialized facilities are required. As yet few synchrotron experiments have been made that directly involve free radicals [38,39].

A monochromator is used to select a band of photons with a narrow energy range from the light source. Because of the high absorbance of air at photon energies in excess of about 5 eV, it is necessary to evacuate the monochromator to a pressure of 10^{-4} Torr or better. The high pressure gas discharge lamp must be separated from the monochromator to obtain this pressure and it is here that a major problem arises. There is no suitable window material that transmits light at energies greater than 12 eV. Thus, for higher energies the entire system must be windowless, which means that high differential pumping systems are required to remove effectively the gas that continuously diffuses through the entrance slit into the monochromator from the light source. Experiments are usually carried out with either a Seya–Namioka or normal incidence type monochromator [34], with typical operating resolutions of approximately 2 Å, corresponding to an energy resolution of 0.016 eV at 10 eV (1240 Å). There can be a problem with contamination by second and higher order radiation from the diffraction grating, particularly when synchrotron radiation is used [37], but this may be eliminated for low energy (< 11.8 eV) radical studies by the use of a LiF filter [39].

The intensity of the light source is not constant and varies with wavelength, so that it is necessary to monitor the photon beam in order to correct the experimental photoion current for photoionization efficiency. One method is to measure the photoelectric yield from a metal surface, such as tungsten, but the data must be calibrated because the yield varies with photon energy [40]. A simpler and more commonly used method of recording photon intensity is to use a photomultiplier with a sodium salicylate phosphor. No calibration is required because the relative fluorescence quantum efficiency remains essentially constant up to about 40 eV [34].

In addition to measurements of the minimum energy required for ion formation, the ionization energy can also be obtained from a study of the energetics of the ionized electron. When a radical is irradiated by photons of a fixed wavelength, electrons that are in accessible energy levels will be ejected with kinetic energies representing the difference in energy between the incident photon and that of the ion formed. The kinetic energy of the ionized electron is measured with an electron energy analyzer and, since the energy of the incident photons is known, the ionization energy for each electronic state of the ion can be determined. The most widely used technique for this is conventional vacuum ultraviolet photoelectron spectroscopy (PES) in which the helium resonance line at 58.4 nm (21.22 eV) is used as the photon source [41]. Other sources of different energies used include the He II resonance line at 40.81 eV and the neon and argon resonance doublets, which have energies in the region of 16.7 and 11.7 eV respectively. For a doublet light source the photoelectron spectrum will consist of two superimposed spectra shifted in energy by the energy difference of the two emission lines. The plot of ejected photoelectron abundance as a function of its kinetic energy is known as a photoelectron spectrum, and provides information about the localization and bonding character of the ejected electron. Unlike a mass spectrometer, a photoelectron spectrometer does not have any mass filtering capability to remove contaminant species, so that it is necessary to ensure that the sample has a high purity. A wide range of free radicals has been studied using PES, particularly by the research groups of Beauchamp [42–48] and Dyke [49–56].

The two threshold laws for ionization by electron and photon impact are slightly different [8]. To a good approximation it can be shown that the ionization probability above threshold is governed by the conditions for distribution of the excess energy between the emerging electrons [57]. For single ionization by electron impact there are two electrons to share the excess energy, so that if the probability increases as the first power of the energy excess, there will be a linear increase in the ionization efficiency above threshold. Because there is one fewer electron to share in the excess energy partition for photon impact, the threshold law will then be the integral of the equivalent electron impact process, i.e. direct ionization will be observed as a step function (Figure 3.3). Unlike electron impact there is a finite cross section at threshold for photoionization, making it easier to detect the actual ionization onset.

Experiments are not restricted to processes involving positive ions. A large number of negative-ion energetics studies have involved the photo-detachment of an electron from an anion [58] that has been produced by other means, such as a xenon arc lamp together with a diffraction grating monochromator [59] or a tunable dye laser [60]. An ion cyclotron resonance (ICR) spectrometer can be conveniently used to generate, trap and detect the negative ions, and the negative-ion concentration monitored as a function of the photon energy [61]. The threshold for reaction (3.4) provides a direct measure of the electron affinity for the radical $R^{\bullet}$.

FIGURE 3.3 Theoretical ionization efficiency curves for single ionization by electron and photon impact.

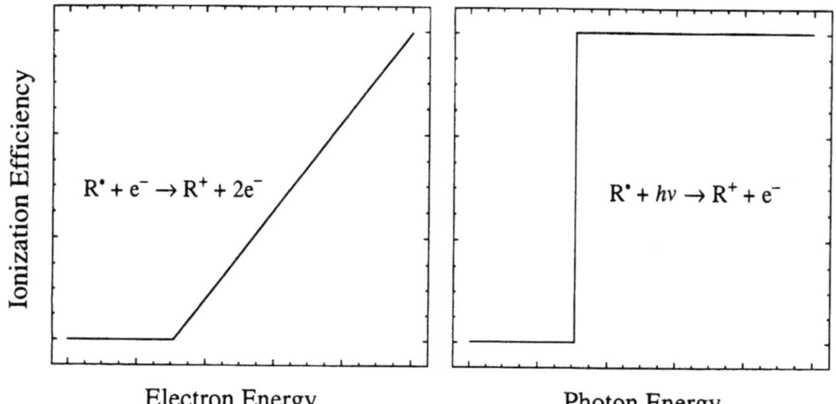

$$R^- + h\nu \rightarrow R^\bullet + e^- \tag{3.4}$$

Negative-ion photoelectron spectroscopy [62] is particularly well suited to the study of neutral free radicals since open-shell systems generally have positive electron affinities [6]. The corresponding anion will therefore be stable with respect to electron detachment. Negative ions have been produced by methods such as DC electrical discharge [63], microwave discharge flowing afterglow [64], electron impact on a molecular beam [65] and electron impact on a molecular beam followed by ion–molecule reaction [66]. The ion beam is accelerated, mass analyzed and then passed into a light-interaction region where photodetachment of the electron occurs. Commonly fixed-frequency lasers, such as an argon-ion laser [63,67] or different harmonics of a pulsed Nd:YAG laser [65], have been used. A small proportion of the photodetached electrons are detected and energy analyzed. Conventional methods, such as the use of a hemispherical electrostatic energy analyzer [63], can be used to determine the electron kinetic energy. It is also possible to employ electron time-of-flight measurements for the energy analysis [65], provided the energy has been calibrated with an ion of known electron affinity [66]. The different experimental techniques for measuring electron affinities have been reviewed in detail elsewhere [58,68].

Appearance Energies

When a neutral molecule is ionized, it is possible that the molecule ion may unimolecularly decompose, producing a fragment ion and a neutral radical. The minimum energy observed for this ionization and fragmentation process is called the appearance energy (AE). This is sometimes called the appearance potential (AP), although use of this terminology is not recommended [69].

$$RX + e^- \rightarrow RX^{+\bullet} + 2e^- \rightarrow R^\bullet + X^+ + 2e^- \tag{3.5}$$

$$RX + h\nu \rightarrow RX^{+\bullet} + e^- \rightarrow R^\bullet + X^+ + e^- \tag{3.6}$$

Provided that the products for these reactions are formed with no excess energy, then the appearance energy can be used to obtain useful thermochemical information for both the cation and the neutral free radical, as it represents the minimum energy for the reaction to occur. Energetic data for the positive ion formed is related to the corresponding radical $X^\bullet$ by its ionization energy.

The determination of the appearance energy from an experimental ionization efficiency curve is not straightforward. For reaction (3.6) at 0 K, and assuming (i) that the total ionization cross section is constant in the

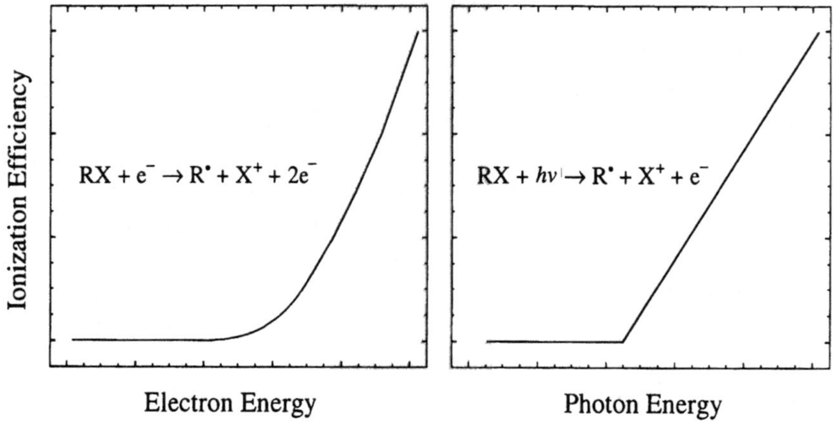

FIGURE 3.4 Theoretical ionization efficiency curves for ionization and fragmentation by electron and photon impact.

immediate threshold region and (ii) that the probability for the molecular ion $RX^{+\bullet}$ producing a fragment ion X^+ (called the breakdown curve for the ion) approximates a step function, then it can be shown that the photoionization efficiency curve should be essentially linear for photon energies greater than the 0 K threshold (E_0) [9]. The corresponding curve for photoionization at temperature T will simply be the convolution of the 0 K curve with the internal thermal energy distribution of the molecule. In this case extrapolation of the linear portion to the energy axis produces an intercept that is lower than E_0 by an amount equivalent to the average effective internal thermal energy of the molecule [70]. The choice of linear section for extrapolation is not always clear. However, some justification can usually be made on the basis of the average thermal energy of the precursor molecule. Because of the different threshold law for ionization by electron impact (Figure 3.4), it is necessary to plot a first differential ionization efficiency curve for reaction (3.5) to obtain a corresponding linear extrapolation in the threshold region [8].

The only way in which the breakdown curve can be experimentally measured is by the threshold technique known as photoelectron–photoion coincidence (PEPICO) [71,72]. The experiment essentially involves selection of the internal energy for a precursor ion by photon impact followed by detection of those mass selected fragment ions formed in coincidence with ejected photoelectrons of zero initial kinetic energy. By measuring the internal energy at which the fractional abundances for both parent and daughter ions are equal, called the crossover energy in a breakdown diagram (Figure 3.5), it is possible to obtain the

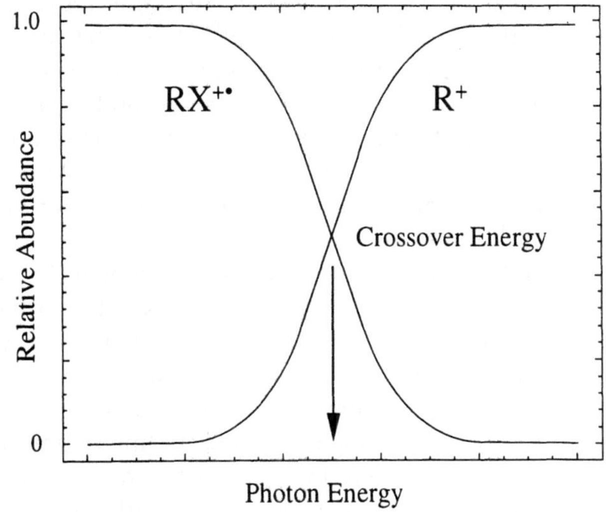

FIGURE 3.5 PEPICO breakdown diagram for RX in the vicinity of the onset for R^+ formation.

0 K appearance energy for the fragmentation reaction (3.6) [73]. The energy difference between the crossover energy and the 0 K threshold will depend on the internal thermal energy of the neutral precursor molecules.

When time-of-flight (TOF) mass analysis is used in a PEPICO experiment, the shape of the coincidence TOF distribution for a selected fragment ion can be compared to the theoretical shape predicted from statistical (quasi-equilibrium/RRKM) theory [74–76]. Different parameters, such as the activation energy for dissociation and vibrational frequencies for the molecular ion and the transition state, are varied, and when the shapes of the theoretical and experimental curves are the same, then it is assumed that the predicted activation energy and transition state shape are correct [77]. The TOF distribution can also provide an estimate of the average kinetic energy release for a selected internal energy [77]. Although these coincidence techniques are desirable for free radical studies they are extremely difficult to perform due to the relatively low densities of such species in the target region. A pulsed PEPICO study of SO radicals produced in a microwave discharge has been reported by Norwood and Ng [78].

The thermochemical threshold energy corresponds to the energy difference between the reactants and the products. Sometimes there is an extra energy barrier that must be overcome before the fragmentation can proceed. This is known as the reverse activation energy (E_{rev}) and is the energy barrier between the products and reactants represented by the transition state ($[RX^{+\bullet}]\ddagger$, Figure 3.6) [79]. It is often assumed to be negligible for simple bond cleavage reactions but can be estimated from kinetic energy release measurements of corresponding metastable ion decompositions or PEPICO experiments. If the average kinetic energy release accompanying the fragmentation is small ($<2\ \text{kJ mol}^{-1}$) then it may be reasonably assumed that the process occurs at, or very close to, the thermochemical limit [80].

Another reason the appearance energy may be higher than the thermochemical threshold energy is the operation of a 'kinetic shift' [81]. This is the energy in excess of the thermochemical threshold energy, including any reverse activation energy, required for the ion to be detected by the instrument (E_{kin}). It can be minimized by increasing the instrument detection sensitivity [82] or by increasing the amount of time available for fragmentation in the ion source [83]. A kinetic shift will be associated with a rate constant that only rises slowly with increasing internal energy. In this case the rate of production of the ion is low for excess energies just above the threshold.

A 'competitive shift' can also cause an increase in the appearance energy. In this situation a particular fragmentation may become thermochemically possible at a particular energy, but is not observed because the precursor molecule is fragmenting via a more favorable competing

FIGURE 3.6 Energy level diagram for a reaction involving a reverse activation energy barrier and a kinetic shift.

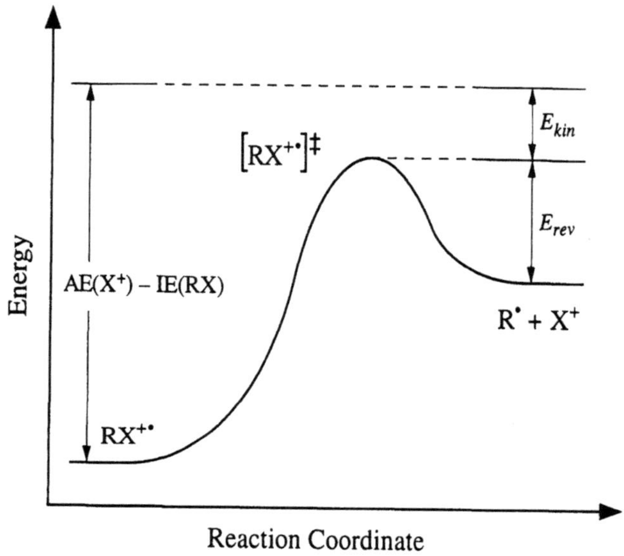

pathway that has become accessible at or below the thermochemical threshold energy of the ion of interest [8].

Appearance energy studies of negative ions are much less common than those involving positive ions. This is due to negative ions being generally formed in lower abundances when compared to their positive ion counterparts. The two main modes of fragment negative-ion formation by electron impact are dissociative electron attachment (dissociative resonance capture)

$$AB + e^- \rightarrow A + B^- \tag{3.7}$$

and ion pair formation

$$AB + e^- \rightarrow A^+ + B^- + e^- \tag{3.8}$$

The appearance energies for both of these processes can be used to derive thermochemical information, such as the bond dissociation energy of AB or the electron affinity of B, provided that the amount of any excess energy is known. It is possible to estimate the excess kinetic energy involved using time-of-flight mass spectrometry [84,85]. Alternatively, studies can be made in which only those ions with essentially thermal energies are collected [86]. Few studies have been made of negative ions produced directly by photon impact [87].

Translational Energy Releases

When two particles collide the collision can be either elastic, in which case there is no change in the internal energy of either of the colliding species, or inelastic, where there is a change in the internal state of the system. These changes are manifested by the interconversion of translational and internal energy. If the collision reaction is endothermic, there is a net conversion of translational energy to internal energy, a situation that may be studied by energy-loss spectroscopy [88].

Charge-inversion energy-loss, or double-charge transfer, is a particular case in which a fast-moving positive ion inelastically collides with a neutral target, M, resulting in a double-electron transfer reaction (3.9) or two successive single-transfer reactions from the neutral to the reactant ion (3.10).

$$R^+ + M \rightarrow R^- + M^{2+} \tag{3.9}$$

$$R^+ + M \rightarrow R^{\bullet} + M^+ \tag{3.10a}$$

$$R^{\bullet} + M \rightarrow R^- + M^+ \tag{3.10b}$$

The energy necessary for these reactions is provided by the incident relative translational energies of R^+ and M and is given by the difference between the initial and final translational energies of the incident ion. Under appropriate experimental conditions, the energy loss, ΔE, will be a direct measurement of the overall change in the internal energy of the colliding system [88].

A reverse-geometry double-focusing mass spectrometer may be conveniently used to study these processes [89]. The incident ion, R^+, is generated by electron impact or chemical ionization and accelerated to several keV before being mass selected and passed into a collision cell containing the neutral target gas, such as xenon, where charge inversion takes place. The resultant negative ion, R^-, is then transmitted through the electric sector to the detector where the negative-ion energy-loss spectrum is measured. Calibration of the energy-loss scale is achieved by using incident ions of known ionization energy and electron affinity [89]. Reactions (3.9) and (3.10) for the double-charge-transfer process can be distinguished by measuring the intensity of the negative-ion current as a function of the target gas pressure. Process (3.9) is linearly dependent whereas process (3.10) is quadratically dependent on the target gas pressure [89].

Since internal energy has been converted to translational energy the energy loss can be calculated from the processes occurring.

$$\Delta E_{(3.9)} = \mathrm{IE}_2(\mathrm{M}) - \mathrm{IE}_1(\mathrm{R}^\bullet) - \mathrm{EA}(\mathrm{R}^\bullet) \tag{3.11}$$

$$\Delta E_{(3.10)} = 2 \times \mathrm{IE}_1(\mathrm{M}) - \mathrm{IE}_1(\mathrm{R}^\bullet) - \mathrm{EA}(\mathrm{R}^\bullet) \tag{3.12}$$

Equations (3.11) and (3.12) have been simplified as they should include terms relating to the recoil energies of the target. However, these will be negligible when the initial energy is large and the geometry of the apparatus is arranged such that the secondary ions are observed at small angles [88]. It is also assumed here that M^+ is formed in the same state for both reactions (3.10a) and (3.10b).

As the double ionization energy for the target, $\mathrm{IE}_2(\mathrm{M})$, is greater than $2 \times \mathrm{IE}_1(\mathrm{M})$, the energy loss for process (3.9) will be greater than the energy loss for process (3.10). If the ionization energies of the target, M, are known then the sum of the ionization energy and the electron affinity for the radical $R^\bullet$ can be calculated. A knowledge of either one of these quantities then enables the other to be calculated. It should be noted, however, that the energy losses measured will be vertical energies, by virtue of the Franck–Condon principle, in which case any derived ionization energy or electron affinity can only represent an upper limit to the adiabatic value [88].

The accuracy of thermochemical information derived from double-charge-transfer experiments is limited by the energy spread of the incident R^+ ion beam, with errors in reaction energies being typically ≅0.5 eV. Also, interpretation of the energy-loss spectra may be complicated by the presence of overlapping peaks due to ionization of different electronic states of the target gas. The presence of excited states of the initial positive ion can also affect the spectrum, although this may be reduced by using ionizing electrons of lower energy [89]. It has been suggested [90] that the vibrational excitation of R^+ and R^- is likely to be similar, tending to compensate any effect on the observed energy loss.

The neutralization of cations provides a convenient means of preparing unusual radical species [91]. Fast beams of readily produced positive ions can be converted into the corresponding radical by electron transfer from an appropriate target gas in a collision cell. For example, the hypervalent radicals $^{\bullet}CH_5$, $^{\bullet}NH_4$ and $H_3O^{\bullet}$ can all be formed by neutralization of the corresponding $CH_5{}^+$, $NH_4{}^+$ and H_3O^+ cations. Various experiments, including translational energy measurements, have been performed on these radicals to obtain useful thermochemical information [91,92]. Reionization of neutral species for subsequent mass analysis can also be achieved by collision with another suitably chosen target gas in a second collision cell. This combined technique is referred to as neutralization–reionization mass spectrometry (NRMS) [93–95].

Ion–Molecule Reactions

It is possible to investigate reactions between ions and neutral free radical or molecular species.

$$A^+ + B^{\bullet} \rightleftharpoons C^+ + D^{\bullet} \tag{3.13}$$

These ion-neutral processes often form the basis of kinetic and thermochemical studies. They may also be used to generate unusual species that are difficult to obtain by other means. In this case, the ions or radicals that have been produced can then be studied by other techniques. Radicals and ions generally need to be prepared *in situ*. Ions are commonly produced by electron or photon impact, whereas radicals can be generated by homolytic fission of a suitable precursor. It is also possible for both species to be produced via other ion-neutral reactions.

The flowing afterglow plasma method for studying ion–molecule reactions, developed by Ferguson and coworkers [96], belongs to a group of techniques that involves the upstream production of an ion that later reacts with one or more neutral reagent gases. The plasma of electrons, ions and excited neutrals is commonly produced by conventional electron impact or microwave discharge acting on a carrier gas. Reactant ions are either formed by introduction of reactant gas with the carrier or by addition downstream from the plasma source region. During the time taken for these reactant ions to traverse the flow tube to the reaction region, relaxation of excited states can occur as a result of collisions between the ions and other species in the flowing afterglow. This thermalization allows the reactant ions to attain low internal energies by the time they reach the reaction zone. Some control of the internal energy of the reactant species is possible by heating the flow tube wall and by heating or cooling the carrier gas. Various neutral reactant gases can be added and reactions allowed to occur as the species are carried downstream. At the end of the reaction zone, reactant or product ions are sampled by a mass spectrometer and the ion signal recorded as a function of the reagent gas flow rate. From this information, rate coefficients can be determined and ionized products identified. Optical spectroscopic methods can also be used to monitor ion–molecule reactions that take place in a flowing afterglow [97].

The presence of ions derived from the source gas or by reaction with species derived from the source gas can complicate the identification of the product ion. Also, reaction of these additional species with the product ion, or in competition with the product ion for another neutral, can result in erroneous rate coefficients. These problems may be prevented by remotely producing the ion of interest, mass selecting it and then injecting it into a flowing carrier gas stream, thereby allowing time to thermalize before reaction with various neutral reactant gases. This apparatus is commonly known as a selected ion flow tube (SIFT) [98]. The various flow reactor techniques for studying gas-phase ion chemistry have been reviewed recently [99,100].

One of the most successful methods for studying ion–molecule equilibria is the technique of pulsed electron beam, high pressure mass spectrometry (HPMS), which has been largely developed by Kebarle and coworkers [101]. Reaction gas mixtures are introduced into an ion source at pressures of several Torr. A pulsed electron gun is then used to produce ions over a short period ($\cong 10\ \mu s$) and the subsequent changes in reactive ion concentrations following the pulse ($\cong 10$ ms) are monitored. By using relatively high pressures the ions are effectively trapped in the ion source, with the effect that their lifetimes, and therefore scale of rate measurements, are significantly increased. The frequency of non-reactive collisions with the bulk gas is such that thermal equilibrium is usually established very rapidly on the time scale of the experiment [102]. High capacity vacuum pumps are used to maintain the large pressure differential between the ion source and mass analyzer. Ion detection is synchronized with the pulsed electron beam.

HPMS is essentially a kinetic technique involving the measurement of the rates at which ion–molecule reactions, particularly those involving proton and electron transfer processes, approach equilibrium. Various thermochemical parameters for species involved in the reactions may then be derived. Because this technique can be used to study reactions over wide temperature ranges, it is possible to measure directly entropy changes that otherwise would have to be estimated.

The technique of ion cyclotron resonance spectrometry (ICR) can also be used to study ion–molecule equilibria and hence obtain equilibrium constants and Gibbs energy changes, simply by observing the relative abundances of reactant and product ions [103]. When ions are injected into a magnetic field, they follow circular paths with frequencies that are mass dependent. A mass spectrum is readily obtained by exciting these cyclotron motions with an alternating electric field and measuring the absorption of RF energy as a function of frequency. A particular feature of ICR is that ions can be trapped for relatively long periods, making it suitable for the investigation of ion–molecule reactions. Drift ICR experiments involve the continuous formation of ions and passage through the ICR cell. Studies are usually made of the effect of pressure for selected reaction times. In contrast, trapped ICR experiments, including Fourier Transform mass spectrometry (FTMS) [104], are used to study ion–molecule reactions as a function of time at fixed low ($<10^{-5}$ Torr) pressures. Double resonance techniques enable the selective removal of ions from the reaction cell, although separate ion formation and reaction regions, as in tandem ICR, may also be used. However, unlike the SIFT technique, reactant ions do not have thermal energies [103]. The quadrupole ion trap [105], which is closely related to the ICR and FTMS, provides another convenient means of studying the thermochemistry of ion–molecule reactions.

Reaction Kinetics

The rates of dissociation and ion-neutral reactions, as well as the equilibrium concentrations of the various species, can be used to derive thermochemical information about the reactant and product molecules involved. It is possible to monitor a reaction by measuring the change in concentration with time of one or more of the reactants or products. It is common in kinetic studies to follow the pseudo-first-order decay of one of the reactant species in a known excess of the other reactants. The selection of the method used to observe the concentration of a particular species depends on its properties.

Positively or negatively charged ions can be conveniently monitored by using a mass spectrometer to measure the mass selected ion current as a function of time. For the ion-neutral reaction (3.13), the absolute rate constants for either the forward or reverse reaction can be obtained by following the change in concentration with time for one of the charged species. However, neutral species, such as free radicals, can only be detected by mass spectrometry if they are first converted to ions. Generally this is accomplished by using ionization techniques such as photon or electron impact, although some ions may be generated by a combination of electron impact and chemical ionization.

Considerable care must be taken to ensure that any ion to be monitored forms directly from the radical and not from another species present in the reaction system. Fortunately the energy required for direct production of an ion from a free radical is invariably lower than that from a unimolecular fragmentation process [6], so that it is usually possible to employ a selected ionizing energy to exclude any ions formed from alternative processes. Low energy electron beams can be used in such experiments but, due to their poorly defined upper energy cutoff, it is preferable to use wavelength-selected photons produced from gas resonance lamps.

Gutman and co-workers [106] have developed a photoionization mass spectrometer coupled to a heated tubular quartz reactor for studying the kinetics of polyatomic free radical reactions. A seeded gas containing the free radical source flows through the reactor tube where it is irradiated by a pulsed laser. Part of the reaction mixture is continuously extracted via a sampling hole and collimated into a beam before being ionized and mass analyzed. Depending on the species being detected, different photon sources, usually atomic resonance lamps, are used for the photoionization. Ion detection is synchronized with each laser pulse. Various polyatomic free radical reactions, such as H-atom abstraction from hydrogen halides, have been studied by this technique [107–110].

Molecular modulation mass spectrometry [111,112] is a technique that can be used to study free radicals via their corresponding ions without having to use an energy-selected ionizing beam. If the free radical beam is modulated, such as by using a rotating disk chopper, then a phase sensitive detector may be used to detect selectively the corresponding ion signal in the presence of other unmodulated ions of the same mass. The technique was developed by Wu and Johnston [111] who studied the kinetics of the photolysis of Cl_2 in the presence of O_2 by monitoring the change in concentration of intermediate $ClO^{\bullet}$ and $ClOO^{\bullet}$ radicals. A photolyzing light source was turned on and off at a regular controlled rate, leading to a periodic variation in the concentration of the reactants, products and radical intermediates. The use of a 'skimmer' enabled the formation of a molecular beam that entered the mass spectrometer where ionization, mass separation and counting of the ionized radicals occurred. From an analysis of the phase shift between photolyzing light and radical half-life, the rate constant could then be determined.

Another kinetic technique in which reactants and products are ionized prior to detection is very low pressure pyrolysis (VLPP) [113]. A flow of reactive gas is introduced into a heated, low-pressure (0.1–10 mTorr) Knudsen cell, and the reaction mixture sampled as a modulated effusive molecular beam using conventional electron-impact mass spectrometry. Temperatures up to 1900 K have been used in conjunction with an alumina reactor [113]. Because of the low pressures used, complications due to secondary processes are negligible. However, reactions will not be at their high-pressure limits, so that Arrhenius parameters cannot be independently determined [4]. For unimolecular processes, RRKM theory is used to convert VLPP experimental rate constants to the appropriate high-pressure values. Both unimolecular and bimolecular reactions involving free radicals have been studied. The radicals are commonly produced in the very low pressure reactor (VLPR) via pyrolysis [113], photolysis [114] or microwave discharge [115]. Photochemical, rather than thermal, activation can also be investigated by the addition of an irradiated section to the Knudsen cell [116]. This modified technique is referred to as very low pressure photolysis (VLPΦ).

THERMOCHEMICAL DATA FROM ION STUDIES

Probably the most important piece of thermochemical information that can be derived from ion studies is the enthalpy of formation for the cation or anion, which is related to the radical enthalpy of formation by either the ionization energy (cation) or electron affinity (anion). In some systems it is also possible to use the experimental data to derive the enthalpy of formation for the radical formed in a fragmentation process [117]. However, in both cases it is essential to know the enthalpies of formation for all of the other reactants and products, including the ionized electron. For many species this information can be obtained from extensive evaluated compilations of experimental data [6,118], otherwise it is necessary to make estimates based on group equivalence schemes for neutral [119,120] and ionic species [121].

Convention for Cationic Enthalpies of Formation

For any gas-phase reaction involving a cation, it is necessary to adopt a standard state convention for the electron ejected in the ionization process. This can be readily demonstrated by considering the standard gas-phase enthalpy of formation for a proton, $\Delta H^\circ_f\,(H^+)$, which refers to the enthalpy change for reaction (3.14).

$$\tfrac{1}{2}H_2\,(g) \rightarrow H^+\,(g) + e^-(g) \tag{3.14}$$

At 0 K, the enthalpy of reaction corresponds to the formation of a proton and an electron with zero translational energy, i.e. at rest. However, for the reaction at a temperature T, it is possible to consider $\Delta H^\circ_f(H^+)$ with reference to the production of a proton with $\tfrac{3}{2}RT$ translational energy and an electron either with $\tfrac{3}{2}RT$ or zero translational energy. These have been called the 'thermal electron' and the 'stationary electron' conventions, respectively [122]. At 298 K there will be a difference of 6.2 kJ/mol between these two differently defined enthalpies of formation. It is therefore essential to specify clearly which particular convention has been used. The stationary electron convention, i.e. $\Delta H^\circ_f(e^-) = 0$, is most commonly used in mass spectrometric data and is used exclusively throughout this chapter. Some widely used compilations, such as the JANAF Thermochemical Tables [123], do however adopt the thermal electron convention.

3.3.2 Ionization and Appearance Energy Measurements

For the ionization reactions (3.1) and (3.2), the enthalpy of formation of the radical $R^\bullet$ is simply given by

$$\Delta H^\circ_f(R^\bullet) = \Delta H^\circ_f(R^+) - \text{IE}\,(R^\bullet) \tag{3.15}$$

where the ionization energy corresponds to the adiabatic transition. For some radicals, such as $^\bullet CH_3$ [43,49], there is minimal geometry change following ionization and the adiabatic transition is the most probable. However, if there is a large difference in geometries, such as for $^\bullet C_2H_5$ where the geometry changes to a non-classical hydrogen-bridged structure [124], the adiabatic onset will be difficult to detect and may not be observed [43,53]. In this case, the experimental IE will be overestimated, resulting in a derived $\Delta H^\circ_f(R^\bullet)$ that is too low. Conversely, hot band structure due to ionization of thermally excited radicals may be erroneously assigned to the adiabatic transition, leading to an underestimation of the IE and a correspondingly high value for $\Delta H^\circ_f(R^\bullet)$. This appears to be the case for the *t*-butyl radical PES [43,125].

In applying equation (3.15) it is necessary to have access to an accurate measurement for $\Delta H^\circ_f(R^+)$. For many even-electron ions these can be obtained from appearance energy measurements, often involving simple bond cleavages. It is preferable for the neutral fragment to be an atomic species because, in the absence of any possible rotational or vibrational excitation, the thermochemistry associated with the reaction is more clearly defined. Although high-energy resolution techniques are often used to measure the appearance energy, the uncertainties in the enthalpies of formation for the neutral precursor and fragment species are often the major source of error in the derived cationic enthalpy of formation. Extensive compilations of evaluated data have been made for most commonly observed ions [5,6].

Apart from being used to obtain cationic enthalpies of formation, the appearance energies for reactions (3.5) and (3.6) can also be used to derive the enthalpy of formation for the radical formed in the unimolecular decomposition by means of equation (3.16).

$$\Delta H^\circ_f(R^\bullet) = \text{AE}\ (X^+) + \Delta H^\circ_f(RX) - \Delta H^\circ_f(X^+) \tag{3.16}$$

The appearance energy for the ion X^+ will only result in an accurate enthalpy of formation if it corresponds to the thermochemical threshold, i.e. provided there is no excess energy due to a kinetic shift, competitive shift or reverse activation energy as discussed earlier. Otherwise, the result obtained will represent an upper limit. This method has been successfully applied to a wide range of free radicals by Holmes and Lossing [117,126–128].

It should be noted that equation (3.16) is only strictly correct at 0 K. To obtain an accurate enthalpy of formation at any other temperature, the internal thermal energy of the dissociating molecule must be allowed for [9]. Because the unimolecular fragmentation occurs in isolation there is no means for the products to thermalize with the surroundings and so they will not be formed at any well-defined equilibrium thermodynamic temperature, i.e. they may be considered to be formed at some 'quasi-temperature' (denoted by * in Figure 3.7) [122]. Thus, to extract useful thermodynamic information from the appearance energy it is necessary to correct the data to a particular temperature T, usually 298 K.

For reaction (3.6) the threshold products are formed with 0 K translational energy with respect to the centre of mass, with essentially 0 K vibrational and rotational energy, and with a centre-of-mass translational energy equivalent to that of the precursor molecule (RX) at the temperature of the experiment. To a good approximation all of the internal

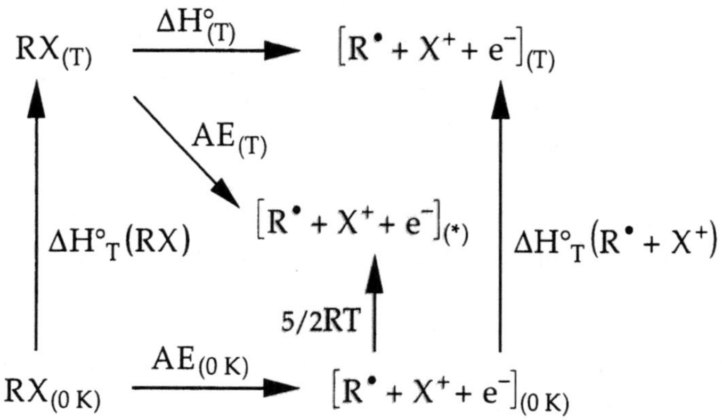

FIGURE 3.7 Thermochemical cycle for unimolecular decomposition of RX.

thermal energy of the precursor is available for dissociation. In this case the correction needed to convert the experimental $AE_{(T)}$ to a true thermodynamic $\Delta H°$ at temperature T is shown to be [122]

$$\Delta H°_T = AE_T + \Delta H_{cor} \tag{3.17}$$

where

$$\Delta H_{cor} = \Delta H°_T\,(R^\bullet) + \Delta H°_T\,(X^+) - \tfrac{5}{2}RT \tag{3.18}$$

The thermochemical cycle for unimolecular decomposition of RX to form $R^\bullet + X^+ + e^-$ is shown in Figure 3.7.

In equation (3.18), $\Delta H°_T(X)$ represents the enthalpy change $H°_T - H°_0$ for the particular ion or neutral and can be readily derived from statistical mechanical calculations [123]. Although the vibrational frequencies needed to calculate $\Delta H°_T(X)$ are frequently unavailable from experimental information, particularly where the ionic fragment is concerned, high-level *ab initio* molecular orbital calculations can usually produce reliable estimates [129]. Depending on the nature of the fragments, the thermal energy correction term, ΔH_{cor}, is typically in the range of 10–30 kJ/mol. Equation (3.16) should therefore be modified to

$$\Delta H°_f(R^\bullet) = AE_T\,(X^+) + \Delta H°_f\,(RX) - \Delta H°_f\,(X^+) + \Delta H_{cor} \tag{3.19}$$

The application of equation (3.19) to electron impact appearance energy data has been questioned by Holmes [80], who has suggested that not all, nor even a major fraction, of the internal thermal energy of the precursor necessarily participates in the activation process. If such a correction term was required then the radical enthalpies of formation derived from a series of monoenergetic electron impact studies would be as much as 20 kJ/mol too high [80]. However, the agreement of such data with accepted values may be the result of the actual threshold being overestimated due to possible kinetic and competitive effects, together with the less favorable threshold ionization probability for electron impact compared to the corresponding photon impact process. The fact that no hot band structure was observed in a similar monoenergetic electron impact study of ionization energies [130], even at high sensitivities, strongly suggests that there will be difficulties with detecting the true threshold in an electron impact fragmentation process.

When the enthalpy of formation of a radical is derived from the appearance energy of an ion, it is important that the correct radical has been identified. It would seem probable that the most likely fragmentation would be a simple bond cleavage, but it is possible that there are alternative reaction pathways in which a different combination of neutrals has been lost but which still results in the same product ion. Several factors need to be considered. The neutral loss should be capable of producing chemically sensible radicals and molecules, in which case the normal mass spectrum should be examined for the presence of the intermediate ion. An examination of the metastable ion mass spectrum is useful as it can confirm parent–daughter relationships [131]. Similarly, collisional activation and collision-induced dissociative ionization experiments can be used to help identify the structure of the radical [79]. Finally, if an alternative pathway is to be thermochemically competitive, the product energies should be lower and any energy barrier for the alternative pathway should not exceed the observed threshold for the desired reaction, i.e. the appearance energy for the competitive process must be lower.

3.3.3 Proton Affinity Measurements

Proton affinity (PA) is defined as the negative of the enthalpy change for the gas-phase protonation reaction

$$M + H^+ \rightarrow MH^+ \tag{3.20}$$

whereas the gas phase basicity (GB) is the negative of the Gibbs energy change for the same reaction [132]. Provided the enthalpies of formation for both M and MH^+ are accurately known, the proton affinity can be calculated directly from the equation

$$PA = -\Delta H°(3.20) = \Delta H°_f(M) + \Delta H°_f(H^+) - \Delta H°_f(MH^+) \tag{3.21}$$

Most commonly $\Delta H^\circ_f(MH^+)$ is derived from the energetics of a unimolecular fragmentation process or from the ionization of a neutral MH precursor. Since these species are often unable to be studied experimentally, absolute proton affinity values are not available for the majority of compounds. In these cases an absolute proton affinity must be assigned by relative reference to a standard whose proton affinity has been accurately determined. There are a number of primary standards used to quantify the thermodynamic proton affinity ladder (eg. isobutene, propene, ethene, CO, CO_2 and O_2), together with several secondary standards whose proton affinities relative to the primary standards are also well known [132]. The evaluation of a large number of relative proton affinity values has resulted in series of overlapping thermodynamic ladders [133]. Although they have been assigned absolute values by comparison with the reference standards, if the value of the standard is redefined then all of the relative values change as well. High-level *ab initio* molecular orbital calculations can also be used to estimate absolute proton affinities, often to an accuracy of only a few kJ/mol [134], providing confirmation of experimentally measured values.

The most common way of measuring proton affinities is via the ion–molecule proton transfer reaction

$$MH^+ + N \rightleftharpoons NH^+ + M \quad (3.22)$$

If the equilibrium constant, K_{eq}, can be accurately determined, the Gibbs energy change may be readily derived using the relationship

$$\Delta G^\circ = -RT \ln K_{eq} \quad (3.23)$$

This method gives the gas phase basicity relative to the species N, and will allow determination of the relative PA using equation (3.24), provided that the entropy change, ΔS°, for reaction (3.22) can also be determined.

$$\Delta G^\circ = \Delta H^\circ - T\,\Delta S^\circ \quad (3.24)$$

Combination of equations (3.23) and (3.24) leads to

$$\ln K_{eq} = \frac{\Delta S^\circ}{R} - \frac{\Delta H^\circ}{RT} \quad (3.25)$$

Thus, if the equilibrium constant can be measured over a range of temperatures, then both ΔH° and ΔS° can be directly obtained from a van't Hoff plot of ln K_{eq} versus T^{-1} [133]. Although the reaction entropy change can be evaluated experimentally in this way, it is commonly estimated by assuming that it may be approximated from changes in rotational symmetry numbers between products and reactants [132]. This simple entropy estimate may however be subject to error when strong intramolecular hydrogen bonding occurs in the ion.

When free radicals are involved in the proton transfer reaction it will be difficult to directly evaluate the equilibrium constant because of the low radical concentrations involved. The proton affinity can then be estimated by 'bracketing' it between two known values [135,136]. This involves reacting the species with a series of bases of known proton affinity. If the proton transfer reaction (3.22) occurs, then the species M is assumed to have a lower proton affinity than the base N. However, if no proton transfer reaction occurs, then M is assumed to have a higher proton affinity than N. The actual value is taken as the average of the proton affinities for the two different bracketing bases involved, with the error being half the difference between them. This means that the smaller the gap between the bracketing bases, the smaller the error in the unknown PA value. Since ΔH° is equal to the difference between PA(M) and PA(N), if proton transfer does occur then the reaction must be exothermic.

This method is less reliable than others because it relies on the assumption that occurrence of the proton transfer reaction depends only on the relative PA values, which is not always the case. For example, an exothermic proton transfer reaction may not be observed if there is another energetically favorable channel available to the reactants. Alternatively, an endothermic proton transfer may occur if there is a negative Gibbs energy change in almost thermoneutral conditions. There may also be other interfering proton transfer reactions taking place to further complicate the thermochemical analysis [132].

Proton affinity bracketing is confined to exothermic or thermoneutral reactions in which the product is able to be detected. The process of 'threshold bracketing' is applicable to both exothermic and endothermic reactions [137]. Ion-molecule reactions between MH^+ and several bases, N, of known proton affinity are performed according to reaction (3.22) and the concentration of the product NH^+ is measured as a function of the ion kinetic energy. A different shaped curve is obtained if the reaction is exothermic or endothermic. It is assumed that the proton affinity of the unknown species, M, will be greater than PA(N) for any endothermic reaction and less than PA(N) for any exothermic reaction, enabling a bracketed PA value to be allocated.

Electron Affinity Measurements

The adiabatic electron affinity corresponds to the minimum energy for the process

$$R^{\bullet} + e^{-} \rightarrow R^{-} \quad (3.26)$$

where both $R^{\bullet}$ and R^{-} are in their ground rovibronic states and the electron has no translational energy, i.e. is at rest. It is defined as the negative of the 0 K enthalpy (or Gibbs energy) change for the electron attachment reaction (3.26).

$$EA(R^{\bullet}) = \Delta H^{\circ}_{f}(R^{\bullet}) - \Delta H^{\circ}_{f}(R^{-}) \quad (3.27)$$

Care must be taken to correctly specify the temperatures relating to the thermochemical terms in equation (3.27). If the enthalpies of formation are expressed at 298 K then the electron affinity will also be a 298 K value, although this will be a reasonable approximation to the 0 K value for a small geometry difference between the radical and anion [6].

Various techniques can be used to measure the electron affinity for radical species [68]. A number of widely used methods involve the study of gas-phase ion–molecule equilibria by either HPMS, ICR or flowing afterglow techniques [138]. Relative electron affinities determined in this way will generally represent an adiabatic value because there is sufficient time during the collisions for the reactants to achieve their equilibrium geometries. Thus, the usual Franck–Condon restrictions associated with other spectroscopic techniques, such as photodetachment and appearance energy measurements, will not apply.

Consider the following equilibrium involving proton transfer to R^{-}.

$$R^{-} + AH \rightleftharpoons A^{-} + RH \quad (3.28)$$

By studying the energetics of reaction (3.28) for a series of acids AH, it is possible to use bracketing to obtain $PA(R^{-})$, which represents the enthalpy change for reaction (3.29).

$$RH \rightarrow R^{-} + H^{+} \quad (3.29)$$

Similarly, using proton-transfer bracketing reactions involving different bases B and the protonated radical,

$$RH^{+\bullet} + B \rightarrow BH^{+} + R^{\bullet} \quad (3.30)$$

enables the proton affinity for the radical $R^{\bullet}$ to be derived, which corresponds to the enthalpy change for reaction (3.31).

$$RH^{+\bullet} \rightarrow R^{\bullet} + H^{+} \tag{3.31}$$

Now, the ionization energy for RH is given by the enthalpy change for reaction (3.32),

$$RH \rightarrow RH^{+\bullet} + e^{-} \tag{3.32}$$

so that the electron affinity for the radical $R^{\bullet}$ can be obtained from equation (3.33).

$$\begin{aligned} EA(R^{\bullet}) &= -\Delta H°(3.26) = \Delta H°(3.32) + \Delta H°(3.31) - \Delta H°(3.29) \\ &= IE(RH) + PA(R^{\bullet}) - PA(R^{-}) \end{aligned} \tag{3.33}$$

The relationships between the various quantities in equation (3.33) are shown in the thermochemical cycle of Figure 3.8. Thus, provided IE(RH) is available, $EA(R^{\bullet})$ can be readily derived. Again it is important that all enthalpy changes are corrected to 0 K values. For example, this particular method has been used with ICR measurements to obtain the electron affinity for the amidogen radical [135].

The gas-phase acidity, $\Delta G°_{acid}$, for a molecule RH is given by the Gibbs energy change for reaction (3.29). At 0 K the enthalpy change for this reaction, $\Delta H°_{acid}$, which is simply the proton affinity for the anion, is related to $EA(R^{\bullet})$ by equation (3.34). The thermochemical cycle is shown in Figure 3.9. Thus, homolytic bond strengths, BDE(R—H), can be derived from electron affinities and gas-phase acidities [7]. Extensive interlocking ladders of relative gas-phase acidities have been obtained from various thermochemical studies of ion–molecule reactions [6,139,140].

FIGURE 3.8 Thermochemical cycle for $EA(R^{\bullet})$.

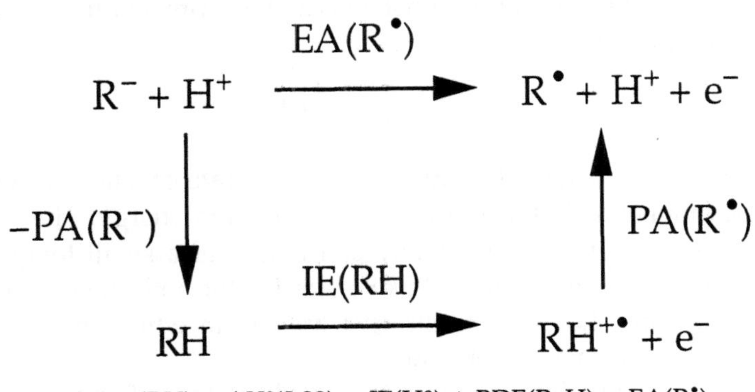

$$\Delta H°_{acid}(RH) = \Delta H°(3.29) = IE(H°) + BDE(R\text{–}H) - EA(R^{\bullet}) \tag{3.34}$$

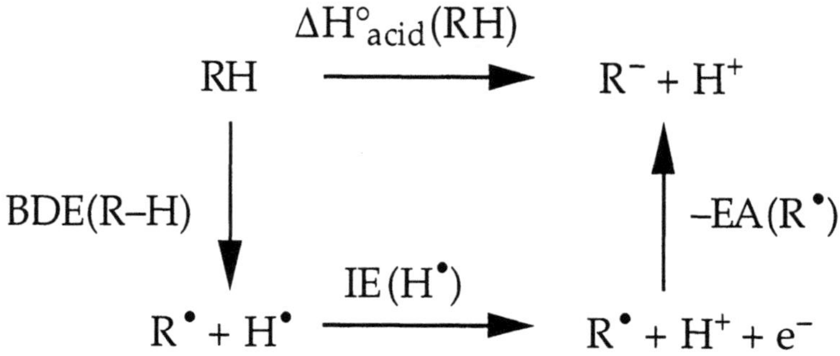

FIGURE 3.9 Thermochemical cycle for $\Delta H°_{acid}$(RH)

The gas-phase acidities for several C_2–C_5 alkanes have only been recently measured for the first time by DePuy and coworkers [141].

Kinetic Measurements

For the bimolecular reaction

$$R^{\bullet} + AB \underset{k_r}{\overset{k_f}{\rightleftharpoons}} RA + B^{\bullet} \tag{3.35}$$

the equilibrium constant is given by the ratio of the rate constants for the forward and reverse reactions [119].

$$K_{eq} = \frac{k_f}{k_r} \tag{3.36}$$

Thus, in combination with equation (3.23), it is possible to use the kinetics for reaction (3.35) to obtain the Gibbs energy change.

Most chemical reaction rate constants can be expressed in terms of the Arrhenius equation.

$$k = A\, e_q - E_a/RT \tag{3.37}$$

This describes the effect of temperature on the rate constant, where A is the pre-exponential factor and E_a is the activation energy. The A-factor and activation energy are generally constant over a small temperature range (~100 K), so that plots of ln k versus T^{-1} for both the forward and reverse reactions should yield straight lines from which the respective activation energies may be obtained.

The reaction enthalpy change corresponds to the difference between the two activation energies [119], i.e.

$$\Delta H^\circ = E_f - E_r \tag{3.38}$$

where the reaction temperature is usually selected as the midpoint of the temperature range used for the Arrhenius plots. The reaction entropy change at this temperature may then be obtained from equation (3.24). It is necessary to use heat capacity data for the reactants and products [119] to convert ΔH° and ΔS° to their corresponding values at 298 K. This method of using variable-temperature experiments to obtain enthalpies of reaction is known as a 'second-law' calculation [7,123].

A more reliable method for estimating reaction enthalpies, which is based on a knowledge of the absolute entropy of the reactants and products, is the 'third-law' method [7,123]. It is particularly useful when experimental data is only available at a single temperature. The Gibbs energy change of reaction may be corrected to a 298 K value from a knowledge of the reactant and product heat capacities and their temperature dependencies [119]. Thus, ΔH°_{298} can be calculated using equation (3.24), where ΔS°_{298} is obtained from the known entropies for the reactants and products. A combination of both second- and third-law calculations is particularly useful as it can indicate whether the experimental data and thermochemical values used in the calculations are mutually consistent [7,123].

ENTHALPIES OF FORMATION FOR SELECTED RADICALS

Although thermochemical information for free radicals has been obtained from a wide range of different ion studies, the following discussion is restricted to photoionization measurements involving cation formation. The data presented is not intended to be a comprehensive review, but rather to give an indication of how they compare with information generated from reactions involving only neutral species.

To facilitate comparisons all data have been corrected to a temperature of 298 K. This necessarily involves the use of $H^\circ_{298} - H^\circ_0$ values, summarized in Table 3.1. Where experimental frequencies are unavailable, which is the case for most cations, high-level *ab initio* calculations have been used to derive the respective values. Appearance energies obtained at 298 K have been converted to enthalpies of reaction at 298 K by using equation (3.17). The 298 K enthalpies of formation for neutral species involved in appearance energy studies are given in Table 3.2. Where

TABLE 3.1 H°_{298}–H°_{0} values (kJ/mol) for selected species in the gas phase.

Species	$H^{\circ}_{298}-H^{\circ}_{0}$	Reference
CHO	10.0	123
CHO^+	9.0	134
CH_3	10.4	123
CH_3^+	10.0	122
CH_2OH	11.3	186
CH_2OH^+	10.2	134
CH_4	10.0	123
CH_3OH	11.4	195
CH_3CO	12.4	192
CH_3CO^+	12.0	134
C_2H_5	13.0	196
$C_2H_5^+$	11.3	134
C_2H_5I	13.9	152
C_3H_5	12.2	169
$C_3H_5^+$	11.3	169
i-C_3H_7	15.7	197
i-$C_3H_7^+$	15.5	134
i-C_3H_7Br	17.2	152
i-C_3H_7I	17.2	152
C_3H_8	14.7	198
t-C_4H_9	19.5	199
t-$C_4H_9^+$	19.5	134
t-C_4H_9I	20.9	200
OH	9.2	123
Monatomic gas	6.2	123

necessary, an energy conversion factor of 1 eV = 96.4846 kJ/mol has been applied to the calculations [123].

Methyl

The methyl radical has been widely studied by a variety of techniques, with most experimentally determined enthalpies of formation agreeing within ± 1 kJ/mol. McMillen and Golden [4] recommended a value of 146.9 ± 0.6 kJ/mol, based on a VLPP study of the reaction between $^{\bullet}Cl$ and CH_4 [115]. This has been supported by the kinetic experiments of Gutman and coworkers for the reactions of $^{\bullet}CH_3$ + HI (146 ± 1 kJ/mol) [107], $^{\bullet}CH_3$ + HBr (148 ± 3 kJ/mol) [142], and $^{\bullet}CH_3$ + HCl (145.6 ± 1.3 kJ/mol) [143]. A kinetic study of the pyrolysis of ethane yielded a value of $\Delta H^{\circ}_{f,298}(^{\bullet}CH_3)$ = 147 ± 2 kJ/mol [144].

TABLE 3.2 Enthalpies of formation (kJ/mol) for selected species in the gas phase.

Species	$\Delta H^\circ_{f,298}$	Reference
Br	111.86 ± 0.06	123
HCHO	− 108.6 ± 0.5	118
HCOOH	− 378.7 ± 0.6	118
CH_3	145.7 ± 0.8	123
CH_3Br	− 35.5 ± 1.1	118
CH_3I	14.7 ± 1.3	118
CH_4	− 74.4 ± 0.4	118
CH_3OH	− 201.5 ± 0.3	118
CO	− 110.53 ± 0.17	123
CH_2CO	− 47.5 ± 1.6	118
C_2H_4	52.5 ± 0.4	118
CH_3CHO	− 166.1 ± 0.5	118
$CH_3(CO)OH$	− 432.8 ± 1.5	118
$H(CO)OCH_3$	− 355.5 ± 0.8	118
C_2H_5	120.2 ± 0.9	150
C_2H_5Br	− 61.9 ± 1.7	118
C_2H_5Cl	− 112.1 ± 1.1	118
C_2H_5I	− 7.5 ± 1.7	118
C_2H_6	− 83.8 ± 0.4	118
C_3H_5Br	43.5 ± 2.7	118[a]
C_3H_6	20.0 ± 0.8	118
cyclo-C_3H_6	53.3 ± 0.6	118
$CH_3(CO)CH_3$	− 217.3 ± 0.7	118
C_3H_8	− 104.7 ± 0.5	118
i-C_3H_7Br	− 99.4 ± 2.5	118
i-C_3H_7Cl	− 144.9 ± 1.3	118
i-C_3H_7I	− 40.3 ± 3.8	118
$CH_3(CO)_2CH_3$	− 327.1 ± 1.3	118
1-C_4H_8	0.1 ± 1.0	118
(E)-2-C_4H_8	− 11.4 ± 1.0	118
(Z)-2-C_4H_8	− 7.1 ± 1.0	118
i-C_4H_8	− 16.9 ± 0.9	118
$CH_3(CO)C_2H_5$	− 238.7 ± 0.8	118
t-C_4H_9I	− 72.0 ± 3.3	118
i-C_4H_{10}	− 134.2 ± 0.7	118
neo-C_5H_{12}	− 168.1 ± 0.8	118
CO	− 110.53 ± 0.17	123
Cl	121.302 ± 0.008	123
H	217.999 ± 0.006	123
H^+	1530.05 ± 0.04	123
OH	39.3 ± 1.2	123[b]
I	106.76 ± 0.04	123

[a] Empirically estimated from $\Delta H^\circ_{f,298}$(allyl X) − $\Delta H^\circ_{f,298}$(ethyl X) = 105.4 ± 2.1 kJ/mol [120].
[b] Obtained from the recommended 0 K value of 9.261 ± 0.29 kcal mol^{-1} = 38.7 ± 1.2 kJ/mol [123].

Photoelectron spectra show that the adiabatic ionization energy corresponds to a vertical transition so that this will be well defined. Values of 9.840 ± 0.005 eV [49], and 9.84 ± 0.02 eV [43] have been obtained from studies of $^{\bullet}CH_3$ produced by pyrolysis of azomethane, both in excellent agreement with the value of 9.843 ± 0.001 eV (949.7 ± 0.1 kJ/mol) derived from vacuum ultraviolet spectroscopic measurements [145]. The corresponding 298 K ionization energy then becomes 949.7 + (10.0 − 10.4) = 949.3 ± 0.1 kJ/mol.

Several ionization and fragmentation processes have been used to obtain the enthalpy of formation for CH_3^+.

(a) $$CH_4 + h\nu \rightarrow CH_3^+ + H^{\bullet} + e^- \qquad (3.39)$$

$$\begin{aligned} \text{AE (0 K)} &= 14.324 \pm 0.003 \text{ eV [146]} = 1382.0 \pm 0.3 \text{ kJ/mol} \\ \Delta H^{\circ}_{298}(3.39) &= 1382.0 + (10.0 + 6.2 - 10.0) = 1388.2 \pm 0.3 \text{ kJ/mol} \\ \Delta H^{\circ}_{f,298}(CH_3^+) &= 1388.2 - 218.0 + (-74.4) = \underline{1095.8 \pm 0.5} \text{ kJ/mol} \end{aligned}$$

(b) $$CH_3Br + h\nu \rightarrow CH_3^+ + Br^{\bullet} + e^- \qquad (3.40)$$

$$\begin{aligned} \text{AE (298 K)} &= 12.77 \pm 0.01 \text{ eV [122]} = 1232.1 \pm 1.0 \text{ kJ/mol} \\ \Delta H^{\circ}_{298}(3.40) &= 1232.1 + (10.0 + 6.2 - 6.2) = 1242.1 \pm 1.0 \text{ kJ/mol} \\ \Delta H^{\circ}_{f,298}(CH_3^+) &= 1242.1 - 111.9 + (-35.5) = \underline{1094.7 \pm 1.5} \text{ kJ/mol} \end{aligned}$$

(c) $$CH_3I + h\nu \rightarrow CH_3^+ + I^{\bullet} + e^- \qquad (3.41)$$

$$\begin{aligned} \text{AE (298 K)} &= 12.18 \pm 0.01 \text{ eV [122]} = 1175.2 \pm 1.0 \text{ kJ/mol} \\ \Delta H^{\circ}_{298}(3.41) &= 1175.2 + (10.0 + 6.2 - 6.2) = 1185.2 \pm 1.0 \text{ kJ/mol} \\ \Delta H^{\circ}_{f,298}(CH_3^+) &= 1185.2 - 106.8 + 14.7 = \underline{1093.1 \pm 1.6} \text{ kJ/mol} \end{aligned}$$

Taking the average of these three values gives $\Delta H^{\circ}_{f,298}(CH_3^+)$ = 1094.5 ± 1.4 kJ/mol, which, when combined with the 298 K ionization energy for $^{\bullet}CH_3$, leads to $\Delta H^{\circ}_{f,298}(^{\bullet}CH_3)$ = 145.2 ± 1.4 kJ/mol, in good agreement with the above kinetic data.

Ethyl

There has been substantial change to the enthalpy of formation for the ethyl radical over the past decade, although current experimental data seem to be essentially in accord. For many years the accepted value was 108.4 ± 4.2 kJ/mol [147], which was also that recommended by McMillen and Golden [4]. However, this was challenged by Tsang [148] who proposed a higher value of 117 − 121 kJ/mol. A subsequent review of kinetic data by Cao and Back [149] supported the upward revision and suggested that 117.2 kJ/mol was an appropriate value; this particular value was also adopted by Lias *et al.* [6] in their comprehensive evaluation of gas-phase ion and neutral thermochemistry. The value obtained by Gutman and coworkers [110] from a kinetic study of both the reactions of $^{\bullet}C_2H_5$ + HBr and $^{\bullet}Br + C_2H_6$ was 121.0 ± 1.5 kJ/mol. More recently, Pilling and coworkers [150] have obtained a value of 120.2 ± 0.9 kJ/mol based on the $^{\bullet}H + C_2H_4$ system. They also reviewed other kinetic data, from which they reported a weighted mean of 119 ± 3 kJ/mol for $\Delta H^{\circ}_{f,298}(^{\bullet}C_2H_5)$ [150].

The adiabatic and vertical ionization energies for the ethyl radical are not the same [43,53]. This is due to the small Franck–Condon factors associated with the transition from neutral $^{\bullet}C_2H_5$ to the lowest energy state of $C_2H_5^+$, which *ab initio* calculations show has a nonclassical H-bridged structure [124]. As a result there is some doubt about the actual adiabatic ionization energy, with values of 8.39 ± 0.02 eV [43] and 8.26 ± 0.02 eV [53] being assigned from photoelectron spectra. A detailed theoretical and experimental study of the ionization energy by photoionization mass spectrometry indicated that both these estimates were too high and proposed a value of 8.117 ± 0.008 eV [151]. This converts to 783.2 + (11.3 − 13.0) = 781.5 ± 0.8 kJ/mol at 298 K, using enthalpy correction data from Table 3.1.

Appearance energy measurements for four different fragmentation processes can be used to derive the enthalpy of formation for $C_2H_5^+$.

(a)
$$C_2H_6 + h\nu \rightarrow C_2H_5^+ + H^{\bullet} + e^- \tag{3.42}$$

$$\begin{aligned}
\text{AE (298 K)} &= 12.40 \pm 0.03 \text{ eV [122]} = 1196.4 \pm 2.9 \text{ kJ/mol}\\
\Delta H^{\circ}_{298}(3.42) &= 1196.4 + (11.3 + 6.2 - 6.2) = 1207.7 \pm 2.9 \text{ kJ/mol}\\
\Delta H^{\circ}_{f,298}(C_2H_5^+) &= 1207.7 - 218.0 + (-83.8) = \underline{905.9 \pm 2.9} \text{ kJ/mol}
\end{aligned}$$

(b)
$$C_2H_5Cl + h\nu \rightarrow C_2H_5^+ + Cl^{\bullet} + e^- \tag{3.43}$$

$$\begin{aligned}
\text{AE (298 K)} &= 11.67 \pm 0.02 \text{ eV [122]} = 1126.0 \pm 1.9 \text{ kJ/mol}\\
\Delta H^{\circ}_{298}(3.43) &= 1126.0 + (11.3 + 6.2 - 6.2) = 1137.3 \pm 1.9 \text{ kJ/mol}\\
\Delta H^{\circ}_{f,298}(C_2H_5^+) &= 1137.3 - 121.3 + (-112.1) = \underline{903.9 \pm 2.2} \text{ kJ/mol}
\end{aligned}$$

(c)
$$C_2H_5Br + h\nu \rightarrow C_2H_5^+ + Br^{\bullet} + e^- \tag{3.44}$$

$$\begin{aligned}
\text{AE (298 K)} &= 11.06 \pm 0.01 \text{ eV [122]} = 1067.1 \pm 1.0 \text{ kJ/mol}\\
\Delta H^{\circ}_{298}(3.44) &= 1067.1 + (11.3 + 6.2 - 6.2) = 1078.4 \pm 1.0 \text{ kJ/mol}\\
\Delta H^{\circ}_{f,298}(C_2H_5^+) &= 1078.4 - 111.9 + (-61.9) = \underline{904.6 \pm 2.0} \text{ kJ/mol}
\end{aligned}$$

(d)
$$C_2H_5I + h\nu \rightarrow C_2H_5^+ + I^{\bullet} + e^- \tag{3.45}$$

$$\begin{aligned}
\text{AE (0 K)} &= 10.52 \pm 0.01 \text{ eV [152]} = 1015.0 \pm 1.0 \text{ kJ/mol}\\
\Delta H^{\circ}_{298}(3.45) &= 1015.0 + (11.3 + 6.2 - 13.9) = 1018.6 \pm 1.0 \text{ kJ/mol}\\
\text{AE (298 K)} &= 10.44 \pm 0.01 \text{ eV [122]} = 1007.3 \pm 1.0 \text{ kJ/mol}\\
\Delta H^{\circ}_{298}(3.45) &= 1007.3 + (11.3 + 6.2 - 6.2) = 1018.6 \pm 1.0 \text{ kJ/mol}\\
\Delta H^{\circ}_{f,298}(C_2H_5^+) &= 1018.6 - 106.8 + (-7.5) = \underline{904.3 \pm 2.0} \text{ kJ/mol}
\end{aligned}$$

The average of these leads to $\Delta H^\circ_{f,298}(C_2H_5^+) = 904.7 \pm 0.9$ kJ/mol, which corresponds to $\Delta H^\circ_{f,298}(^\bullet C_2H_5) = 123.2 \pm 1.2$ kJ/mol when combined with the 298 K ionization energy of 781.5 ± 0.8 kJ/mol. The small discrepancy of 3 to 4 kJ/mol with the above kinetically derived values could be due to an underestimated ionization energy and/or an overestimated cationic enthalpy of formation. There is some evidence for the latter following recent theoretical [134] and experimental [133] measurements for the proton affinity of ethene, from which respective values of 681.9 kJ/mol and 680.3 kJ/mol were obtained. The corresponding proton affinity calculated using the above $\Delta H^\circ_{f,298}(C_2H_5^+)$ and data from Table 3.2 is $PA(C_2H_4) = 1530.1 + 52.5 - 904.7 = 677.9 \pm 1.0$ kJ/mol, which would be in better agreement if a slightly lower value for the cationic enthalpy of formation was used. Further work is required to investigate whether there is $\cong 3$ kJ/mol excess energy associated with the fragmentation products in each of the reactions (3.42)–(3.45).

Isopropyl

As for the ethyl radical, the McMillen and Golden [4] recommended value for the enthalpy of formation for the isopropyl radical was also based on an earlier review by Golden and Benson [147]. Again, Tsang [148] proposed that this should be increased from 76.1 ± 4.2 kJ/mol to 81–88 kJ/mol. His subsequent studies further increased this to 93.3 ± 2.5 kJ/mol [153] although this was later modified to 87.9 ± 2.1 kJ/mol by Tschuikow-Roux and Chen [154]. From a study of the reaction kinetics for isopropyl with HBr, Gutman and coworkers [142] obtained a value of 88 ± 3 kJ/mol. They also reworked the original experimental data cited by Golden and Benson [147] to derive a value of 88–93 kJ/mol in support of their measurements. A more recent comprehensive study of the isopropyl/HBr system has produced a value of 90.0 ± 1.7 kJ/mol for $\Delta H^\circ_{f,298}$(isopropyl) [110].

Photoelectron spectra of the isopropyl radical are in essential agreement concerning the adiabatic ionization energy assignment. From the pyrolysis of 2-methylpropyl nitrite Houle and Beauchamp [43] obtained 7.36 ± 0.02 eV, whereas Dyke *et al.* [155] measured it to be 7.37 ± 0.02 eV using both azoisopropane pyrolysis and the reaction between $F^\bullet$ and propane as the isopropyl radical source. Although there are appreciable changes in the CCC and CCH bond angles following ionization, producing a 0.3 eV higher vertical ionization energy, the adiabatic transition appears to have been observed because lower energy structure in the photoelectron spectrum could be identified as hot bands [155]. Taking the adiabatic ionization energy as 7.365 eV gives a corresponding 298 K ionization energy of $710.6 + (15.5 - 15.7) = 710.4 \pm 1.9$ kJ/mol using data from Table 3.1.

Appearance energies for the following gas-phase reactions can be used to obtain a measure of the enthalpy of formation for the isopropyl cation.

(a) $$C_3H_8 + h\nu \rightarrow i\text{-}C_3H_7^+ + H^\bullet + e^- \quad (3.46)$$

$$\begin{aligned} \text{AE (0 K)} &= 11.59 \pm 0.01 \text{ eV [156]} = 1118.3 \pm 1.0 \text{ kJ/mol} \\ \Delta H^\circ_{298}(3.46) &= 1118.3 + (15.5 + 6.2 - 14.7) = 1125.3 \pm 1.0 \text{ kJ/mol} \\ \Delta H^\circ_{f,298}(i\text{-}C_3H_7^+) &= 1125.3 - 218.0 + (-104.7) = \underline{802.6 \pm 1.1} \text{ kJ/mol} \end{aligned}$$

(b) $$i\text{-}C_3H_7Cl + h\nu \rightarrow i\text{-}C_3H_7^+ + Cl^\bullet + e^- \quad (3.47)$$

$$\begin{aligned} \text{AE (298 K)} &= 10.92 \pm 0.01 \text{ eV [157]} = 1053.6 \pm 1.0 \text{ kJ/mol} \\ \Delta H^\circ_{298}(3.47) &= 1053.6 + (15.5 + 6.2 - 6.2) = 1069.1 \pm 1.0 \text{ kJ/mol} \\ \Delta H^\circ_{f,298}(i\text{-}C_3H_7^+) &= 1069.1 - 121.3 + (-144.9) = \underline{802.9 \pm 1.6} \text{ kJ/mol} \end{aligned}$$

(c) $$i\text{-}C_3H_7Br + h\nu \rightarrow i\text{-}C_3H_7^+ + Br^\bullet + e^- \quad (3.48)$$

$$\begin{aligned} \text{AE (0 K)} &= 10.42 \pm 0.01 \text{ [152]} = 1005.4 \pm 1.0 \text{ kJ/mol} \\ \Delta H^\circ_{298}(3.48) &= 1005.4 + (15.5 + 6.2 - 17.2) = 1009.9 \pm 1.0 \text{ kJ/mol} \\ \text{AE (298 K)} &= 10.33 \pm 0.01 \text{ eV [157]} = 996.7 \pm 1.0 \text{ kJ/mol} \\ \Delta H^\circ_{298}(3.48) &= 996.7 + (15.5 + 6.2 - 6.2) = 1012.2 \pm 1.0 \text{ kJ/mol} \\ \text{Average } \Delta H^\circ_{298}(3.48) &= 1011.1 \pm 1.2 \text{ kJ/mol} \\ \Delta H^\circ_{f,298}(i\text{-}C_3H_7^+) &= 1011.1 - 111.9 + (-99.4) = \underline{799.8 \pm 2.8} \text{ kJ/mol} \end{aligned}$$

(d) $$i\text{-}C_3H_7I + h\nu \rightarrow i\text{-}C_3H_7^+ + I^\bullet + e^- \quad (3.49)$$

$$\begin{aligned} \text{AE (0 K)} &= 9.77 \pm 0.02 \text{ eV [152]} = 942.7 \pm 1.9 \text{ kJ/mol} \\ \Delta H^\circ_{298}(3.49) &= 942.7 + (15.5 + 6.2 - 17.2) = 947.2 \pm 1.9 \text{ kJ/mol} \\ \text{AE (298 K)} &= 9.70 \pm 0.01 \text{ eV [157]} = 935.9 \pm 1.0 \text{ kJ/mol} \\ \Delta H^\circ_{298}(3.49) &= 935.9 + (15.5 + 6.2 - 6.2) = 951.4 \pm 1.0 \text{ kJ/mol} \\ \text{Average } \Delta H^\circ_{298}(3.49) &= 949.3 \pm 2.1 \text{ kJ/mol} \\ \Delta H^\circ_{f,298}(i\text{-}C_3H_7^+) &= 949.3 - 106.8 + (-40.3) = \underline{802.2 \pm 4.3} \text{ kJ/mol} \end{aligned}$$

The average of these determinations gives $\Delta H^\circ_{f,298}(i\text{-}C_3H_7^+) = 801.9 \pm 1.4$ kJ/mol, which when combined with IE_{298}(isopropyl) $= 710.4 \pm 1.9$ kJ/mol, gives a 298 K enthalpy of formation for the isopropyl radical of 91.5 ± 2.4 kJ/mol, in good agreement with the above kinetically determined values. Although the proton affinity for propene, calculated here to be $1530.1 + 20.0 - 801.9 = 748.2 \pm 1.6$ kJ/mol, is slightly higher than recent theoretical and experimental values [133,134], any lowering of the present proton affinity would only lead to a higher value for the cationic enthalpy of formation and hence that of the free radical. Similarly, a lower adiabatic ionization energy would also lead to a higher value for the isopropyl enthalpy of formation.

t-Butyl

The enthalpy of formation for the *t*-butyl radical continues to be the center of both theoretical and experimental investigations with recent determinations varying by as much as 12 kJ/mol [158]. McMillen and Golden [4] recommended a value of 36.4 ± 4.2 kJ/mol. However, the kinetic data of Tsang [148] implied that this was too low and that it should be in the range of 46–54 kJ/mol. A later measurement by Tsang [153] produced a value of 51.7 ± 2.2 kJ/mol. Although different VLPR experiments were in accord [114,159] they consistently gave lower values, with the most recent being 38.5 ± 2.1 kJ/mol [114]. In a comprehensive study of the kinetics for both of the reactions $t\text{-}C_4H_9$ + HBr and $i\text{-}C_4H_{10}$ + Br, the research groups of Pilling and Gutman [110] obtained a value of 51.3 ± 1.8 kJ/mol, supporting the Tsang data. In recognition of this unresolved problem, Cohen [160] has proposed a 'fence-straddling' value of 46.0 ± 6.3 kJ/mol.

The adiabatic ionization energy for the *t*-butyl radical has been measured by photoelectron spectroscopy. Dyke *et al.* [161] used the pyrolysis of 2,2′-azoisobutane to obtain a value of 6.58 ± 0.01 eV whereas Houle and Beauchamp [43] identified the adiabatic transition at 6.70 ± 0.03 eV, based on the lower temperature pyrolysis of 2,2-dimethylpropylnitrite. This latter result was also in good agreement with an earlier study by Koenig *et al.* [162]. It was suggested that interfering hot band structure was the reason for the lower assignment [43]. If an adiabatic value of 6.70 eV is assumed, the 298 K ionization energy becomes 646.4 + (19.5 − 19.5) = 646.4 kJ/mol.

The choice of ionization and fragmentation processes for measurement of $\Delta H°_{f,298}(t\text{-}C_4H_9^+)$ is rather limited. Although the *t*-butyl cation is often observed in mass spectra, often there is no molecular ion, indicating that any appearance energy will simply reflect the ionization energy for the precursor molecule, i.e. the thermochemical threshold occurs at a lower energy [163]. There are however three processes for which this does not appear to be the case.

(a)
$$i\text{-}C_4H_{10} + h\nu \rightarrow t\text{-}C_4H_9^+ + H^\bullet + e^- \qquad (3.50)$$

$$\begin{aligned} \text{AE (298 K)} &= 10.821 \pm 0.008\ \text{eV [125]} = 1044.1 \pm 0.8\ \text{kJ/mol} \\ \Delta H°_{298}(3.50) &= 1044.1 + (19.5 + 6.2 - 6.2) = 1063.6 \pm 0.8\ \text{kJ/mol} \\ \Delta H°_{f,298}(t\text{-}C_4H_9^+) &= 1063.6 - 218.0 + (-134.2) = \underline{711.4 \pm 1.1}\ \text{kJ/mol} \end{aligned}$$

(b)
$$t\text{-}C_4H_9I + h\nu \rightarrow t\text{-}C_4H_9^+ + I^\bullet + e^- \qquad (3.51)$$

$$\begin{aligned} \text{AE (0 K)} &= 9.180 \pm 0.015\ \text{eV [164]} = 885.7 \pm 1.4\ \text{kJ/mol} \\ \Delta H°_{298}(3.51) &= 885.7 + (19.5 + 6.2 - 20.9) = 890.5 \pm 1.4\ \text{kJ/mol} \\ \Delta H°_{f,298}(t\text{-}C_4H_9^+) &= 890.5 - 106.8 + (-72.0) = \underline{711.7 \pm 3.6}\ \text{kJ/mol} \end{aligned}$$

(c) $$neo\text{-}C_5H_{12} + h\nu \rightarrow t\text{-}C_4H_9^+ + {}^\bullet CH_3 + e^- \qquad (3.52)$$

$$\begin{aligned} \text{AE (298 K)} &= 10.38 \pm 0.02\ \text{eV [125]} = 1001.5 \pm 1.9\ \text{kJ/mol} \\ \Delta H^\circ_{298}(3.52) &= 1001.5 + (19.5 + 10.4 - 6.2) = 1025.2 \pm 1.9\ \text{kJ/mol} \\ \Delta H^\circ_{f,298}(t\text{-}C_4H_9^+) &= 1025.2 - 145.7 + (-168.1) = \underline{711.4 \pm 2.2}\ \text{kJ/mol} \end{aligned}$$

The average of these three values is $\Delta H^\circ_{f,298}(t\text{-}C_4H_9^+) =$ 711.5 ± 0.2 kJ/mol, from which a proton affinity for isobutene of 1530.1 + (−16.9) − 711.5 = 801.7 ± 0.9 kJ/mol may be derived. This is in excellent agreement with the value of 802.1 kJ/mol obtained by both experimental [133] and theoretical studies [134]. The enthalpy of formation for the *t*-butyl radical is then calculated to be 711.5 − 646.4 = 65.1 ± 2.9 kJ/mol, that is significantly greater than the highest values obtained via kinetic studies. Thus, it would appear that even the higher adiabatic ionization energy of 6.70 eV is too low, presumably as a result of hot band interference. Additional experiments are required to resolve this discrepancy.

Allyl

There is much less controversy concerning the enthalpy of formation for the allyl radical, although the actual value is still not well defined. Following VLPP studies involving the recombination of allyl radicals to form 1,5-hexadiene, and the reaction of $C_3H_5^\bullet$ with HI to form $C_3H_6 + I^\bullet$, Golden and coworkers obtained values of 163.6 ± 6.3 kJ/mol [165], and 164.8 ± 6.3 kJ/mol [166], respectively. The former result was that recommended in the 1982 review of Golden and McMillen [4]. In a subsequent shock tube study of allyl bromide, Tsang [167] obtained a value of 173.6 kJ/mol, although this was based on an enthalpy of formation for allyl bromide of 49.4 kJ/mol [168]. As discussed by Traeger [169], there is considerable doubt surrounding the experimentally measured values for all of the allyl halides. Using a value of $\Delta H^\circ_{f,298}$ (allyl bromide) = 43.5 kJ/mol, empirically estimated from data for ethyl bromide [120], would lower Tsang's [167] enthalpy of formation for the allyl radical to 167.7 kJ/mol, in much better agreement with the previous results. More recently, a shock tube study of 1,5-hexadiene produced a value of 166.9 ± 2.1 kJ/mol using both second-law and third-law calculations [170].

The photoelectron spectrum for the allyl radical shows that the adiabatic and vertical ionization energies are the same [42], i.e. there is minimal geometry change following ionization. The value of 8.13 ± 0.02 eV obtained from the pyrolysis of 3-butenyl nitrite converts to a 298 K value of 784.4 + (11.3 − 12.2) = 783.5 ± 1.9 kJ/mol.

There are five different photoionization fragmentation processes that can be used to derive a reliable estimate for $\Delta H^{\circ}_{f,298}(C_3H_5^+)$. Although the $C_3H_5^+$ appearance energies for the allyl halides are well defined, they have not been used here because of the uncertainty associated with the enthalpies of formation for the neutral precursors. The AE for formation of $C_3H_5^+$ from propene has also been excluded as it appears to be affected by a kinetic shift [169].

(a)
$$\text{cyclo-}C_3H_6 + h\nu \rightarrow C_3H_5^+ + H^{\bullet} + e^- \qquad (3.53)$$

$$\begin{aligned} \text{AE (298 K)} &= 11.43 \pm 0.01 \text{ eV [169]} = 1102.8 \pm 1.0 \text{ kJ/mol} \\ \Delta H^{\circ}_{298}(3.53) &= 1102.8 + (11.3 + 6.2 - 6.2) = 1114.1 - 1.0 \text{ kJ/mol} \\ \Delta H^{\circ}_{f,298}(C_3H_5^+) &= 1114.1 - 218.0 + 53.3 = \underline{949.4 \pm 1.2} \text{ kJ/mol} \end{aligned}$$

(b)
$$1\text{-}C_4H_8 + h\nu \rightarrow C_3H_5^+ + {}^{\bullet}CH_3 + e^- \qquad (3.54)$$

$$\begin{aligned} \text{AE (298 K)} &= 11.20 \pm 0.02 \text{ eV [169]} = 1080.6 \pm 1.9 \text{ kJ/mol} \\ \Delta H^{\circ}_{298}(3.54) &= 1080.6 + (11.3 + 10.4 - 6.2) = 1096.1 \pm 1.9 \text{ kJ/mol} \\ \Delta H^{\circ}_{f,298}(C_3H_5^+) &= 1096.1 - 145.7 + 0.1 = \underline{950.5 \pm 2.3} \text{ kJ/mol} \end{aligned}$$

(c)
$$(E)\text{-}2\text{-}C_4H_8 + h\nu \rightarrow C_3H_5^+ + {}^{\bullet}CH_3 + e^- \qquad (3.55)$$

$$\begin{aligned} \text{AE (298 K)} &= 11.30 \pm 0.02 \text{ eV [169]} = 1090.3 \pm 1.9 \text{ kJ/mol} \\ \Delta H^{\circ}_{298}(3.55) &= 1090.3 + (11.3 + 10.4 - 6.2) = 1105.8 \pm 1.9 \text{ kJ/mol} \\ \Delta H^{\circ}_{f,298}(C_3H_5^+) &= 1105.8 - 145.7 + (-11.4) = \underline{948.7 \pm 2.3} \text{ kJ/mol} \end{aligned}$$

(d)
$$(Z)\text{-}2\text{-}C_4H_8 + h\nu \rightarrow C_3H_5^+ + {}^{\bullet}CH_3 + e^- \qquad (3.56)$$

$$\begin{aligned} \text{AE (298 K)} &= 11.25 \pm 0.02 \text{ eV [169]} = 1085.5 \pm 1.9 \text{ kJ/mol} \\ \Delta H^{\circ}_{298}(3.56) &= 1085.5 + (11.3 + 10.4 - 6.2) = 1101.0 \pm 1.9 \text{ kJ/mol} \\ \Delta H^{\circ}_{f,298}(C_3H_5^+) &= 1101.0 - 145.7 + (-7.1) = \underline{948.2 \pm 2.3} \text{ kJ/mol} \end{aligned}$$

(e)
$$i\text{-}C_4H_8 + h\nu \rightarrow C_3H_5^+ + {}^{\bullet}CH_3 + e^- \qquad (3.57)$$

$$\begin{aligned} \text{AE (298 K)} &= 11.33 \pm 0.02 \text{ eV [169]} = 1093.2 \pm 1.9 \text{ kJ/mol} \\ \Delta H^{\circ}_{298}(3.57) &= 1093.2 + (11.3 + 10.4 - 6.2) = 1108.7 \pm 1.9 \text{ kJ/mol} \\ \Delta H^{\circ}_{f,298}(C_3H_5^+) &= 1108.7 - 145.7 + (-16.9) = \underline{946.1 \pm 2.2} \text{ kJ/mol} \end{aligned}$$

The average of these determinations is 948.6 ± 1.6 kJ/mol, which leads to a value of 165.1 ± 2.5 kJ/mol for $\Delta H^{\circ}_{f,298}$(allyl) when combined with the above 298 K ionization energy. This in good agreement with the recent value of 166.9 ± 2.1 kJ/mol obtained by Roth *et al.* [170].

Formyl

Although $H^{\bullet}CO$ is one of the simplest radical species, its enthalpy of formation is still not firmly established. The value of 37.2 ± 5.0 kJ/mol recommended by McMillen and Golden [4] was based on the gas-phase kinetic studies of Walsh and Benson [171]. The most recent JANAF value is 43.5 ± 8 kJ/mol [123], derived from thermochemical information for glyoxal [172,173]. Lias *et al.* [6] however chose to use the spectroscopic measurement of Becker *et al.* [174] (44.8 ± 4 kJ/mol in their extensive thermochemical compilation. The most accurately determined experimental value to date resulted from a fluorescence experiment by Chuang *et al.* [175] in which a value of 41.8 ± 0.8 kJ/mol was obtained.

The formyl radical undergoes a significant change in geometry following ionization, so that there is a large difference between the adiabatic and vertical ionization energies [54]. Indeed, it appears that the adiabatic transition has not been directly observed in photoelectron spectroscopic measurements. From an analysis of the vibrational structure in the photoelectron spectrum, Dyke [54] has proposed that the unobserved adiabatic ionization energy should occur at 8.14 ± 0.04 eV. This would correspond to a 298 K ionization energy of 785.4 + (9.0 − 10.0) = 784.4 ± 3.9 kJ/mol.

Although photoionization HCO^+ appearance energies are available for a number of compounds it has been found that, because of competitive and kinetic effects, only formic acid can produce a reliable estimate for $\Delta H^\circ_{f,298}(HCO^+)$ [176].

(a) $$HCOOH + h\nu \rightarrow HCO^+ + {}^{\bullet}OH + e^- \tag{3.58}$$

$$\begin{aligned} \text{AE (298 K)} &= 12.76 \pm 0.01\ \text{eV [176]} = 1231.1 \pm 1.0\ \text{kJ/mol} \\ \Delta H^\circ_{298}(3.58) &= 1231.1 + (9.0 + 9.2 - 6.2) = 1243.1 \pm 1.0\ \text{kJ/mol} \\ \Delta H^\circ_{f,298}(HCO^+) &= 1243.1 - 39.3 + (-378.7) = \underline{825.1 \pm 1.7}\ \text{kJ/mol} \end{aligned}$$

This cationic enthalpy of formation corresponds to a proton affinity for carbon monoxide of 1530.1 + (−110.5) − 825.1 = 594.5 ± 1.7 kJ/mol, in excellent agreement with the values of 593.3 kJ/mol [177], 593.0 kJ/mol [134] and 593.7 kJ/mol [178] obtained from recent high-level *ab initio* calculations. It is slightly higher than the 591.6 ± 4.2 kJ/mol determined by Adams *et al.* [179] from variable temperature SIFT studies of relative proton affinities. However, any downward correction to the present proton affinity would require a corresponding increase in the experimental appearance energy for reaction (3.58), which is highly unlikely.

Combining $\Delta H^\circ_{f,298}(HCO^+)$ = 825.1 kJ/mol with the above $IE_{298}(H^{\bullet}CO)$ = 784.4 kJ/mol gives a value of 40.7 ± 4.3 kJ/mol for $\Delta H^\circ_{f,298}(H^{\bullet}CO)$. Although this is lower than the value obtained from spectroscopic measurements [174,175], it is probably a reflection of the uncertainty associated with the unobserved adiabatic ionization transition. Because of this, it is appropriate to employ an alternative means of estimating the enthalpy of formation for the formyl radical.

Methyl formate readily fragments following photoionization to produce CH_2OH^+ and $H^{\bullet}CO$ [176]. Since the enthalpy of formation for this cation is well defined (see section 3.4.7), the corresponding value for the formyl radical may be derived from the CH_2OH^+ appearance energy.

(b) $$H(CO)OCH_3 + h\nu \rightarrow CH_2OH^+ + H^{\bullet}CO + e^- \qquad (3.59)$$

$$\begin{aligned} \text{AE (298 K)} &= 11.32 \pm 0.01 \text{ eV [176]} = 1092.2 \pm 1.0 \text{ kJ/mol} \\ \Delta H^{\circ}_{298}\,(3.59) &= 1092.2 + (10.2 + 10.0 - 6.2) = 1106.2 \pm 1.0 \text{ kJ/mol} \\ \Delta H^{\circ}_{f,298}(H^{\bullet}CO) &= 1106.2 - 708.6 + (-355.5) = \underline{42.1 \pm 1.6} \text{ kJ/mol} \end{aligned}$$

Although this is in excellent agreement with the recent result of Chuang *et al.* [175] it can only be regarded as an upper limit due to the possibility of excess energy being associated with an appearance energy.

Hydroxymethyl

Recently there has been considerable interest in the enthalpy of formation for the hydroxymethyl radical, with several experimental [108,180–182] and theoretical [183,184] studies being made. It is interesting to note that each of the experimental investigations involved ions. Previously McMillen and Golden [4] had recommended a value of -25.9 ± 6.3 kJ/mol, which was based on an earlier review by Golden and Benson [147]. Tsang's [185] shock tube investigations of the thermal stability of some branched alcohols however indicated a higher value of -17.6 ± 8 kJ/mol. From a photoionization study of $^{\bullet}CD_2OH$ and $^{\bullet}CD_3O$, Ruscic and Berkowitz [31] proposed a value of -15.5 kJ/mol. This was subsequently revised to $-16.6 + 0.9$ kJ/mol following a detailed photoionization study of the formation of CH_2OH^+ from methanol [181]. A similar investigation by Traeger and Holmes [180] yielded a value of -18.9 ± 1.0 kJ/mol. The essential difference between these two results is the way in which the fragment ion threshold was extracted from the experimental photoionization efficiency curve. Earlier, a kinetic study of the reaction between $^{\bullet}CH_2OH$ and both HBr and HI had produced a value of -8.9 ± 1.8 kJ/mol [108]. It was subsequently shown that this was too high due to an incorrect value of S°_{298} for the hydroxymethyl radical [180]. Depending on the vibrational analysis used for the statistical mechanical calculation, the entropy can be either 240.2 J/K/mol or 245.6 J/K/mol, resulting in corresponding values of -14.7 or -12.2 kJ/mol for $\Delta H^{\circ}_{f,298}(^{\bullet}CH_2OH)$ [181]. High level *ab initio* calculations have produced corresponding values of -15.6 ± 1.5 kJ/mol [183] and -15.2 ± 3.5 kJ/mol [184] in support of the earlier Ruscic and Berkowitz [31] measurement. However, Klemm and coworkers [182] have recently obtained a value of -20.4 kJ/mol for $\Delta H^\circ_{f,298}(^\bullet CH_2OH)$, from a photoionization study of the hydroxymethyl radical using synchrotron radiation, which agrees well with the Traeger and Holmes [180] value.

The adiabatic ionization energy for the hydroxymethyl radical has been measured by photoelectron spectroscopy to be 7.55 ± 0.01 eV [54]. Photoionization mass spectrometry experiments have obtained similar values of 7.549 ± 0.006 eV [31] and 7.56 ± 0.02 eV [39], in excellent agreement. Adopting a value of 7.55 ± 0.01 eV results in a 298 K ionization energy of 728.5 + (10.2 − 11.3) = 727.4 ± 1.0 kJ/mol.

Unlike other possible precursor molecules, the formation of CH_2OH^+ from methanol is not subject to competition from a lower energy dissociation channel and should provide a thermochemical threshold free from any competitive shift [180]. Furthermore, high-level G2 *ab initio* molecular orbital calculations do not predict any excess energy involved in the fragmentation process [186].

(a)
$$CH_3OH + h\nu \rightarrow CH_2OH^+ + H^\bullet + e^- \qquad (3.60)$$

$$\begin{aligned}
\text{AE (298 K)} &= 11.578 \pm 0.007\ \text{eV [180]} = 1117.1 \pm 0.7\ \text{kJ/mol}\\
\Delta H^\circ_{298}(3.60) &= 1117.1 + (10.2 + 6.2 - 6.2) = 1127.3 \pm 0.7\ \text{kJ/mol}\\
\text{AE (0 K)} &= 11.649 \pm 0.003\ \text{eV [181]} = 1123.9 \pm 0.3\ \text{kJ/mol}\\
\Delta H^\circ_{298}(3.60) &= 1123.9 + (10.2 + 6.2 - 11.4) = 1128.9 \pm 0.3\ \text{kJ/mol}\\
\text{Average } \Delta H^\circ_{298}(3.60) &= 1128.1 \pm 0.8\ \text{kJ/mol}\\
\Delta H^\circ_{f,298}(CH_2OH^+) &= 1128.1 - 218.0 + (-201.5) = \underline{708.6 \pm 0.9}\ \text{kJ/mol}
\end{aligned}$$

The proton affinity for HCHO has been calculated by Smith and Radom [134] to be 711.8 kJ/mol, whereas the value adopted by Lias *et al.* [132] in their extensive compilation of evaluated gas-phase basicities and proton affinities is 718 kJ/mol. However, Lias *et al.* [132] used $PA(H_2O)$ = 697 ± 8 kJ/mol from an earlier photoionization study of the water dimer [187] as their reference point. If the recent value of 690.8 ± 2.9 kJ/mol, obtained from a variable temperature SIFT study [188], is used instead, the proton affinity for formaldehyde is then lowered to 712 kJ/mol. This is in excellent agreement with both the Smith and Radom [134] G2 *ab initio* calculation and the value of PA(HCHO) = 1530.1 + (−108.6) − 708.6 = 712.9 ± 1.0 kJ/mol, supporting the present value for $\Delta H^\circ_{f,298}(CH_2OH^+)$. Combination with the above 298 K ionization energy of 727.4 ± 1.0 kJ/mol then leads to $\Delta H^\circ_{f,298}(^\bullet CH_2OH) = -18.8 \pm 1.3$ kJ/mol.

Acetyl

Although the acetyl radical has been studied in many chemical systems its enthalpy of formation is still not firmly established. The McMillen and

Golden [4] recommended value of -24.3 ± 1.7 kJ/mol was based on kinetic data from the earlier Golden and Benson [147] review. However, from a study of the reaction kinetics involved in acetyl radical chemistry over a wide pressure and temperature range, Anastasi and Maw [189] concluded that a higher value of -16 kJ/mol was more likely. This was supported by Tsang [190] who, from a shock-tube study of a series of ketones, proposed a value of -13.8 kJ/mol. More recently, Gutman and coworkers [109] obtained a significantly higher value of -10.0 ± 1.2 kJ/mol, based on a kinetic study of the $CH_3{}^{\bullet}CO + HBr$ reaction. Two high-level molecular orbital calculations have produced respective values of -17.6 ± 8.4 kJ/mol [191] and -9.2 ± 2.9 kJ/mol [192].

The acetyl cation enthalpy of formation may be derived from the appearance energies for several different unimolecular fragmentation processes.

(a) $$CH_3CHO + h\nu \rightarrow CH_3CO^+ + H^{\bullet} + e^- \tag{3.61}$$

$$\begin{aligned} \text{AE (298 K)} &= 10.67 \pm 0.01 \text{ eV [193]} = 1029.5 \pm 1.0 \text{ kJ/mol} \\ \Delta H^\circ_{298}(3.61) &= 1029.5 + (12.0 + 6.2 - 6.2) = 1041.5 \pm 1.0 \text{ kJ/mol} \\ \Delta H^\circ_{f,298}(CH_3CO^+) &= 1041.5 - 218.0 + (-166.1) = \underline{657.4 \pm 1.1} \text{ kJ/mol} \end{aligned}$$

(b) $$CH_3(CO)CH_3 + h\nu \rightarrow CH_3CO^+ + {}^{\bullet}CH_3 + e^- \tag{3.62}$$

$$\begin{aligned} \text{AE (298 K)} &= 10.38 \pm 0.01 \text{ eV [193]} = 1001.5 \pm 1.0 \text{ kJ/mol} \\ \Delta H^\circ_{298}(3.62) &= 1001.5 + (12.0 + 10.4 - 6.2) = 1017.7 \pm 1.0 \text{ kJ/mol} \\ \Delta H^\circ_{f,298}(CH_3CO^+) &= 1017.7 - 145.7 + (-217.3) = \underline{654.7 \pm 1.5} \text{ kJ/mol} \end{aligned}$$

(c) $$CH_3(CO)C_2H_5 + h\nu \rightarrow CH_3CO^+ + {}^{\bullet}C_2H_5 + e^- \tag{3.63}$$

$$\begin{aligned} \text{AE (298 K)} &= 10.32 + 0.01 \text{ eV [193]} = 995.7 \pm 1.0 \text{ kJ/mol} \\ \Delta H^\circ_{298}(3.63) &= 995.7 + (12.0 + 13.0 - 6.2) = 1014.5 + 1.0 \text{ kJ/mol} \\ \Delta H^\circ_{f,298}(CH_3CO^+) &= 1014.5 - 120.2 + (-238.7) = \underline{655.6 \pm 1.6} \text{ kJ/mol} \end{aligned}$$

(d) $$CH_3(CO)OH + h\nu \rightarrow CH_3CO^+ + {}^{\bullet}OH + e^- \tag{3.64}$$

$$\begin{aligned} \text{AE (298 K)} &= 11.54 \pm 0.01 \text{ eV [193]} = 1113.4 \pm 1.0 \text{ kJ/mol} \\ \Delta H^\circ_{f,298}(3.64) &= 1113.4 + (12.0 + 9.2 - 6.2) = 1128.4 \pm 1.0 \text{ kJ/mol} \\ \Delta H^\circ_{f,298}(CH_3CO^+) &= 1128.4 - 39.3 + (-432.8) = \underline{656.3 \pm 2.2} \text{ kJ/mol} \end{aligned}$$

From a theoretical study of the acetyl cation, Smith and Radom [194] concluded that, unlike reaction (3.62), there was a reverse activation energy associated with reaction (3.61). The slightly higher $\Delta H^\circ_{f,298}(CH_3CO^+)$ obtained above for process (3.61) is consistent with this, although not to the extent of the calculated 9.8 kJ/mol [194]. If the appearance energy measurement for acetaldehyde is excluded from the average, is obtained a value of 655.5 ± 0.8 kJ/mol from the present data. This corresponds to a proton affinity for ketene of 1530.1 + (−47.5) − 655.5 = 827.1 ± 1.8 kJ/mol, in good agreement with both a recent high-level *ab initio* calculation of 825.0 kJ/mol [134], and the experimentally determined value of 828.0 kJ/mol [132].

Unfortunately there is no reliable experimental value available for the adiabatic ionization energy of the acetyl radical [6], so that the cationic enthalpy of formation cannot be used directly to obtain $\Delta H^{\circ}_{f,298}(CH_3{}^{\bullet}CO)$. Because of the large difference between the adiabatic (7.21 ± 0.05 eV) and vertical (8.39 ± 0.05 eV) ionization energies measured by photoelectron spectroscopy [56] it is most unlikely that the actual adiabatic transition has been observed. However, the appearance energy for the formation of $CH_3CO^+ + CH_3{}^{\bullet}CO$ from biacetyl should provide a good estimate as this process is free from any competitive shift. Similarly, any kinetic shift should be small because of the simple bond cleavage involved [193].

(e) $$CH_3(CO)_2CH_3 + h\nu \rightarrow CH_3CO^+ + CH_3{}^{\bullet}CO + e^- \qquad (3.65)$$

$$\begin{aligned} \text{AE (298 K)} &= 9.88 \pm 0.01 \text{ eV [193]} = 953.3 \pm 1.0 \text{ kJ/mol} \\ \Delta H^{\circ}_{298}(3.65) &= 953.3 + (12.0 + 12.4 - 6.2) = 971.5 \pm 1.0 \text{ kJ/mol} \\ \Delta H^{\circ}_{f,298}(CH_3{}^{\bullet}CO) &= 971.5 - 655.5 + (-327.1) = \underline{-11.1 \pm 1.8} \text{ kJ/mol} \end{aligned}$$

Although this value for the acetyl radical enthalpy of formation is in good agreement with the experiments of Gutman and coworkers [109] it can only be regarded as an upper limit. Any reverse activation energy associated with the fragmentation would result in the products of reaction (3.65) being formed with excess energy, i.e. not at the thermochemical threshold, and a lower value for $\Delta H^{\circ}_{f,298}(CH_3{}^{\bullet}CO)$. Further work is required to resolve this issue, along with many others concerning the energetics of organic free radicals.

REFERENCES

1. NONHEBEL, D. C. and WALTON, J. C. (1974) *Free-Radical Chemistry*. Cambridge University Press: London.
2. WERTZ, J. E. and BOLTON, J. R. (1972) *Electron Spin Resonance: Elementary Theory and Practical Applications*. McGraw-Hill: New York.
3. FOSTER, S. C. and MILLER, T. A. (1989) *J. Phys. Chem.* **93,** 5986.
4. MCMILLEN, D. F. and GOLDEN, D. M. (1982) *Annu. Rev. Phys. Chem.* **33,** 493.
5. ROSENSTOCK, H. M., DRAXL, K., STEINER, B. W. and HERRON, J. T. (1977) *J. Phys. Chem. Ref. Data, Suppl. 1,* 6.

6. Lias, S. G., Bartmess, J. E., Liebman, J. F., Holmes, J. L., Levin, R. D. and Mallard, W. G. (1988) *J. Phys. Chem. Ref. Data, Suppl. 1*, 17.
7. Berkowitz, J., Ellison, G. B. and Gutman, D. (1994) *J. Phys. Chem.*, **98,** 2744.
8. Rosenstock, H. M. (1976) *Int. J. Mass Spectrom. Ion Phys.* **20,** 139.
9. Chupka, W. A. (1971) *J. Chem. Phys.* **54,** 1936.
10. Levy, D. H. (1980) *Annu. Rev. Phys. Chem.* **31,** 197.
11. Engelking, P. C. (1991) *Chem. Rev.* **91,** 399.
12. Hipple, J. A. and Stevenson, D. P. (1943) *Phys. Rev.* **63,** 121.
13. Lossing, F. P. (1963) In *Mass Spectrometry*, McDowell, C. A. (ed.); McGraw-Hill: New York, pp. 442–505.
14. Marmet, P. and Morrison, J. D. (1962) *J. Chem. Phys.* **36,** 1238.
15. Maeda, K., Semeluk, G. P. and Lossing, F. P. (1968) *Int. J. Mass Spectrom. Ion Phys.* **1,** 395.
16. Simpson, J. A. (1964) *Rev. Sci. Inst.*, **35,** 1698.
17. Lossing, F. P and Semeluk, G. P. (1970) *Can. J. Chem.* **48,** 955.
18. Lossing, F. P. (1971) *Can. J. Chem.* **49,** 357.
19. Lossing, F. P. (1972) *Can. J. Chem.* **50,** 3973.
20. Lossing, F. P. (1972) *Bull. Soc. Chim. Belges* **81,** 125.
21. Lossing, F. P. and Traeger, J. C. (1976) *Int. J. Mass Spectrom. Ion. Phys.* **19,** 9.
22. Lossing, F. P. and Maccoll, A. (1976) *Can. J. Chem.* **54,** 990.
23. Berkowitz, J. (1989) *Acc. Chem. Res.* **22,** 413.
24. Berkowitz, J. and Ruscic, B. (1991) In *Vacuum Ultraviolet Photoionization and Photodissociation of Molecules and Clusters*; Ng, C. Y., (ed.); World Scientific: Singapore, pp. 1–41.
25. Elder, F. A., Giese, C., Steiner, B. and Inghram, M. (1962) *J. Chem. Phys.* **36,** 3292.
26. Lifshitz, C. and Chupka, W. A. (1967) *J. Chem. Phys.* **47,** 3439.
27. Elder, F. A. and Parr, A. C. (1969) *J. Chem. Phys.* **50,** 1027.
28. Gibson, S. T., Greene, J. P. and Berkowitz, J. (1985) *J. Chem. Phys.* **83,** 4319.
29. Berkowitz, J., Mayhew, C. A. and Ruscic, B. (1988) *J. Chem. Phys.* **88,** 7396.
30. Ruscic, B., Schwarz, M. and Berkowitz, J. (1989) *J. Chem. Phys.*, **91,** 6780.
31. Ruscic, B. and Berkowitz, J. (1991) *J. Chem. Phys.* **95,** 4033.
32. Ruscic, B. and Berkowitz, J. (1993) *J. Chem. Phys.* **98,** 2568.
33. Reid, N. W. (1971) *Int. J. Mass Spectrom. Ion Phys.* **6,** 1.
34. Samson, J. A. R. (1967) *Techniques of Vacuum Ultraviolet Spectroscopy*. Wiley: New York.
35. Nicholson, A. J. C. (1965) *J. Chem. Phys.* **43,** 1171.
36. Parr, G. R. and Taylor, J. W. (1973) *Rev. Sci. Instr.* **44,** 1578.

37. SHIROMARU, H., ACHIBA, Y., KIMURA, K. and LEE, Y. T. (1987) *J. Phys. Chem.* **91,** 17.
38. NESBITT, F. L., MARSTON, G., STIEF, L. J., WICKRAMAARATCHI, M. A., TAO, W. and KLEMM, R. B. (1991) *J. Phys. Chem.* **95,** 7613.
39. TAO, W., KLEMM, R. B., NESBITT, F. L. and STIEF, L. J. (1992) *J. Phys. Chem.* **96,** 104.
40. CAIRNS, R. B. and SAMSON, J. A. R. (1966) *J. Opt. Soc. Am.* **56,** 1568.
41. RABALAIS, J. W. (1977) *Principles of Ultraviolet Photoelectron Spectroscopy.* Wiley-Interscience: New York.
42. HOULE, F. A. and BEAUCHAMP, J. L. (1978) *J. Am. Chem. Soc.* **100,** 3290.
43. HOULE, F. A. and BEAUCHAMP, J. L. (1979) *J. Am. Chem. Soc.* **101,** 4067.
44. HOULE, F. A. and BEAUCHAMP, J. L. (1981) *J. Phys. Chem.* **85,** 3456.
45. SCHULTZ, J. C., HOULE, F. A. and BEAUCHAMP, J. L. (1984) *J. Am. Chem. Soc.* **106,** 3917.
46. SCHULTZ, J. C., HOULE, F. A. and BEAUCHAMP, J. L. (1984) *J. Am. Chem. Soc.* **106,** 7336.
47. DEARDEN, D. V. and BEAUCHAMP, J. L. (1985) *J. Phys. Chem.* **89,** 5359.
48. HAYASHIBARA, K., KRUPPA, G. H. and BEAUCHAMP, J. L. (1986) *J. Am. Chem. Soc.* **108,** 5441.
49. DYKE, J. M., JONATHAN, N., LEE, E. and MORRIS, A. (1976) *J. Chem. Soc., Faraday Trans. 2* **72,** 1385.
50. DYKE, J. M., JONATHAN, N. and MORRIS, A. (1982) *Int. Rev. Phys. Chem.* **2,** 3.
51. ANDREWS, L., DYKE, J. M., JONATHAN, N., KEDDAR, N. and MORRIS, A. (1984) *J. Phys. Chem.* **88,** 1950.
52. ANDREWS, L., DYKE, J. M., JONATHAN, N., KEDDAR, N., MORRIS, A. and RIDHA, A. (1984) *J. Phys. Chem.* **88,** 2364.
53. DYKE, J. M., ELLIS, A. R., KEDDAR, N. and MORRIS, A. (1984) *J. Phys. Chem.* **88,** 2565.
54. DYKE, J. M. (1987) *J. Chem. Soc., Faraday Trans. 2* **83,** 69.
55. DYKE, J. M., LEE, E. P. F. and ZAMANPOUR NIAVARAN, M. H. (1989) *Int. J. Mass Spectrom. Ion Processes* **94,** 221.
56. COCKETT, M. C. R., DYKE, J. M. and ZAMANPOUR, H. (1991) In *Vacuum Ultraviolet Photoionization and Photodissociation of Molecules and Clusters.* Ng, C. Y., (ed.); World Scientific: Singapore, pp. 43–99.
57. WANNIER, C. H. (1955) *Phys. Rev.* **100,** 1180.
58. DRZAIC, P. S., MARKS, J. and BRAUMAN, J. I. (1984) In *Gas-Phase Ion Chemistry.* Bowers, M. T. (ed.); Academic Press: New York, **3,** 167.
59. RICHARDSON, J. H., STEPHENSON, L. M. and BRAUMAN, J. I. (1973) *J. Chem. Phys.* **59,** 5068.
60. JANOUSEK, B. K. and BRAUMAN, J. I. (1980) *J. Chem. Phys.* **72,** 694.
61. WETZEL, D. M. and BRAUMAN, J. I. (1987) *Chem. Rev.* **87,** 607.

62. Mead, R. D., Stevens, A. E. and Lineberger, W. C. (1984) In *Gas-Phase Ion Chemistry*. Bowers, M. T., (ed.); Academic Press: New York, **3,** 213.

63. Ellis, H. B. and Ellison, G. B. (1983) *J. Chem. Phys.* **78,** 6541.

64. Leopold, D. G., Murray, K. K., Stevens Miller, A. E. and Lineberger, W. C. (1985) *J. Chem. Phys.* **83,** 4849.

65. Metz, R. B., Weaver, A., Bradforth, S. E., Kitsopoulos, T. N. and Neumark, D. M. (1990) *J. Phys. Chem.* **94,** 1377.

66. Bradforth, S. E., Kim, E. H., Arnold, D. W. and Neumark, D. M. (1993) *J. Chem. Phys.* **98,** 800.

67. Ervin, K. M., Ho, J. and Lineberger, W. C. (1989) *J. Chem. Phys.* **91,** 5974.

68. Christodoulides, A. A., McCorkle, D. L. and Christophorou, L. G. (1984) In *Electron–Molecule Interactions and their Applications*. Christophorou, L. G., (ed.); Academic Press: Orlando, **2,** 423.

69. Price, P. (1991) *J. Am. Soc. Mass Spectrom.* **2,** 336.

70. Guyon, P. M. and Berkowitz, J. (1971) *J. Chem. Phys.* **54,** 1814.

71. Dannacher, J. (1984) *Org. Mass Spectrom.* **19,** 253.

72. Baer, T., Booze, J. and Weitzel, K.-M. (1991) In *Vacuum Ultraviolet Photoionization and Photodissociation of Molecules and Clusters*. Ng, C. Y. (ed.); World Scientific: Singapore, pp. 259–296.

73. Oliveira, M. C., Baer, T., Olesik, S. and Almoster Ferreira, M. A. (1988) *Int. J. Mass Spectrom. Ion Processes* **82,** 299.

74. Rosenstock, H. M. and Krauss, M. (1963) *Adv. Mass Spectrom.* **2,** 251.

75. Lifshitz, C. (1989) *Adv. Mass Spectrom.* **11,** 713.

76. Lifshitz, C. (1992) *Int. J. Mass Spectrom. Ion Processes* **118/119,** 315.

77. Baer, T. (1979) In *Gas-Phase Ion Chemistry*. Bowers, M. T., (ed.); Academic Press: New York, **1,** 153.

78. Norwood, K. and Ng, C. Y. (1989) *Chem. Phys. Lett.* **156,** 145.

79. Holmes, J. L. (1985) *Org. Mass Spectrom.* **20,** 169.

80. Holmes, J. L. (1992) *Int. J. Mass Spectrom. Ion Processes* **118/119,** 381.

81. Chupka, W. A. (1959) *J. Chem. Phys.* **30,** 191.

82. Traeger, J. C. and McLoughlin, R. G. (1978) *Int. J. Mass Spectrom. Ion. Phys.* **27,** 319.

83. Lifshitz, C. (1982) *Mass Spectrom. Rev.* **1,** 309.

84. Harland, P. W., Franklin, J. L. and Carter, D. E. (1973) *J. Chem. Phys.* **58,** 1430.

85. Hacaloglu, J., Gokmen, A. and Suzer, S. (1989) *J. Phys. Chem.* **93,** 3418.

86. Hubin-Franskin, M.-J., Katihabwa, J. and Collin, J. E. (1976) *Int. J. Mass Spectrom. Ion Phys.* **20,** 285.

87. Ruscic, B. and Berkowitz, J. (1990) *J. Chem. Phys.* **93,** 5586.

88. Cooks, R. G. (ed.) (1978) *Collision Spectroscopy*. Plenum Press: New York.

89. GRIFFITHS, W. J., HARRIS, F. M. and BARTON, J. D. (1989) *Rapid Commun. Mass Spectrom.* **3,** 384.
90. LANGFORD, M. L., ALMEIDA, D. P. and HARRIS, F. M. (1990) *Int. J. Mass Spectrom. Ion Processes* **98,** 147.
91. HOLMES, J. L. (1989) *Mass Spectrom. Rev.* **8,** 513.
92. GRIFFITHS, W. J., HARRIS, F. M., BRENTON, A. G. and BEYNON, J. H. (1986) *Int. J. Mass Spectrom. Ion Processes* **74,** 317.
93. WESDEMIOTIS, C. and MCLAFFERTY, F. W. (1987) *Chem. Rev.* **87,** 485.
94. TERLOUW, J. K. (1989) *Adv. Mass Spectrom.* **11,** 984.
95. MCLAFFERTY, F. W. (1992) *Int. J. Mass Spectrom. Ion Processes* **118/119,** 221.
96. FERGUSON, E. E., FEHSENFELD, F. C. and SCHMELTEKOPF, A. L. (1969) *Adv. At. Mol. Phys.* **5,** 1.
97. TSUJI, M. (1988) In *Techniques of Chemistry.* FARRAR, J. M. and SAUNDERS, W. H., (eds); Wiley-Interscience: New York, **20,** 489.
98. SMITH, D. and ADAMS, N. G. (1988) *Adv. At. Mol. Phys.* **24,** 1.
99. ADAMS, N. G. and SMITH, D. In *Techniques of Chemistry.* FARRAR, J. M. and SAUNDERS, W. H., (eds); Wiley-Interscience: New York, **20,** 165.
100. GRAUL, S. T. and SQUIRES, R. R. (1988) *Mass Spectrom. Rev.* **7,** 263.
101. KEBARLE, P. (1988) In *Techniques of Chemistry.* FARRAR, J. M. and SAUNDERS, W. H. (eds); Wiley-Interscience: New York, **20,** 221.
102. SZULEJKO, J. E. and MCMAHON, T. B. (1991) *Int. J. Mass Spectrom. Ion Processes* **109,** 279.
103. KEMPER, P. R. and BOWERS, M. T. (1988) In *Techniques of Chemistry.* FARRAR, J. M. and SAUNDERS, W. H., (eds); Wiley-Interscience: New York, **20,** 1.
104. FREISER, B. S. (1988) In *Techniques of Chemistry.* FARRAR, J. M. and SAUNDERS, W. H., (eds); Wiley-Interscience: New York, **20,** 61.
105. MARCH, R. E. and HUGHES, R. J. (1989) In *Chemical Analysis.* WINEFORDNER, J. D., (ed.); Wiley-Interscience: New York, 102.
106. SLAGLE, I. R., YAMADA, F. and GUTMAN, D. J. (1981) *Am. Chem. Soc.* **103,** 149.
107. SEETULA, J. A., RUSSELL, J. J. and GUTMAN, D. J. (1990) *Am. Chem. Soc.* **112,** 1347.
108. SEETULA, J. A. and GUTMAN D. (1992) *J. Phys. Chem.* **96,** 5401.
109. NIIRANEN, J. T., GUTMAN, D. and KRASNOPEROV, L. N. (1992) *J. Phys. Chem.* **96,** 5881.
110. SEAKINS, P. W., PILLING, M. J., NIIRANEN, J. T., GUTMAN, D. and KRASNOPEROV, L. N. (1992) *J. Phys. Chem.* **96,** 9847.
111. WU, C-H AND JOHNSTON, H. (1972) *Bull. Soc. Chim. Belges* **81,** 135.
112. FITE, W. L. (1975) *Int. J. Mass Spectrom. Ion. Phys.* **16,** 109.
113. GOLDEN, D. M., SPOKES, G. N. and BENSON, S. W. (1973) *Angew. Chem. Int. Ed.* **12,** 534.

114. Müller-Markgraf, W., Rossi, M. J. and Golden, D. M. (1989) *J. Am. Chem. Soc.* **111,** 956.

115. Baghal-Vayjooee, M. H., Colussi, A. J. and Benson, S. W. (1979) *Int. J. Chem. Kinet.* **11,** 147.

116. Golden, D. M., Rossi, M. J., Baldwin, A. C. and Barker, J. R. (1981) *Acc. Chem. Res.* **14,** 56.

117. Holmes, J. L. and Lossing, F. P. (1984) *Int. J. Mass Spectrom. Ion Processes,* **58,** 113.

118. Pedley, J. B., Naylor, R. D. and Kirby, S. P. (1986) *Thermochemical Data of Organic Compounds,* 2nd edn; Chapman and Hall: London.

119. Benson, S. W. (1976) *Thermochemical Kinetics,* 2nd edn; Wiley-Interscience: New York.

120. Cohen, N. and Benson, S. W. (1993) *Chem. Rev.* **93,** 2419.

121. Holmes, J. L., Fingas, M. and Lossing, F. P. (1981) *Can. J. Chem.* **59,** 80.

122. Traeger, J. C. and McLoughlin, R. G. (1981) *J. Am. Chem. Soc.* **103,** 3647.

123. Chase, M. W., Davies, C. A., Downey, J. R., Frurip, D. J., McDonald, R. A. and Syverud, A. N. (1985) *J. Phys. Chem. Ref. Data* **14,** Suppl. 1 (*JANAF Thermochemical Tables,* 3rd edn).

124. Raghavachari, K., Whiteside, R. A., Pople, J. A. and Schleyer, P. V. R. (1981) *J. Am. Chem. Soc.* **103,** 5649.

125. Traeger, J. C. To be published.

126. Holmes, J. L., Lossing, F. P. and Maccoll, A. (1988) *J. Am. Chem. Soc.* **110,** 7339.

127. Holmes, J. L. and Lossing, F. P. (1988) *J. Am. Chem. Soc.* **110,** 7343.

128. Holmes, J. L., Lossing, F. P. and Mayer, P. M. (1991) *J. Am. Chem. Soc.* **113,** 9723.

129. Pople, J. A., Schlegel, H. B., Krishnan, R., DeFrees, D. J., Binkley, J. S., Frisch, M. J., Whiteside, R. A., Hout, R. F. and Hehre, W. J. (1981) *Int. J. Quantum Chem. Quantum Chem. Symp.* **15,** 269.

130. Holmes, J. L. and Lossing, F. P. (1991) *Org. Mass Spectrom.* **26,** 537.

131. Holmes, J. L. and Terlouw, J. K. (1980) *Org. Mass Spectrom.* **15,** 383.

132. Lias, S. G., Liebman, J. F. and Levin, R. D. (1984) *J. Phys. Chem. Ref. Data* **13,** 695.

133. Szulejko, J. E. and McMahon, T. B. (1993) *J. Am. Chem. Soc.* **115,** 7839.

134. Smith, B. J. and Radom, L. (1993) *J. Am. Chem. Soc.* **115,** 4885.

135. DeFrees, D. J., Hehre, W. J., McIver, R. T. and McDaniel, D. H. (1979) *J. Phys. Chem.* **83,** 232.

136. DeFrees, D. J.; McIver, R. T. and Hehre, W. J. (1980) *J. Am. Chem. Soc.* **102,** 3334.

137. Kinter, M. T. and Bursey, M. M. (1987) *Org. Mass Spectrom.* **22,** 775.

138. Kebarle, P. and Chowdhury, S. (1987) *Chem. Rev.* **87,** 513.

139. BARTMESS, J. E. and MCIVER, R. T. (1979) In *Gas-Phase Ion Chemistry*. BOWERS, M. T. (ed.); Academic Press: New York, **2,** 87.
140. KOPPEL, I. A., TAFT, R. W., ANVIA, F., ZHU, S-Z., HU, L-Q., SUNG, K-S., DESMARTEAU, D. D., YAGUPOLSKII, L. M., YAGUPOLSKII, Y. L., IGNAT'EV, N. V., KONDRATENKO, N. V., VOLKONSKII, A. Y., VLASOV, V. M., NOTARIO, R. and MARIA, P-C. (1994) *J. Am. Chem. Soc.* **116,** 3047.
141. DEPUY, C. H., GRONERT, S., BARLOW, S. E., BIERBAUM, V. M. and DAMRAUER, R. (1989) *J. Am. Chem. Soc.* **111,** 1968.
142. RUSSELL, J. J., SEETULA, J. A. and GUTMAN, D. (1988) *J. Am. Chem. Soc.* **110,** 3092.
143. RUSSELL, J. J., SEETULA, J. A., SENKAN, S. M. and GUTMAN, D. (1988) *Int. J. Chem. Kinet.* **20,** 759.
144. PACEY, P. D. and WIMALASENA, J. H. (1984) *J. Phys. Chem.* **88,** 5657.
145. HERZBERG, G. (1961) *Proc. Roy. Soc.* **A262,** 291.
146. MCCULLOH, K. E. and DIBELER, V. H. (1976) *J. Chem. Phys.* **64,** 4445.
147. GOLDEN, D. M. and BENSON, S. W. (1969) *Chem. Rev.* **69,** 125.
148. TSANG, W. (1978) *Int. J. Chem. Kinet.* **10,** 821.
149. CAO, J-R. and BACK, M. H. (1984) *Int. J. Chem. Kinet.* **16,** 961.
150. HANNING-LEE, M. A., GREEN, N. J. B., PILLING, M. J. and ROBERTSON, S. H. (1993) *J. Phys. Chem.* **97,** 860.
151. RUSCIC, B., BERKOWITZ, J., CURTISS, L. A. and POPLE, J. A. (1989) *J. Chem. Phys.* **91,** 114.
152. ROSENSTOCK, H. M., BUFF, R., FERREIRA, M. A. A., LIAS, S. G., PARR, A. C., STOCKBAUER, R. L. and HOLMES, J. L. (1982) *J. Am. Chem. Soc.* **104,** 2337.
153. TSANG, W. (1985) *J. Am. Chem. Soc.* **107,** 2872.
154. TSCHUIKOW-ROUX, E. and CHEN, Y. (1989) *J. Am. Chem. Soc.* **111,** 9030.
155. DYKE, J., ELLIS, A., JONATHAN, N. and MORRIS, A. (1985) *J. Chem. Soc., Faraday Trans. 2* **81,** 1573.
156. CHUPKA, W. A. and BERKOWITZ, J. (1967) *J. Chem. Phys.* **47,** 2921.
157. TRAEGER, J. C. (1980) *Int. J. Mass Spectrom. Ion Phys.* **32,** 309.
158. GUTMAN, D. (1990) *Acc Chem. Res.* **23,** 375.
159. ISLAM, T. S. A. and BENSON, S. W. (1984) *Int. J. Chem. Kinet.* **16,** 995.
160. COHEN, N. (1992) *J. Phys. Chem.* **96,** 9052.
161. DYKE, J., JONATHAN, N., LEE, E., MORRIS, A. and WINTER, M. (1977) *Physica Scripta* **16,** 197.
162. KOENIG, T., BALLE, T. and SNELL, W. (1975) *J. Am. Chem. Soc.* **97,** 662.
163. MCLOUGHLIN, R. G. and TRAEGER, J. C. (1979) *J. Am. Chem. Soc.* **101,** 5791.
164. KEISTER, J. W., RILEY, J. S. and BAER, T. (1993) *J. Am. Chem. Soc.* **115,** 12613.
165. ROSSI, M., KING, K. D. and GOLDEN, D. M. (1979) *J. Am. Chem. Soc.* **101,** 1223.

166. Rossi, M. and Golden, D. M. (1979) *J. Am. Chem. Soc.* **101,** 1230.
167. Tsang, W. (1984) *J. Phys. Chem.* **88,** 2812.
168. Stull, D. R., Westrum, E. F. and Sinke, G. C. (1969) *The Chemical Thermodynamics of Organic Compounds.* Wiley: New York.
169. Traeger, J. C. (1984) *Int. J. Mass Spectrom. Ion Processes* **58,** 259.
170. Roth, W., Bauer, F., Beitat, A., Ebbrecht, T. and Wüstefeld, M. (1991) *Chem. Ber.* **124,** 1453.
171. Walsh, R. and Benson, S. W. (1966) *J. Am. Chem. Soc.* **88,** 4570.
172. Hartley, D. B. (1967) *J. Chem. Soc., Chem. Commun.* 1281.
173. Fletcher, R. A. and Pilcher, G. (1970) *Trans. Faraday Soc.* **66,** 794.
174. Becker, K. H., Lippmann, H. and Schurath, U. (1977) *Ber. Bunsenges. Phys. Chem.* **81,** 567.
175. Chuang, M-C., Foltz, M. F. and Moore, C. B. (1987) *J. Chem. Phys.* **87,** 3855.
176. Traeger, J. C. (1985) *Int. J. Mass Spectrom. Ion Processes* **66,** 271.
177. Komornicki, A. and Dixon, D. A. (1992) *J. Chem. Phys.* **97,** 1087.
178. Martin, J. M. L., Taylor, P. R. and Lee, T. J. (1993) *J. Chem. Phys.* **99,** 286.
179. Adams, N. G., Smith, D., Tichy, M., Javahery, G., Twiddy, N. D. and Ferguson, E. E. (1989) *J. Chem. Phys.* **91,** 4037.
180. Traeger, J. C. and Holmes, J. L. (1993) *J. Phys. Chem.* **97,** 3453.
181. Ruscic, B. and Berkowitz, J. (1993) *J. Phys. Chem.* **97,** 11451.
182. Kuo, S-C., Zhang, Z., Klemm, R. B., Liebman, J. F., Stief, L. J. and Nesbitt, F. L. (1994) *J. Phys. Chem.*, **98,** 4026.
183. Espinosa-Garcia, J. and Olivares del Valle, F. J. (1993) *J. Phys. Chem.* **97,** 3377.
184. Bauschlicher, C. W. and Partridge, H. (1994) *J. Phys. Chem.* **98,** 1826.
185. Tsang, W. (1976) *Int. J. Chem. Kinet.* **8,** 173.
186. Ma, N. L., Smith, B. J., Pople, J. A. and Radom, L. (1991) *J. Am. Chem. Soc.* **113,** 7903.
187. Ng, C. Y., Trevor, D. J., Tiedemann, P. W., Ceyer, S. T., Kronebusch, P. L., Mahan, B. H. and Lee, Y. T. (1977) *J. Chem. Phys.* **67,** 4235.
188. McIntosh, B. J., Adams, N. G. and Smith, D. (1988) *Chem. Phys. Lett.* **148,** 142.
189. Anastasi, C. and Maw, P. R. (1982) *J. Chem. Soc., Faraday Trans. 1* **78,** 2423.
190. Tsang, W. (1984) *Int. J. Chem. Kinet.* **16,** 1543.
191. Francisco, J. S. and Abersold, N. J. (1991) *Chem. Phys. Lett.* **187,** 354.
192. Bauschlicher, C. W. (1994) *J. Phys. Chem.* **98,** 2564.
193. Traeger, J. C., McLoughlin, R. G. and Nicholson, A. J. C. (1982) *J. Am. Chem. Soc.* **104,** 5318.
194. Smith, B. J. and Radom, L. (1990) *Int. J. Mass Spectrom. Ion Processes* **101,** 209.

195. Chen, S. S., Wilhoit, R. C. and Zwolinski, B. J. (1977) *J. Phys. Chem. Ref. Data* **6,** 105.

196. Heuts, H., Gilbert, R. G. and Radom, L. To be published.

197. Chen, Y., Rauk, A. and Tschuikow-Roux, E. (1990) *J. Phys. Chem.* **94,** 2775.

198. Chao, J., Wilhoit, R. C. and Zwolinski, B. J. (1973) *J. Phys. Chem. Ref. Data* **2,** 429.

199. Smith, B. J. and Radom, L. To be published.

200. Radom, L. Private communication.

4

Theoretical Studies of the Energetics of Radicals

Joseph S. Francisco and
John A. Montgomery, Jr.

INTRODUCTION

It is becoming more apparent that many reactions involving oxidation, polymerization, combustion, and atmospheric processes generally involve short-lived reactive intermediates such as radicals. Such species have also been found to play significant roles in biological processes. Understanding the role that radicals play in these processes require that the mechanisms by which the reactions take place be understood. Mechanistic understanding of chemical reactions requires knowledge of their fundamental thermochemical properties such as enthalpies, entropies, and heat capacities. These properties can be used with simple thermodynamic principles to provide insight into those chemical transformations which are energetically permitted. Therefore the availability of accurate thermochemical data on radicals is essential for understanding issues of their reactivity. O'Neal and Benson [1] noted that accurate data for heats of formation of radicals have been elusive, and that few heats of formations that were generally accepted prior to 1960 are still accepted today. For example, consider the heat of formation of acetyl radical, CH_3CO. In 1955, Szwarc and Taylor [2] measured the heat of formation to be -10.8 kcal mol^{-1}; by 1969 the accepted value [3] was -5.8 ± 0.4 kcal mol^{-1}. Presently, the accepted value [4] is -2.35 ± 0.29 kcal mol^{-1}. This refinement and readjustment in heat of formation values is not unusual. Improvement in measurement techniques and the introduction of new techniques that overcome intrusive limitations all contribute to reducing errors in the measurements. However, there are experimental limitations that make it almost impossible to measure certain heats of formation. In these cases, what tools are available to provide accurate energetics for radical systems? *Ab initio* molecular orbital methods have developed to the point where accurate energies can be obtained and thereby provide results for experimentally elusive species. In this chapter the present state of the art in computational methodology is described. The value of such calculations in the guidance of assessing the feasibility of postulated reaction mechanisms and intermediates is illustrated.

THERMOCHEMISTRY AND KINETICS

Chemical kinetics is the study of the change of chemical composition with time [5]. The rate of change of a reaction mixture with time is known as the rate of reaction. The rate of reaction is an experimentally derived quantity which is determined by measuring the concentration of one or more reactants or products in the reaction mixture. The mathematical expression that describes how the rate of reaction depends on the concentration of species in the mixture is given by

$$R = k \prod_{i=1}^{j} (C_i^{n_i}) \tag{4.1}$$

where R is the rate of reaction, C_i is the concentration of each of the components of the reaction, n_i is the stoichiometric coefficient of each of the components of the reaction, and k is the rate constant. The rate constant does not depend on the concentration of the species in the reaction mixture. The rate constant is also independent of time. However, the rate does depend on temperature. Arrhenius [6] found that the rate constant can be expressed in two parts: (1) a pre-exponential term (A-factor); (2) an activation energy term.

$$k(T) = A \exp(-E_a/RT) \tag{4.2}$$

Most elementary chemical reactions have been found to follow the Arrhenius expression. Moreover, it should be noted that the expression is empirical and derived solely from experimental observations. However, the two parts of the Arrhenius expression give insight into the microscopic details of why reactions occur. Thermochemical properties of reactions can also be obtained from the Arrhenius expression. The key quantity in the Arrhenius expression that can provide this information is the activation energy, E_a. This quantity gives the amount of energy

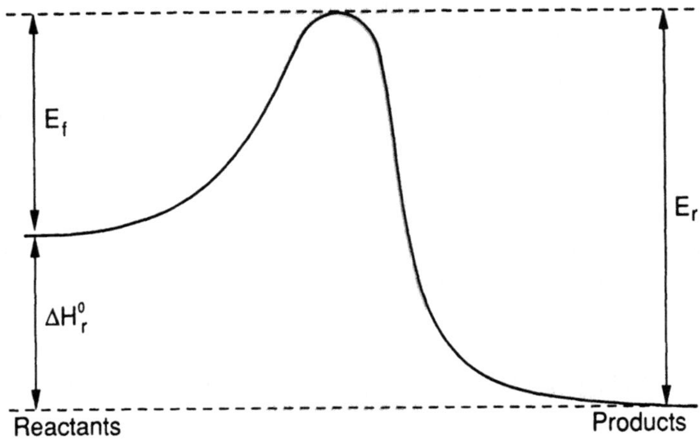

FIGURE 4.1 Pictorial relationship between forward and reverse activation energies and the thermochemical enthalpies.

necessary to initiate the reaction. Moreover, this quantity is also related to thermochemical enthalpies as shown in Figure 4.1. The enthalpy of a reaction is the difference between the foward (E_f) and reverse (E_r) activation energies for a given reaction

$$E_f - E_r = \Delta H_r^0 \tag{4.3}$$

The heat of reaction, ΔH_r^0, is the energy difference between reactants and products and is the difference in their heats of formation

$$\Delta H_r^0 = \Delta H_f^0\,(\text{products}) - \Delta H_f^0\,(\text{reactants}) \tag{4.4}$$

Therefore, if the Arrhenius terms of the forward and reverse reaction are known, then the thermochemical properties of the reactants and products can be determined. Conversely, if the Arrhenius terms of either the forward and reverse reactions are known, the Arrhenius terms of the other can be determined from the known thermochemical properties of the reactants and products. Thus, knowing one gives an estimate of the other, if the heat of formation data are known.

From the relationship between the kinetics and thermochemistry, it is clear for some reactions that measuring the kinetics is equivalent to measuring the heat of formation of the components of the reaction. Such is the case for kinetic measurements of bond-dissociation energies (BDE) in molecules [7]. For the molecule RX, the bond rupture gives two radicals (R• and X•). The bond-dissociation energy is just the enthalpy of reaction for the bond rupture process.

$$RX \rightleftarrows R\bullet + X\bullet \tag{4.5}$$

$$BDE = \Delta H_r^0\,(R\bullet) + \Delta H_f^0\,(X\bullet) - \Delta H_f^0\,(RX) \tag{4.6}$$

Therefore from measuring the kinetics of the bond-rupture processes one can obtain a direct measurement of the heats of formation of the resulting radical products.

Another means of determining the heats of formation of radicals experimentally is by thermal pyrolysis studies. In these studies, the forward and reverse reaction rate constants are measured for which the forward and reversed activation energy are determined. This allows the enthalpy of the reaction to be determined.

$$RR \rightleftarrows R\bullet + R\bullet \tag{4.7}$$

$$\Delta H_r^0 = E_f - E_f = 2\Delta H_f^0\,(R\bullet) - \Delta H_f^0\,(RR) \tag{4.8}$$

What also makes this process attractive is that the radical–radical recombination activation energy usually is zero or nearly zero for most radicals (i.e. $E_r \approx 0$). There is direct experimental evidence to justify the assertion that E_r is zero [8–10]. Therefore only the forward rate needs to be measured and if the enthalpy of the reactant is known, good values of the heat of formation of the radical (R•) can be obtained. The experimental problem is to determine the activation energy of the forward reaction accurately. In most pyrolytic systems the reaction of interest may be complicated by other side reactions which arise from wall reactions or radical chain reactions. Wall reactions can be minimized, but radical chain reactions are not so easily avoided experimentally. To measure the rate of decomposition and hence the activation energy in the presence of complex radical chain reactions, it is common practice to add a radical scavenger to the pyrolysis system. This in effect suppresses or inhibits the chain reactions. The technique is known as the toluene-carrier flow system [11].

Another common method involves radical or atom metatheses kinetics [12–16]. These reactions involve the reaction of a halogen atom X, usually chlorine, bromine, or iodine, with a substrate RY (equation 4.9).

$$X\bullet + RY \rightleftarrows R\bullet + XY \tag{4.9}$$

If R radicals are generated in the presence of XY, the reverse rate can be determined, and if X atoms are produced in the presence of RY, the forward rate can be measured. From the temperature dependent rates, measured over the widest range of temperatures possible, the activation energies of the forward and reverse reactions can be determined, from which ΔH_r^0 can be determined directly from

$$\Delta H_r^0 = E_f - E_r \tag{4.10}$$

It should be noted that the heat of reaction can be determined from the equilibrium constant, K_{eq}, as determined from the measured rate constant at a given temperature by determining the ΔG_r^0, from which ΔH^0 can be extracted. If the heat of formation for XY, X and RY are known, then the heat of formation for the radical R can be determined.

For both the radical kinetics and thermal pyrolysis methods there are various techniques for producing the radical and there are a vast array of sensitive detection methods for monitoring the reactions. Two widely used methods in monitoring the kinetics are the atomic resonance fluorescence method [17–20] and photoionization mass spectroscopy [19,21]. These methods have permitted a wide range of studies to determine heats of formation of radicals.

Photoionization mass spectrometry is another commonly used method for determining the heats of formation of radicals. The principles are fundamentally different from those by which the radical kinetics and atom metatheses methods follow. In the photoionization mass spectrometric (PIMS) method, the heat of formation of the radical is determined from the energetics of ion formation. Firstly, the appearance potential of the dissociative process is measured for RX

$$\mathrm{RX} + h\upsilon \rightarrow \mathrm{R}^+ + \mathrm{X}\bullet + \mathrm{e}^- \qquad E_{\mathrm{AP}}\,(\mathrm{RX}) \tag{4.11}$$

The appearence potential, $E_{\mathrm{AP}}(\mathrm{RX})$, is defined as the minimum energy required to dissociate RX in its ground state into the positive ion R^+ and the neutral radical X•. In the literature, the appearence potential is sometimes referred to as the appearence energy. The quantities are the same. The ionization potential of the radical R• is defined as the minimum energy required to remove an electron from the radical,

$$\mathrm{R}\bullet + h\upsilon \rightarrow \mathrm{R}^+ + \mathrm{e}^- \qquad \mathrm{IP}\,(\mathrm{R}) \tag{4.12}$$

If the two processes are combined

$$\mathrm{RX} + h\upsilon \rightarrow \mathrm{R}^+ + \mathrm{X}\bullet + \mathrm{e}^- \tag{4.13}$$

$$\mathrm{R}^+ + \mathrm{e}^- \rightarrow \mathrm{R}\bullet + h\upsilon \tag{4.14}$$

$$\mathrm{RX} \rightarrow \mathrm{R}\bullet + \mathrm{X}\bullet \tag{4.15}$$

Then the combination allows the appearance and ionization potential to be related to the bond dissociation energy (BDE) by

$$\mathrm{BDE} = E_{\mathrm{AP}}\,(\mathrm{RX}) - \mathrm{IP}\,(\mathrm{R}) \tag{4.16}$$

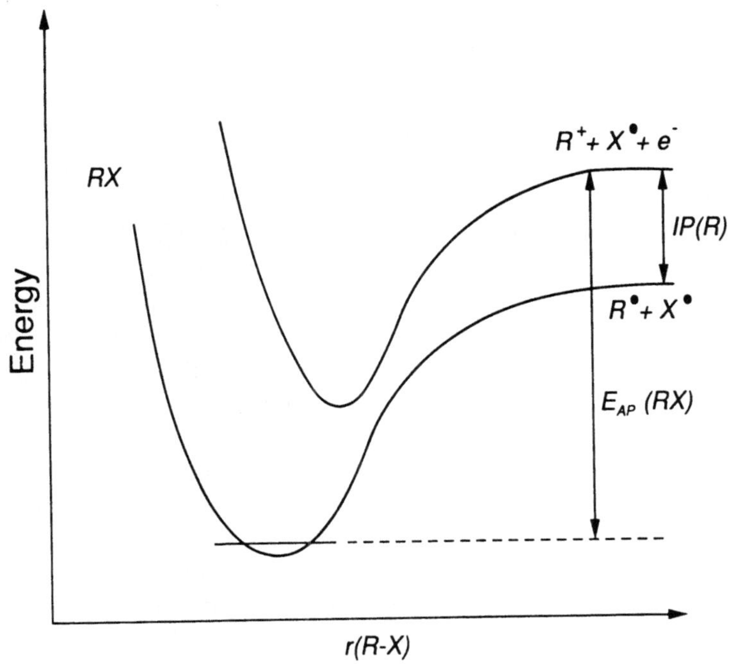

FIGURE 4.2 Pictorial relationship between the appearance and ionization potentials and bond dissociation energy.

This is represented in Figure 4.2. The approach depends upon two measurements: (1) the appearance potential for RX, along with (2) an independent measurement of the ionization potential of the radical, R. These measurements provide the most direct measure of the bond dissociation energies. To illustrate just how direct the method can be, let us take an example, hydrogen fluoride, HF [22]. The appearance potential of H^+ from HF is 19.445 ± 0.01 eV (where 1eV = 23.062 kcal mol^{-1}) and the ionization potential of H• is 13.5984 eV. Using equation (4.16), the bond dissociation energy for HF is D_0(HF) = 5.85 ± 0.01 eV (or 134.9 ± 0.2 kcal mol^{-1}). This compares well with the D_0 (HF) value from JANAF tables of 135.3 ± 0.3 kcal mol^{-1}. Although the range of uncertainty is small for the hydrogen fluoride case, the photoionization mass spectrometric method is usually known to have an accuracy of better than ±1 kcal mol^{-1}. However, there are disadvantages of the method that can introduce sources of errors in the measurement. Firstly, obtaining an accurate measurement of the ionization potential can be difficult, and usually there is considerable spread in the ionization potential values reported from most radicals. This spread results from the shift in the molecular geometry in going from the neutral to the ion. In some cases if the unimolecular decay is slow a kinetic energy shift may occur. This shift can be as small as one eV as in the case of $C_6H_5^+$ from benzene, and greater for larger molecules [23,24].

The above discussion is not intended to provide an extensive overview of all the experimental methods used in determining the energetics of radical systems. This has been done before [25,26]. Moreover, other techniques which utilize negative ion cycles are not discussed, but are reviewed by Berkowitz *et al.* [23]. Nevertheless, the intended scope of this section has been to provide an overview of the basic kinetic and thermochemical connections of how the energetics of radicals are determined experimentally and to provide a very brief overview of the uncertainties in the results of the experimental methods that are used to make the measurements. There are, however, cases where it is difficult to get an accurate measurement of the energetics. In these cases, *ab initio* molecular orbital methods can provide some assistance to the experimentalist. The energetics of radicals can be predicted to an accuracy of ± 2 kcal mol^{-1}. To date, however, decisions on which methods are useful and at what level of theory results are accurate enough is often difficult. In the next section these issues are discussed.

PROCEDURES FOR PREDICTING THE ENERGETICS OF RADICALS

Most procedures for calculation of the energetics of radicals are the same or only slightly different than those applicable to stable closed shell molecules. In empirical methods the lack of accurate experimental data on radicals can limit the accuracy compared to closed shells. In *ab initio* methods, difficulties can arise in obtaining a proper open shell spin state (i.e. eigenvalue of $\langle S^2 \rangle$; the normal value for radicals is 0.75) which do not occur in closed shell molecules, and from the inherent difference in correlation energy between paired and unpaired electrons. Largely due to computational difficulties, *ab initio* calculations traditionally have been done 'one molecule at a time', often with different modifications to the methods made between molecules. This makes it difficult, especially for the non-specialist, to assess the performance of the theory on arbitrary systems. Methods will therefore be emphasized that have been applied systematically to large sets of molecules and for which detailed comparisons with experimental thermochemical data have been made.

The energetics of radicals have been successfully calculated by both *ab initio* and semi-empirical methods. Traditionally, *ab initio* methods have only taken from experiment the fundamental constants and obtained calculated results from first principles (i.e. the Schrödinger equation), whereas semi-empirical theories have introduced parameters that have been determined, for example, from thermochemical data (i.e. heats of formation). Interestingly, the difference between *ab initio* and semi-empirical methods is becoming less distinct. Recently, *ab initio* methods have been developed that include semi-empirical corrections; these hybrid methods are capable of higher accuracy than either *ab initio* or semi-empirical methods alone.

Terminology

In order to help the reader with some of the complexity of the terminology used in the molecular orbital theory, a brief summary is provided here in order to aid the readability of the following section. For more detailed accounts, the reader is referred to several excellent reviews [27,28,29].

The two essential ingredients in an *ab initio* molecular orbital calculation are the basis set and the level of theory used in the calculation. Basis sets that are normally used are at least double zeta plus polarization (DZP) quality [30]. Such basis sets contain two functions for each formal atomic orbital (e.g. 1s, 3d, etc.) on each atom plus one set of functions (polarization functions) on each atom with one quantum number higher than required in the isolated atoms. There are two versions of this basis set type, those due to Dunning and Hay [30] and the other developed by Pople and coworkers [31,32]. The DZP and the larger triple zeta plus polarization (TZP) basis sets are Dunning and Hay type basis sets, while the 6–31G(*d*) and 6–31G(*d,p*) basis sets are considered as Pople basis set types. The larger Pople basis set [33,34] is the 6–311G(*d,p*). These basis sets can be expanded by the addition of more *d* polarization functions or the addition of higher angular momentum functions such as *f*-functions. More recently larger basis sets that are suited for electron correlation have been developed. They are correlated consistent basis sets (cc-PVTZ and cc-PVQZ) [35]. The PVQZ basis set employs *g* polarization functions.

Electron correlation can be incorporated into a calculation in several ways. One of the more common ways is using many body perturbation theory [29]. Pople and co-workers [29] have developed levels of correlation within the Møller–Plesset (MP) formation. They are MP2, which is second order; MP3, third-order, and MP4, fourth order. The MP4 calculation can be performed with single, double and quadruple excitations (SDQ) or with the inclusion of triple excitations (SDTQ). Usually triple can be quite important for the prediction of energetics, but they tend to be computationally expensive. The other form of electron correlation is coupled cluster theory [36,37]. This theory is related to many body perturbation theory. It is commonly formulated with the inclusion of single and double excitation (CCSD), single, double and triple excitations (CCSD(T)), or through quadruple excitations (CCSDTQ). Quadratic configuration (QCISD) is an approximation to CCSD methods and usually give similar results [38]. There are other methods accounting for electron correlation; however these are the more popular approaches.

The choice of basis set and electron correlation treatment determines the quality of a given *ab initio* molecular orbital calculation. In most cases, increasing the size of the basis set or improving the treatment of electron correlation improves the result.

Semi-Empirical Methods

Two different approaches are considered to the semi-empirical calculation of molecular thermochemistry. The first approach is based on the simple chemical idea of group additivity, the second on semi-empirical approximations to molecular orbital theory.

Group additivity methods. The idea that the properties of a molecule may be derived from the properties of its constituent atoms or functional groups has been successfully used by chemists for some time. The systematic application of this principle to molecular energetics is largely due to Benson and coworkers [39–47]. Its applicability is due to the local nature of chemical forces. Atoms in general have significant interactions (chemical bonds) only with their nearest neighbors, and those differences between non-partner interactions are small.

Benson and Buss [39] demonstrated in 1958 that a hierarchy of such schemes could be constructed, with the first approximation consisting of atom equivalents, the next level consisting of bond equivalents, and the third of group equivalents. As simple atom equivalent schemes cannot distinguish between systems conserving the total number of atoms (i.e. reactants and products in a chemical reaction), and bond equivalent schemes cannot distinguish isomeric species, the lowest generally useful level of approximation is that using group equivalents. Benson has tabulated numerical values for the group contribution to ΔH_f, S, and C_p for many commonly encountered functional groups. To illustrate the method, let us consider simple linear (unbranched) alkanes. The group equivalents required to estimate the thermochemical properties of these compounds are, in Benson's notation, [C—(C)(H)$_3$] and [C—(C)$_2$(H)$_2$]. For example, the heat of formation of propane ($CH_3CH_2CH_3$) is calculated as

$$\Delta H_f\,(\text{propane}) = 2\,[\text{C—(C)(H)}_3] + [\text{C—(C)}_2\text{(H)}_2] \tag{4.17}$$

$$= 2\,(-10.2) - 4.93 = -24.33\ \text{kcal mol}^{-1} \tag{4.18}$$

$$\Delta H_f\,(\text{exp}) = -25.0 \pm 0.1\ \text{kcal mol}^{-1} \tag{4.19}$$

using the numerical values from Table A.1 of [44]. The agreement with experiment is excellent for the example above.

Benson has also worked out corrections for some commonly occurring non-bonded (higher-order) interactions, such as those occurring between different rotameric states of chain molecules (*cis-trans*, *trans-gauche*, etc.) and in ring compounds [44].

The physical basis of the group additivity method does not restrict its validity to closed shell molecules, and it has been extended to permit the estimation of the thermochemical properties of radicals [1]. In practice, the lack of available thermochemical data is severely limiting.

More recently, there have been methods proposed that use group additivity ideas to formulate corrections to *ab initio* calculations. The simplest such scheme is due to Wiberg [48], and has been extended by Ibrahim and Schleyer [49]. Wiberg's original scheme was developed for the prediction of the heats of formation of alkanes and alkenes. Group equivalents for CH_3, CH_2, CH, and C are derived using 6–31G* and 4–31G SCF energies and known hydrocarbon heats of formation. The heat of formation (in kcal mol^{-1}) is calculated as $\Delta H_f = 627.5\,(E(ab\ initio) - \Sigma E\,(\text{groups}))$. For example, the heat of formation of propane is calculated from its 6–31G* SCF energy.

$$\Delta H_f\,(\text{propane}) = E_{scf}\,(\text{propane}) - 2(-\text{CH}_3) - (>\text{CH}_2) \tag{4.20}$$

$$= 627.5\,[-118.26365 - 2(-39.59841) - 39.02671] \tag{4.21}$$

$$= -25.1\ \text{kcal mol}^{-1} \tag{4.22}$$

$$\Delta H_f(\text{exp}) = -25.0\ \text{kcal mol}^{-1} \tag{4.23}$$

Comparisons with thermochemical data indicate that the calculated heats of formation are accurate to ± 2 kcal mol^{-1}. As one might anticipate, this scheme breaks down for benzene, giving a 10.7 kcal mol^{-1} error in the calculated heat of formation. Wiberg gives both olefinic and saturated CH equivalents, neither of which is quite correct for the aromatic benzene CH groups. Subsequently, Wiberg extended the method to include acetylenic and oxygen-containing compounds and found a RMS error of

1.23 kcal mol^{-1} in a comparison of the 6–31G* calculated heats of formation with experiment for 42 molecules [50].

A similar scheme was reported by Ibrahim and Schlelyer [49], who extended it to a wider range of molecules. They report 3–21G and 6–31G* atom (rather than group) equivalents for hydrogen, carbon, nitrogen, oxygen and fluorine in a variety of molecular environments. Comparisons with experimental data for a test set of over 100 small molecules, including fluorocarbons, radicals, carbocations, and protonated species show an average error of approximately 2 kcal mol^{-1}. The results for radicals are very encouraging, if limited. Ibrahim and Schleyer report an average error of 1.5 kcal mol^{-1} with the 6–31G* basis for a set of eight free radicals, a level of accuracy not significantly worse than for closed shells. As Ibrahim and Schleyer have used unrestricted Hartree–Fock (UHF) energies for radicals, it is possible that spin contamination affects the accuracy that may be expected for radicals. It is interesting that the largest error found for the eight free radical cases considered occurs for ethynyl, which also has the largest spin contamination ($\langle S^2 \rangle = 1.187$ at 6–31G*), but more evidence is needed before any conclusions may be drawn.

Both Wiberg and Ibrahim and Schleyer demonstrate that these methods work best for 'classical' molecules. The results obtained from aromatic and strained ring compounds are significantly larger that the 1–2 kcal mol^{-1} errors found for other molecules. It is possible that the use of correlated energy calculations (e.g. MP2/6–31G*) would improve the agreement in these cases.

Another related approach is the BAC-MP4 method developed by Melius and co-workers [51–54]. This method is based on bond-additivity corrections to MP4/6–31G**//UHF/6–31G* *ab initio* energies. As indicated below, this method is really an *ab initio* calculation with empirical methods, similar in spirit, if not detail, to such methods as G1 and G2.

Suppose you have calculated the total atomization enthalpy of methane, CH_4. Since four C—H bonds have been broken, a correction for the error in the calculated energy of a C—H bond could be estimated as

$$\mathrm{BAC(C{-}H)} = \tfrac{1}{4}[\Delta H_{at}\,(\mathrm{exp}) - \Delta H_{at}\,(\mathrm{calc})] \qquad (4.24)$$

and used to correct the calculated atomization of the remaining CH_n species. This is the idea behind the BAC-MP4 method. Of course, accurate experimental heats of formation are necessary to obtain the bond corrections.

The BAC-MP4 energy may be written as

$$\mathrm{E(BAC\text{-}MP4)} = \mathrm{E(MP4)} + \sum_{ij} \mathrm{BAC}(\mathrm{A}_i - \mathrm{A}_j) + \mathrm{BAC}(S^2) + \mathrm{BAC\ (UHF\text{-}I)} \qquad (4.25)$$

E(MP4) is the *ab initio* MP4 (SDTQ)/6–31G** energy, calculated at the 6–31G* SCF optimized geometry. A vibrational zero point correction is made using the 6–31G* harmonic vibrational frequencies, divided by 1.12. BAC(A_i–A_j) is the empirical correction to the A_i–A_j bond. The exact form of BAC (A_i–A_j) depends on the chemical environment; for the three atom molecule A_k–A_i–A_j it is

$$BAC(A_i\text{–}A_j) = f_{ij}\, g_{kij} \tag{4.26}$$

where

$$f_{ij} = A_{ij} \exp(-\alpha_{ij} R_{ij}) \tag{4.27}$$

$$g_{kij} = (1 - h_{ik} h_{ij}) \tag{4.28}$$

$$h_{ik} = B_k \exp[-\alpha_{ik}(R_{ik} - 1.4\ \text{Å})] \tag{4.29}$$

R_{ij} is the optimized 6–31G* SCF bond distance, and α_{ij}, A_{ij} and B_k are empirical parameters. The BAC(S^2) term corrects for UHF spin contamination errors in open shell molecules, and is given by

$$BAC(S^2) = E(\text{UMP3}) - E(\text{PUMP3}) \tag{4.30}$$

E(MP3) is the UMP2/6–31G** energy and E(PMP3) is the PMP3/6–31G** energy, calculated using the method of Schlegel [55]. (The PMP3 refers to the spin corrected third-order Møller–Plesset perturbation energy).

The BAC(UHF-I) term is applied to closed shell molecules that are UHF unstable, that is, a broken symmetry UHF wavefunction exists that has a lower energy than the RHF solution. It is given by

$$BAC(\text{UHF-I}) = K_{\text{UHF-I}}\, S(S + 1) \tag{4.31}$$

where $K_{\text{UHF-I}}$ is taken as 10.0 kcal mol^{-1}.

BAC-MP4 energies, calculated as described above, are then used to calculate heats of formation from calculated atomization energies, and known atomic heats of formation. Corrections to 298 K are made from standard formulae of statistical mechanics, using the calculated vibrational frequencies.

Melius has found an average error of 1.3 kcal mol^{-1} from the comparison of approximately 90 calculated BAC-MP4 heats of formation [54]. Good agreement is found for radicals as well as stable species. The method has been successfully applied by Melius and coworkers to a number of problems in the combustion of hydrocarbons and energetic materials such as fuels and explosives.

MINDO, MNDO, AM1 and PM3 methods. Semi-empirical molecular orbital theories, in particular Huckel theory and its extensions, have long been useful for qualitative molecular orbital studies. The serious use of semi-empirical methods for thermochemical calculations was pioneered by Dewar and his coworkers, beginning in 1975 with the MINDO/3 method [56]. Further development resulted in the MNDO [57], AM1 [58] and PM3 [59] methods.

All of these methods are based on parameterized approximations to the Hartree–Fock self consistent field equations. Core electrons are neglected, and the valence molecular orbitals are expanded in a minimum basis of atomic orbitals (LCAO method). Overlap approximations (e.g. INDO, NDDO) are used to simplify and/or eliminate the remaining integral terms and at this point explicit parameterization is introduced. For a detailed account of the resulting semi-empirical SCF equations, the reader is referred to the literature [60,61].

MINDO/3 (modified intermediate neglect of differential overlap, version 3) derives, as the name suggests, from the INDO (intermediate neglect of differential overlap) theory of Pople *et al.* [62]. An important difference is that the parameters in the theory are taken as adjustable parameters, rather than obtained from spectral data or theory. The accuracy of the MINDO/3 method (a mean absolute error of 11.0 kcal mol^{-1}) was found for the calculated heats of formation of a large set of known organic molecules [63]. MINDO/3 had particular difficulties with systems containing lone pairs, a consequence of the INDO approximation. A computational difficulty was created by the use of diatomic as well as atomic parameters; this made the determination of parameterization for new elements difficult. Thus Dewar and Thiel were led to the use of the NDDO (neglect of diatomic differential overlap) approximation, due originally to Pople *et al.* [64]. The result was the MNDO (modified neglect of diatomic overlap) method, which was formulated entirely in terms of atomic parameters. Its accuracy was much improved over MINDO/3. Dewar and Thiel reported an absolute error of 6.3 kcal mol^{-1} for the test set mentioned above [63]. These successes, and the subsequent distribution of the MOPAC computer program [65], have resulted in widespread use of these methods.

Subsequent developments by Dewar and coworkers have taken place within the MNDO framework. Experience with MNDO suggested that many of its largest errors resulted from an excess repulsion between non-bonded atoms (i.e. at van der Waals distances). To correct this problem, Gaussian terms were added to the core–core interaction in the Fock matrix; the resulting method is known as AM1 (Austin method 1). Comparisons of MINDO/3, MNDO, and AM1 found mean absolute errors of 10.8, 6.3 and 5.5 kcal mol^{-1} respectively, in the calculated heats of formation for a test set of 138 organic molecules. For a set of 14 neutral radicals, the mean absolute errors in the calculated heats of formation are found to be 8.7 and 8.0 kcal mol^{-1} for MNDO and AM1, respectively [58].

One should note that much of the effort required to develop these methods was consumed by the numerical determination of the empirical parameters. MNDO required seven parameters per atom; AM1 required up to 19. The numerical difficulties were so great that the original AM1 paper [58] only reported parameters for the four atoms C, H, N, and O. Stewart subsequently devised an improved numerical optimization scheme, which he used to develop the MNDO-PM3 or PM3 (parametric method 3) model [59]. The PM3 Fock matrix is very similar to that used in AM1, differing in the number of parameters. Each element is described by 18 parameters in PM3 (eleven for hydrogen). Fully optimized parameter sets for H, C, N, O, F, A1, Si, P, S, Cl, Br, and I have been reported by Stewart [59]. In an extensive comparison, Stewart finds mean average errors of 7.8 kcal mol^{-1} in the PM3 heats of formation for 657 test molecules. The errors found for MNDO and AM1 are 13.9 and 12.7 kcal mol^{-1}, respectively [66].

It appears that the heats of formation of small to medium size molecules may be calculated with modern semi-empirical theory to an accuracy of roughly 5–10 kcal mol^{-1}. The limited data available on radicals suggests comparable or slightly greater errors should be expected. A great advantage of these methods is their relative computational simplicity, allowing results to be obtained rapidly and for relatively large systems.

Ab Initio Methods

As mentioned earlier, *ab initio* methods, by definition, start from the Schrödinger equation, taking as experimental parameters only the fundamental constants i.e. $\hbar$, m_e, and e. This approach is computationally very demanding, and except for a few simple cases (e.g. the hydrogen molecule), there have not been many results of chemical accuracy (± 1 kcal mol^{-1}) until relatively recently. Quantities of thermochemical interest, such as dissociation energies or heats of reaction, are obtained from *ab initio* calculations as the (relatively small) difference of large numbers, and great care must be taken to ensure an accurate result. For example, the calculation of the dissociation energy of the N_2 molecule into two nitrogen atoms is, even now, an extremely difficult task. The reason for this is the enormous difference between the electronic structure of the closed shell N_2 molecule and two ^4S nitrogen atoms. It has been found, especially when using relatively low levels of theory, that energy differences can be most accurately calculated between systems whose electronic structure is similar.

Most *ab initio* calculations may be analyzed in terms of one-particle (basis set) and N-particle (CI) expansions. Truncation errors in either (or both) usually account for the observed discrepancies with experimental data. For ground state equilibrium systems, modern correlation methods (multi-reference CI and coupled cluster theory) have been found to agree well (less than 1 kcal mol^{-1}) with benchmark full CI calculations [67,68] suggesting that truncation of the one-particle basis is the major remaining source of error in such calculations. However, as one reaches the 1 kcal mol^{-1} level of accuracy in thermochemical calculations, the sources of error become more difficult to pinpoint. Small residual discrepancies may be due to geometry errors, neglect of core–core and core–valence interactions, vibrational zero-point errors, and relativistic effects, in addition to the truncation errors mentioned above. Modern methods, as discussed below, can achieve accuracies of 1–2 kcal mol^{-1} for the thermochemistry of small molecules [69]. Significant improvement will require attention to the areas mentioned above. It is possible, even likely, that in the near future improvements will come in the speed and reliability of these methods, rather than in their absolute accuracy. Conversely, in the comparison of high-level *ab initio* results with experiment, one should realize that large differences (>5 kcal mol^{-1}) are probably meaningful, but small ones may not be.

4.3.4.1 'Conservation Laws'. Attempts have been made to classify, for the purpose of interpreting *ab initio* calculations, the degree of electronic similarity between reactants and products in chemical reactions. Much of this work is due to Pople and co-workers [29].

The dissociation of a closed shell molecule into two radicals results in products which have fewer electron pairs than in the reactants. It was recognized long ago that SCF methods, which by definition neglect electron correlation, will have significant errors in the calculated heats of such reactions. This observation led naturally to the study of isogyric reactions [70], defined as those reactions conserving the number of electron pairs. Thus, the O—H bond dissociation of water,

$$H_2O\ (^1A_1) \rightarrow OH\ (^2\Pi) + H(^2S) \qquad (4.32)$$

is not isogyric, but the reaction

$$H_2O(^1A_1) + H(^2S) \rightarrow OH(^2\Pi) + H_2(^1\Sigma_g) \qquad (4.33)$$

is.

Note that bond dissociations in high spin radicals can be isogyric, as in

$$CH(^4\Sigma_g) \rightarrow C(^3P) + H(^2S) \tag{4.34}$$

but of course, this is not always the case, as in, for example

$$OH(^2\Pi) \rightarrow O(^3P) + H(^2S). \tag{4.35}$$

It is clear that, in the water example, the bond dissociation energy $D_o(H{-}OH)$ may be obtained from the calculated heat of either reaction (4.32) or (4.33), but we expect a cancellation of errors in the correlation energy to occur for the isogyric reaction (4.33). However, note that use of the isogyric reaction to determine $D_o(H{-}OH)$ requires experimental data, in this case the experimentally well known value of $D_o(H_2)$. It is, of course, quite possible to encounter cases where the necessary data is not known, and this is a fundamental limitation of the method.

It is possible to formulate other criteria for the electronic similarity of reactants and products. Consideration of possible similarities in chemical bonding leads to the concept of the isodesmic reaction [29,71], defined as one in which the numbers of each type of bond is conserved, and only the relationships between them differ. For example, the reaction

$$CH_3OCH_3 + H_2O \rightarrow 2CH_3OH \tag{4.36}$$

is isodesmic: there are 6 C—H bonds, 2 C—O bonds, and 2 O—H bonds on each side of the reaction. The number of electron pairs is conserved, so this reaction is also isogyric. One can imagine reactions that are isodesmic, but not isogyric, for example

$$CH_2(^3B_1) + CH_2F_2 \rightarrow CF_2(^1A_1) + CH_4 \tag{4.37}$$

but in practice such cases are seldom encountered (and best avoided).

The classification of reactions as isogyric is exact; not so for isodesmic reactions, although it is nearly exact often enough to be useful. The definition breaks down somewhat in practice when the bonding cannot be cleanly characterized, and caution is advised in such cases. An interesting class of isodesmic reactions is the so-called bond separation reactions [29,71,72]. For a given molecule, consider a particular bond between two non-hydrogens. Replace all other bonds to non-hydrogens in the molecule with bonds to hydrogens. Do this for each bond between two non-hydrogens to form the products, and add water, ammonia, and methane as necessary to ensure stochiometric balance. For example, the bond separation reactions for ketene and carbon tetrafluoride are

$$H_2CCO + CH_4 \rightarrow CH_2CH_2 + H_2CO \qquad (4.38)$$

$$CF_4 + 3CH_4 \rightarrow 4CH_3F \qquad (4.39)$$

For *n*-propyl radical, one finds

$$CH_3CH_2CH_2 + CH_4 \rightarrow CH_3CH_3 + CH_3CH_2 \qquad (4.40)$$

Bond separation reactions give a systematic way of relating the heat of formation of a larger molecule to a group of smaller prototypes. They have also been used to discuss substituent effects on molecular stability. The calculation of bond separation energies can be done with good accuracy (5 kcal mol^{-1} or better) with modest basis SCF calculations. For example, Radom *et al.* found a mean absolute error of 3.1 kcal mol^{-1} in the calculated heats of formation of 21 small molecules using 4-31G SCF bond separation energies [72].

Further generalization of these ideas has been considered. The homodesmotic (or homodesmic) reaction has been proposed by George *et al.* [73,74]. It may be defined as an isodesmic reaction in which the numbers of each type of atom, as defined by the type and number of substituents and hybridization, are conserved. For example, the reaction

$$3CH_3CH_2CH_3 \rightarrow cyclo\text{-}(CH_3)_3 + 3CH_3CH_3 \qquad (4.41)$$

is homodesmic, but the isodesmic reaction

$$3CH_3CH_3 \rightarrow cyclo\text{-}(CH_3)_3 + 3CH_4 \qquad (4.42)$$

is not. Few isodesmic reactions are homodesmic, which limits the practical applicability of the idea.

G1 and G2 methods. The systematic development by Pople and coworkers of accurate methods for the *ab initio* calculation of molecular energies culminated in the Gaussian-1 (Gl) [75,76] and Gaussian-2 (G2) [77] methods. These methods consolidate a number of ideas, some of which have already been mentioned.

The G1 method was developed as a general procedure for the calculation of molecular energies with a target accuracy of ± 2 kcal mol^{-1} for calculated energy differences. The G1 energy may be written as

$$E(\mathrm{G1}) = E(\mathrm{MP4/6\text{–}311G}(d,p)) + \Delta E(+) + \Delta E(2df) + \Delta E(\mathrm{QCI}) + \Delta E(\mathrm{HLC}) + \Delta \mathrm{E(ZPE)} \qquad (4.43)$$

where

$$E(+) = E(\text{MP4/6–311} + \text{G}(d,p)) - E(\text{MP4/6–311G}(d,p)) \quad (4.44)$$

$$E(2df) = E(\text{MP4/6–311G}(2df,p)) - E(\text{MP4/6–311G}(d,p)) \quad (4.45)$$

$$E(\text{QCI}) = E(\text{QCISD(T)/6–311G}(d,p)) - E(\text{MP4/6–311G}(d,p)) \quad (4.46)$$

$$\Delta E(\text{HLC}) = -0.19\ \text{n}_\alpha - 5.95\ \text{n}_\beta \text{ (in millihartrees).} \quad (4.47)$$

The n_α and n_β appearing in ΔE(HLC) are the number of α and β valence electrons. The MP4 and QCISD(T) energies are calculated at the MP2(Full)/6–31G* optimized geometry, using the frozen core approximation. The MP4 energies used are the full fourth-order results, including triples. All calculations are based on RHF wavefunctions for closed shells, UHF for open shells. ΔE(ZPE) is the HF/6–31G* vibrational zero-point energy, scaled by 0.8929.

The G1 energy contains an empirical correction, ΔE(HLC), referred to as a 'higher level correction' by Pople *et al.* [75]. The HLC is intended to approximately correct for remaining basis set deficiencies. The numerical parameters in the HLC were chosen to reproduce the exact known energies of the hydrogen atom and molecule. The form of the HLC evolved naturally from earlier work, [70,78–81] in which MP4 calculations were used to evaluate the heats of reaction for isogyric reactions of the form

$$AH_n + H \rightarrow AH_{n-1} + H_2. \quad (4.48)$$

As the HLC cancels in the calculation of the heats of isogyric reactions, it can be seen that G1 represents a generalization of this idea. The utility of the HLC in the calculation of other processes clearly depends on the relative accuracy of the 6–311 + G(2*df,p*) basis sets remaining roughly constant across the periodic table, which seems to occur in practice.

If the contributions of diffuse functions, higher polarization, and post-MP4 correlation effects are additive, then the G1 energy calculation is effectively at the QCISD(T)/6–311 + G(2*df,p*) level. Comparison with experimental dissociation energies for a set of 31 small molecules gives a mean absolute difference of 1.47 kcal mol^{-1} [75].

G1 theory was found to have difficulties with certain compounds, in particular, ionic molecules (such as LiF), triplets (CH_2 and O_2), some hydrides (NH_3 and N_2H_4), and hypervalent molecules (such as SO_2 and ClO_2). This led to the development of the Gaussian-2 (G2) theory [77]. The G2 energy may be written

$$E(\text{G2}) = E(\text{G1}) + \Delta + 1.14\ n_{\text{pair}} \quad (4.49)$$

$$\Delta = E(\text{MP2}/6-311+\text{G}(3df,2p) - E(\text{MP2}/6-311\text{G}(2df,p)) - E(\text{MP2}/6-311+\text{G}(d,p)) + E(\text{MP2}/6-311\text{G}(d,p)) \quad (4.50)$$

G2 requires only slightly more computational effort than G1, in particular, a single MP2/6 − 311 + G(3*df*, 2*p*) energy calculation. The new term Δ corrects, at the MP2 level, for non-additivity of the contributions of diffuse functions and higher polarization, and adds an additional *d* function on non-hydrogens and an additional *p* on hydrogens. The 1.14 n_{pair} term adjusts the HLC so that there is zero mean deviation from experiment in the calculated atomization energies of their test set (55 molecules).

The G2 atomization energies are found to be significantly improved relative to G1, the mean absolute deviations from experiment being 1.21 kcal mol^{-1} and 1.53 kcal mol^{-1}, respectively [77]. The problems G1 encountered with hydrides and ionic molecules are much improved.

With current (1995) computers, G2 is restricted by CPU and storage limitations to relatively small molecules (5–6 heavy atoms). Realizing this, Curtiss *et al.* [82] devised modified versions, G2 (MP2) and G2 (MP3), which substitute MP2 and MP3 basis set corrections for the MP4 corrections in full G2. These methods were found to have mean average deviations from experiment of 1.58 and 1.52 kcal mol^{-1}, respectively. It was concluded that G2 (MP3) had no significant advantage over G2 (MP2), whose use was recommended in studies for which the full G2 is not possible.

The G2 (MP2) energy may be written

$$E(\text{G2(MP2)}) = E(\text{QCISD(T)}/6-311\text{G}(d,p)) - E(\text{MP2}/6-311\text{G}(d,p)) + E(\text{MP2}/6-311+\text{G}(3df,2p)) + \Delta E(\text{HLC}) + E(\text{ZPE}). \quad (4.51)$$

ΔE(HLC) and Δ*E*(ZPE) are identical to that used in G1 and G2. Note that the most computationally demanding step in G1 and G2, evaluation of the MP4/6 − 311G (2*df*,*p*) energy, has been eliminated.

One might ask to what extent the additivity approximations in G2 contribute to the remaining errors of the method. This question was answered by comparison of energies obtained from exact QCISD(T)/6 − 311 + G(3*df*, 2*d*) calculations with G2 energies. The mean absolute deviation from experiment went from 1.21 kcal mol^{-1} to 1.17 kcal mol^{-1}, indicating that the basis set additivity approximations used in G2 are quite accurate [83].

The G1 and G2 methods have been considered in detail for several reasons:

1. They are defined for any molecular system ('model chemistry').

2. They give results of demonstrated high accuracy, usually better than 2 kcal mol^{-1}.
3. They are easy to use and the necessary computer programs are readily available; many G1 and G2 results have appeared in the literature.

However, calculations of equal or greater accuracy have been done with other methods; some of those are now mentioned.

Other methods. A method closely related to G2 was proposed by Martin [84,85]. Using both atomic natural orbital (ANO) and Dunning's correlation consistent basis sets (cc-pVTZ and cc-pVQZ) to calculate CCSD(T) energies, an empirical correction of the form

$$\Delta E = \alpha N_\sigma + bN_\pi + cN_{\text{pair}} \tag{4.52}$$

is found, gives mean absolute errors less than 1 kcal/mol with *spdf* basis sets and about 0.5 kcal/mol with *spdfg* basis sets for the calculated atomization energies of 13 small molecules. Note that this method, as proposed, is defined only for closed shell molecules. It does, however, raise the interesting possibility of creating G2-like (i.e. composite) methods using other basis sets and more elaborate forms for the empirical correction.

Most modern *ab initio* approaches to the calculation of accurate molecular energies are based on either infinite-order perturbation theory or multi-reference configuration interaction (MRCI). The latter approach has been successfully employed by Bauschlicher and coworkers at the NASA Ames laboratory [67,69]. They have demonstrated in a number of examples using small basis sets (e.g. double zeta) that MRCI calculations based on a complete active space SCF reference can give results that agree closely with benchmark full CI results. When similar MRCI calculations are performed with large ANO basis sets, accurate thermochemical results have been obtained [69,86–89]. An advantage of this approach is that, due to the use of configuration interaction methods, exact spin states are obtained for radical species. Unfortunately, MRCI methods are computationally very expensive. Partly as a result, the Ames group has investigated the use of single reference methods based on the coupled pair functional, and coupled cluster theory [90–93].

Dunning and co-workers [94–97] have used exponential extrapolations of the form

$$E(i) = c_1 + c_2 \exp(-c_3\beta) \tag{4.53}$$

where c_1 is the complete basis set (CBS) limit and β is the number of *d*-functions in the basis set, to estimate the CBS limit of a sequence of *ab initio* (MRCI) calculations performed with their correlation consistent basis sets. Using these methods, they have made accurate estimates of total correlation energies, as well as thermochemical and spectroscopic properties, for several small molecules.

A novel approach has been developed by Petersson and co-workers based on the study of the convergence properties of natural orbital expansions. They have shown [98–101] that the convergence of second-order Møller–Plesset pair correlation energies ${}^{\alpha\beta}e_{ij}^{(2)}(N)$ calculated with N pair natural orbitals goes asymptotically as $1/N$;

$$\lim_{N\to\infty} {}^{\alpha\beta}e_{ij}^{(2)}(\mathrm{N}) = {}^{\alpha\beta}e_{ij}^{(2)}(\infty) + \left(\frac{225}{4608}\right)|S|_{ij}^{2}\frac{1}{N+\delta_{ij}}. \tag{4.54}$$

These asymptotic results have been used to develop practical methods for extrapolating finite basis set calculations to obtain estimates of the complete basis set (CBS) limit [101–103]. It has been demonstrated that CBS extrapolations of MP2 calculations using (14*s*9*p*4*d*2*f*, 6*s*3*p*1*d*)/(6*s*6*p*3*d*2*f*, 4*s*2*p*1*d*) atomic pair natural orbital (APNO) basis sets combined with QCISD(T) calculations of the higher order terms in the Møller–Plesset perturbation series (CBS-QCI/APNO method) gives RMS errors of less than 1 kcal mol^{-1} in the calculated atomization energies of 29 first-row molecules [104]. A recent modification of this method [105] is computationally feasible for species with up to four first-row atoms and has a mean absolute error of 0.53 kcal mol^{-1} for the 64 first-row examples from the G2 test set. The advantage of this approach is that the contribution of the higher angular momentum basis functions is not explicitly calculated, but instead estimated from lower angular momentum terms. However, care must be taken to ensure that the low-lying natural orbitals, upon which the extrapolation is based, are adequately converged to ensure an accurate extrapolation.

Lastly, it is noted that there has been a freshening of interest in density functional theory (DFT) [106] in the last several years. These methods are based on the Hohenberg–Kohn theorem [107], which states that the electronic energy of a system is a unique function of its electron density. Thus it remains only to find the functional form of the electron density, a task that has engaged theorists ever since. Although it is beyond the scope of this article to review density functional theory, recent developments [108] suggest that DFT may soon take a place alongside other standard methods for the accurate calculation of molecular energetics. For example, Johnson *et al.* [109] have recently demonstrated a density functional model using HF/6–31G(*d*) densities that has a mean absolute error of 4.14 kcal mol^{-1} on the G2 test set (excluding electron affinities). Similar results have been obtained by other workers [110–113]. The low computational cost of DFT makes it a very attractive alternative to traditional *ab initio* methods.

EXAMPLES: COMPARISON BETWEEN EXPERIMENT AND THEORY

It is important to recognize that at present there are theoretical methods available for estimating heats of formation that can provide reliable results that could be helpful to the experimentalist. Some of these methods are quite elaborate and computationally demanding, while others are quite simple. Just how useful some of the methodologies are in estimating heats of formation of open shell systems will be illustrated. The goal here is to compare the various methodologies with each other and see if any methods achieve consistent results. More importantly, it is useful to see how the results of various theoretical methods compare directly with experiment. Also a case will be highlighted, for a closed shell system, where discrepancies exist between the theoretical and experimental results.

Ethynyl Radical (C_2H) and Vinyl Radical (C_2H_3)

This section begins with a discussion of two examples from hydrocarbon chemistry, the ethynyl and vinyl radicals, each of which has received considerable recent attention, both experimentally and theoretically. The experimental results are briefly summarized, followed by a comparison with theory.

The experimentally measured value of the C–H bond dissociation energy of acetylene, and consequently the heat of formation of the ethynyl radical, has been the subject of controversy, as recent experimental determinations gave conflicting values near either 127 kcal mol^{-1} or 132 kcal mol^{-1}. Two groups have directly determined $D_0(H{-}CCH)$ from measurements of the translational energy release upon photodissociation of acetylene. Wodtke and Lee [114] found 132 ± 2 kcal mol^{-1}, while the first value [115] reported by Wittig's group was 127 ± 1.5 kcal mol^{-1}. Field and coworkers [116] at nearly the same time, deduced an upper bound of 126.647 ± 0.002 kcal mol^{-1} from Stark anti-crossing spectroscopy. Using a thermodynamic cycle, Ervin *et al.* [117] determined $D_0(H{-}CCH)$ from the gas phase acidity of acetylene and the electron affinity of C_2H, obtaining a value of 131.3 ± 0.7 kcal mol^{-1}. This value is in very good agreement with the earlier, but less precise, result of Janousek *et al.* [118] of 130.4 ± 5 kcal mol^{-1}, obtained from the same thermodynamic cycle. Subsequently, it was discovered that Wittig's value of 127 ± 1.5 kcal mol^{-1} was an artefact of C_2H photolysis, and the most recent result from his laboratory is D_o(HCCH) = 131.8 ± 0.5 kcal mol^{-1} [119]. In another recent measurement, Ruscic and Berkowitz [120] measured the threshold for photoion pair production in acetylene ($C_2H_2 \rightarrow H^+ + C_2H^-$) and from it found D_o(H–CCH) = 131.1 ± 0.7 kcal mol^{-1}. Baldwin *et al.* [121] used photofragment imaging to determine D_o(H–CCH) = 131 ± 1 kcal mol^{-1}. Thus, there is an emerging consensus that the correct value is in the range 131–132 kcal mol^{-1}, and the recent review of Berkowitz *et al.* [23] recommends the 131.3 ± 0.7 kcal mol^{-1} value of Ervin *et al.* [117]. This value is adopted in Table 4.1.

The C—H bond dissociation of ethylene has also been the subject of several recent experimental studies, and again there is some uncertainty in the exact value. The recently recommended [23] experimental value of 109.7 ± 0.8 kcal mol^{-1} was obtained by Ervin *et al.* [117], again from a negative ion thermochemical cycle, using their measured values of the electron affinity of vinyl radical and the gas phase acidity of ethylene. A different ion cycle, involving the heat of formation of $C_2H_3^+$ and the ionization potential of C_2H_3 can be used to determine a lower bound to $D_o(H-C_2H_3)$. Unfortunately, there is a 2–3 kcal mol^{-1} uncertainty in $\Delta H_f(C_2H_3^+)$ [23,122] and a 7.8 kcal mol^{-1} difference in the two recent measurements [123,124] of IP(C_2H_3). Therefore a useful bound is not available from this route [23]. It is noted that the IP of C_2H_3 is somewhat problematic because the ionization is so non-vertical; this may explain some of the discrepancy. Recent kinetic studies [125,126] of the

TABLE 4.1 Calculated C—H bond dissociation energies of acetylene and ethylene

Method	$D_0(H-C_2H)$	$D_0(H-C_2H_3)$
PM3	—	98.8
6-31G* atom eq.	128.3	108.8
BAC-MP4	130.1	110.8
DFT	133.3	108.7
MRCI	130.1	110.4
G1	133.4	110.1
G2	133.4	110.4
CBS-QCI	131.5	109.7
Exp	131.3 ± 0.7	109.7 ± 0.8

$Cl + C_2H_4 \rightleftarrows HCl + C_2H_3$ reaction give $D_0(H\text{-}C_2H_3) = 105.1 \pm 0.3$ kcal mol^{-1}. The reason for the 5 kcal mol^{-1} discrepancy of this result with the ion cycle result of Ervin *et al.* [117] is unclear.

Unlike experiment, theory has provided a consistent picture of the thermochemistry of acetylene and ethylene. Table 4.1 summarizes the relevant calculations of the C-H bond dissociation energies of acetylene and ethylene. It can be seen that only one calculation, the PM3 C—H bond dissociation energy of ethylene, is in error by more than 1 kcal mol^{-1}. All other calculated results agree within their approximate error limits (as discussed above) with the experimental values (from [23]), a significant achievement of modern computational chemistry. The triple bond of acetylene, and the resulting large ($\langle S^2 \rangle = 1.187$, 6-31G*) UHF spin contamination of the ethynyl radical make this a difficult case computationally. As pointed out by Curtiss *et al.*, use of the UMP2/6-31G* geometry for C_2H results in D_0(H-CCH) that is too large (albeit by only 2.1 kcal mol^{-1}). Use of the more accurate QCISD(T)/6-31G* geometries reduced the G2 value for $D_0(H\text{-}C_2H)$ to 132.3 kcal mol^{-1}, an error of only 1.0 kcal mol^{-1} compared to the experimental value. Excepting the PM3 result, the calculated values for the vinyl radical are all within 1.6 kcal mol^{-1} of each other and within 1 kcal mol^{-1} of experiment. Indeed, it seems that the close agreement of the calculated results for $D_0(H\text{-}C_2H_3)$ with the gas phase acidity/electron affinity value [117] offers a strong argument to favor it over the kinetic value [125,126]. These calculations are representative of the level of accuracy that may be achieved with currently available computational methods on small first-row molecules. For a given molecule, the computational cost of these calculations is approximately

$$\text{semi-empirical} \ll \text{SCF} < \text{DFT} \ll \text{CI, coupled cluster, G2}$$

indicating that the SCF based methods, in this case 6-31G* atom equivalents, and DFT are the most cost effective. What one obtains from the more expensive calculations is a demonstrated high level of accuracy (for example, the ± 2 kcal mol^{-1} for G2) that may be relied on in studies of unknown systems, rather than, in many cases, a clearly much more accurate calculation. Of course, in cases such as those considered here, that extra effort is clearly of significant value.

Chlorine Oxide Radicals

Chlorine oxide radicals such as chlorine monoxide, ClO, and chlorine peroxy radical, ClOO, have been implicated in atmospheric chemical mechanisms for polar ozone destruction [127], namely

$$ClO + BrO \rightarrow Br + ClOO \tag{4.55}$$

$$ClOO + M \rightarrow Cl + O_2 + M \tag{4.56}$$

$$Cl + O_3 \rightarrow ClO + O_2 \tag{4.57}$$

$$Br + O_3 \rightarrow BrO + O_2 \tag{4.58}$$

$$\text{net} \quad 2O_3 \rightarrow 3O_2 \tag{4.59}$$

Another mechanism involving ClO radicals is the self-reaction that yields ClOO radical [128] via

$$ClO + ClO \rightarrow Cl + ClOO \tag{4.60}$$

$$ClOO + M \rightarrow Cl + O_2 + M \tag{4.61}$$

$$2\{Cl + O_3 \rightarrow ClO + O_2\} \tag{4.62}$$

$$\text{net} \quad 2O_3 \rightarrow 3O_2 \tag{4.63}$$

The termolecular addition reaction of ClO to form $(ClO)_2$ is the rate-limiting step in a proposed mechanism for polar ozone destruction [128].

Chlorine oxide (ClO) radical and FOCl. ClO radical, the simplest chlorine oxide radical, has long been a classical system for *ab initio* methods. The heat of formation for ClO is determined from the dissociation energy of ClO combined with the enthalpies of formation for Cl atom and O atom at 0 K. Using the G1 method, Pople *et al.* [76] estimated the dissociation energy of ClO to be 60.9 kcal mol^{-1}. Curtiss *et al* [77] redetermined the dissociation energy using G2 theory to be 61.5 kcal mol^{-1}. The dissociation energy of ClO has been determined spectroscopically by Coxem and Ramsay [129] from the convergence limit of the $A \leftarrow X$ sub bands assuming dissociation into the $Cl(^2P_{3/2})$ and $O(^1D)$. The dissociation energy value obtained is 63.427 ± 0.008 kcal mol^{-1}. The deviation between the experimental value and G2 theory is only 1.9 kcal mol^{-1}. Based on the experimental value, the heat of formation for ClO is determined to be 24.15 ± 0.02 kcal mol^{-1} at O K and $24.30 + 0.02$ at 298 K. The heat of formation for ClO is an important quantity for determining ClO reaction enthalpies for various reactions. One example is the reaction of fluorine atoms with ClO radical to yield FOCl. With the introduction of new alternative hydrocarbons which mainly contain fluorine, the release of fluorine from these materials into the atmosphere will introduce new chemistry, which results from the interaction of chlorine oxide and fluorine/fluorine oxides species. The FOCl/FClO system is the simplest of the $FClO_x$ species that results from this interaction. Understanding its energetics is central to characterizing other open and closed shell $FClO_x$ species.

Although FOCl is not a radical, it is a nice example from the recent literature, which highlights the usefulness of theoretical methods in pinpointing experimental discrepancies. There has been, to date, one experimental measurement of the heat of formation of FOCl. Balaev and coworkers measured a value of $-29.8 \pm 3.5\ \text{kcal mol}^{-1}$ using electron-impact mass spectroscopy [130]. DeKock and Jasperse [131], using the MNDO semi-empirical methods, estimated the heat of formation to $27.3\ \text{kcal mol}^{-1}$. The two results show a major discrepancy of $57.1\ \text{kcal mol}^{-1}$! More recently, the heat of formation for FOCl has been re-examined theoretically using several methods, such as isodesmic schemes, G1 and G2, and bond-additivity.

Calculation of the heat of formation of FOCl using an isodesmic scheme relies on the reaction

$$FOCl + H_2O \rightarrow HOF + HOCl \tag{4.64}$$

This isodesmic reaction scheme requires accurate experimental heats of formation for H_2O, HOF, and HOCl. The experimental uncertainties in the heats of formation of these compounds are less than $2\ \text{kcal mol}^{-1}$. However, the larger uncertainty lies in the calculated energies, which are a function of electron correlation. Francisco and Sander [132a] found that small basis sets and modest electron correlation can produce discrepancies of $\pm 5\ \text{kcal mol}^{-1}$ in the results. Using the isodesmic scheme with large basis sets and high electron correlation, i.e. QCISD(T)/6-311 ++ G(3*df*, 3*pd*) level of theory, the heat of formation of FOCl is predicted to $13.4 \pm 2\ \text{kcal mol}^{-1}$ at 298 K.

Using G1 and G2 methods, the heat of formation of FOCl is estimated at 11.0 ± 3 and $12.0 \pm 2\ \text{kcal mol}^{-1}$ respectively. There is convergence and good agreement between G1 and G2 results. The G2 result differs with the isodesmic result by $1.4\ \text{kcal mol}^{-1}$, which is within the estimated uncertainty of the results.

Colussi and Grela [133], using the bond additivity method, determined bond contributions for the heats of formation for various fluorine oxides and chlorine oxide compounds. In FOCl, the major bond contributions result from [F—O] and [Cl—O], whose contributions are 3.0 and $9.8\ \text{kcal mol}^{-1}$, respectively. The resulting heat of formation of FOCl is the sum of the bond contributions which give 12.8 ± 2 kcal mol^{-1}. Comparing the bond additivity result with that from the isodesmic scheme shows a difference of 1.1 kcal mol^{-1}; this too is within the estimated accuracy of ± 2 kcal mol^{-1}.

The isodesmic method, G1 and G2, and the bond additivity methods provide a consistent picture of the heat of formation of FOCl to be close to 13.9 ± 2 kcal mol^{-1}. The MNDO result overestimates the heat of formation. Moreover, studies by Lee [132b] and Francisco and Sander [132c] on the FClO isomer show it to be 4.5 kcal mol^{-1} below that of FOCl. It is therefore unlikely that the species observed in the experimental determination is FClO. Nevertheless, FOCl is a good example where the high level methods point to a serious experimental discrepancy. A summary of these results is shown in Table 4.2.

Chlorine peroxy (ClOO) radical. The identification of ClOO as an important intermediate in polar ozone destruction mechanisms has been inferred from kinetic experiments by Friedl and Sander [134]. In the experiment, ClOO rapidly decomposes to chlorine atoms and molecular oxygen, which suggests that ClOO is weakly bound. The first estimate of the Cl–OO bond dissociation energy was the 8.0 kcal mol^{-1} value of Benson and Buss [135]. Kinetic studies performed by Johnston [136] and Clyne *et al.* [137] led to the conclusion that ClOO has a heat of formation that is 3.1 kcal mol^{-1} lower than that of OClO. Nicovich *et al.* [138] redetermined the heat of formation of ClOO, and derived a value of 23.4 ± 0.5 kcal mol^{-1} at 298 K. Baer *et al.* [139] determined a value of 23.3 kcal mol^{-1} (at 298 K), while Mauldin *et al.* [140] measured a value of 24.0 ± 0.4 kcal mol^{-1} (at 0 K).

Theoretical studies largely corroborate the experimental finding. Gole [141] performed the first SCF calculations on ClOO radical using constrained geometry optimization. This was later followed by full geometry

TABLE 4.2 Summary of the heat of formation (in kcal mol^{-1} of FOCl at 298 K

Theory	
Isodesmic	13.4 ± 2
G1	11.0 ± 3
G2	12.0 ± 2
Bond additivity	12.8 ± 2
MNDO	27.3
Experimental	
Electron impact mass spectroscopy	− 29.8 ± 3.5

optimizations at the SCF level using a large *sp* basis set without polarization functions by Hinchliffe [142]. The first high level optimization of the ClOO radical was done by Jafri [143] using a double-zeta plus polarization basis set and MCSCF-CI wave functions. This work led to the first theoretical estimate of the bond dissociation energy ($Cl + O_2$) of 7.5 ± 4 kcal mol^{-1}. Peterson and Werner (144) in their study of the energetics of the OClO radical also obtained a bond dissociation energy of 2.46 kcal mol^{-1}. Francisco and Sander [145] have re-examined the bond dissociation energy and heat of formation using QCISD(T) methods with large basis sets (6−311++G(*3df,3pd*) basis set) and isodesmic schemes. They estimate the bond dissociation energy to be 4.0 ± 2 kcal mol^{-1}. This is in reasonable agreement with the experimental measurements of Nicovich *et al.* [138] (4.76 ± 0.49 kcal mol^{-1}), Baer *et al.* [139] (4.983 ± 0.005 kcal mol^{-1}), and Mauldin [140] (4.6 ± 0.4 kcal mol^{-1}). Francisco and Sander [145] estimated the heat of formation for ClOO to be 24.2 ± 2 kcal mol^{-1} at the QCISD(T) level as compared to the experimental value of 23.4 ± 0.1 kcal mol^{-1} that was determined by Nicovich *et al.* [138]. The difference can be attributed to the fact that a small basis set and moderate correlation are included. Rathman and Schindler [146] using G1 theory estimated the heat of formation to be 29.0 kcal mol^{-1} Between the G1 and QCISD(T) results there is a 4.8 kcal mol^{-1} difference. Note that the difference is greater than the mean absolute difference of 1.47 kcal mol^{-1} found for G1 theory. This is a result of geometry differences used in the calculations. It is well to remember that G1 theory uses geometries calculated at the MP2/6–31G(*d*) level of theory. At this level the ClO bond length is 1.720 Å, while at the QCISD(T)/6–311G(2*d*) level it is 2.106 Å. Because the ClO bond is so stretched, higher order correlation is needed to properly describe this bond. Since MP2 truncates the higher order terms, the geometry introduces an error in the G1 result.

Acetyl Radical (CH_3CO)

Over the years there have been many experimental determinations of the heat of formation of acetyl (CH_3CO) radical [147–160]. The first experimental determination used the pyrolysis method with acetone as a reagent [147]. The heat of formation obtained was − 10.8 kcal mol^{-1}. The pyrolysis experiment was repeated with acetone in 1962 by O'Neal and Benson [148]. These authors determined a value of − 6.2 kcal mol^{-1}. The heat of formation was redetermined using a different technique (photolysis) with different reagents (azomethane/CO) [149,150]. In 1958 a value of − 3.0 kcal mol^{-1} was obtained [149] while a re-examination [150] of the results in 1965 resulted in a heat of formation of $- 4 \pm 2$ kcal mol^{-1}. A photoionization mass spectroscopic study [151] in 1964 obtained a value of − 6.5 kcal mol^{-1} but electron impact/mass spectroscopic study [158] in 1984 yielded a value of − 4.5 kcal mol^{-1}. Electron impact studies [156] yield a low value for the heat of formation but with a large experimental uncertainty, e.g. -2 ± 5 kcal mol^{-1}. More recent studies [160] yield a heat of formation of -2.39 ± 0.39 kcal mol^{-1} using the kinetics method. Prior to this result, the experimental literature appeared to converge to − 5 kcal mol^{-1} for the heat of formation. The evolution of the experimental values is illustrated in Table 4.3.

Francisco and Abersold [161] using the isodesmic reaction scheme

$$CH_3CO + CH_4 \rightarrow HCO + CH_3CH_3 \tag{4.65}$$

and employing electron correlation at the UMP4SDTQ level and a modest basis set [6−311 + G(*d,p*)] obtained a heat of formation of −4.2 kcal mol^{-1}. Smith *et al.* [162] using G1 obtained a heat of formation of -1.8 ± 2 kcal mol^{-1}. More recently Bauschlicher [93] has determined a

TABLE 4.3 Experimental and theoretical heats of formation of acetyl radical

	Reference	Heat of formation (298 K)	Year
Experiment			
Acetone pyrolysis	[147]	−10.8	1955
Azomethane/CO photolysis	[149]	−3.0	1958
Acetone pyrolysis	[148]	−6.2, −6.3	1962
Photoionization mass spectroscopy	[151]	−6.5	1964
Azomethane/CO photolysis	[150]	−4 ± 2	1965
Spectrophotometric technique	[152]	−6.3 ± 2.0	1966
Spectrophotometry	[153]	−5.4 ± 0.8	1966
Biacetyl pyrolysis	[154]	−5.1 ± 2.0	1969
Calorimetry	[155]	−5.8 ± 0.4	1969
Electron impact	[156]	−2 ± 5	1969
Shock tube studies of ketone	[157]	3.3	1984
Electron impact/mass spectrometry	[158]	−4.5	1984
Photoelectron spectroscopy	[159]	−5.4 ± 2.1	1989
Kinetics	[160]	−2.39 ± 0.29	1992
Theory			
Isodesmic (MP2/MP4)	[161]	−4.2 ± 2	1991
G1	[162]	−1.8 ± 2	1991
G2	[93]	−1.8 ± 2	1994
Isodesmic (MP2/QCISD(T))	[93]	−2.2 ± 0.7	1994

G2 result which is the same value as the G1. Using a different isodesmic reaction,

$$CH_3 + CH_3C(O)H \rightarrow CH_4 + CH_3CO \quad (4.66)$$

which conserves the bonds broken and formed, and employing higher order electron correlation through the use of the quadratic configuration interaction method including a perturbational estimate of the triplet excitation [QCISD(T)] and large basis sets, Bauschlicher obtained a fully converged result of -2.2 ± 0.7 kcal mol^{-1}.

It is interesting to note that G1 and G2 and isodesmic results agree with the more recent experimental kinetic studies value to within 0.2 kcal mol^{-1}. It is also interesting to note that it has taken nearly 37 years for the experimental results to converge, but only 2 years for the theoretical results to converge.

Cyanato (NCO) Radical

The cyanato radical is an important intermediate in combustion processes whose thermochemistry has been difficult to establish. The JANAF value [163] (38 ± 2.5 kcal mol^{-1}) is derived from the photoionization threshold for photodissociation of the H—NCO bond [164]. More recent studies [165] of the photodissociation of HNCO into singlet NH and CO gave a revised value for ΔH_f(HNCO) of $-24.9^{+0.7}_{-2.8}$, resulting in ΔH_f(NCO) = $36.1^{+0.7}_{-2.8}$ kcal mol^{-1}, while photolysis studies of the reaction NCO → CN + O(^{3}P) were used to infer a lower bound [166] ΔH_f(NCO) > 37 kcal mol^{-1}. Very recently, Cyr *et al.* [167] observed production of N(^{2}D) + CO dissociation products 20.3 kcal mol^{-1} above the origin of NCO(B^{2}II) and concluded that ΔH_f(NCO) = 30.5 ± 1 kcal mol^{-1}, a significantly lower value than any previously reported.

In an attempt to resolve this question, an extensive theoretical study was undertaken by East and Allen [168]. They have calculated the heats of the following reactions:

$$CO_2 + NH_3 \rightarrow HNCO + H_2O \quad (4.67)$$

$$CO + NH_3 \rightarrow HCN + H_2O \quad (4.68)$$

$$HNCO \rightarrow H + NCO \quad (4.69)$$

$$HNCO \rightarrow H^+ + NCO^- \quad (4.70)$$

$$N(^4S) + CO \rightarrow NCO \quad (4.71)$$

$$HCN + O \rightarrow H + NCO \quad (4.72)$$

$$NH + CO \rightarrow H + NCO \quad (4.73)$$

at levels of theory up to CCSD(T)/QZ(2*d*1*f*,2*p*1*d*) and MP2/[13*s*8*p*6*d*4*f*, 8*s*6*p*4*d*] and base their final values on composite energies using the large basis through second order, and the smaller (QZ) basis to obtain higher order corrections. This approach is very similar to G2(MP2), although larger basis sets are employed. Calculations on related molecules are used to determine bond additivity corrections used in evaluating the heats of reactions (4.69), (4.71), (4.72), and (4.73). Using known experimental heats of formation, the calculated heats of reaction are used to derive the heats of formation for NCO. Reactions (4.67) and (4.68) are used to make small corrections in the heats of formation of HCN and HNCO, which are subsequently used to calculate the heats of reactions (4.69), (4.70) and (4.72). The values obtained for ΔH_f(NCO) are 31.2, 31.4, 31.6, 31.2, and 31.9 kcal mol^{-1}, using reactions (4.69), (4.70), (4.71), (4.72), and (4.73), respectively. Their final recommendation is 31.4 ± 0.5 kcal mol^{-1}, which is in excellent agreement with the experimental value of Cyr *et al.* [167].

Melius [169] has found ΔH_f(NCO) = 32.1 ± 4.8 kcal mol^{-1} using the previously described BAC-MP4 method. The G1 and G2 atomization energies give values of ΔH_f(NCO) of 29.2 and 29.4 kcal mol^{-1}, respectively [170]. These results are all in agreement, within their respective error limits, with the experimental value of Cyr *et al.* [167] and the calculations of East and Allen [168].

It is noted that none of the reactions used by East and Allen [168] to determine the heat of formation of NCO are isodesmic and only (4.70) is isogyric. Indeed, reaction (4.70) involves the comparison of two isoelectronic species, and if basis sets adequate to describe the diffuse nature of the NCO^- anion are used, then this should be the most attractive scheme computationally due to the electronic similarity between reactants and products. An indication that this is so is given by the 31.4 kcal mol^{-1} value of ΔH_f(NCO), calculated from (4.70) without the use of any bond additivity correction, which is equal to East and Allen's final recommended value.

One can repeat East and Allen's calculation using G2 energies [170], the results for ΔH_f (NCO) are 29.4, 30.7, 30.6, 31.2, 30.0 and 31.5 kcal mol^{-1} using their five reactions (4.69)–(4.73). The average of these values is 30.8 kcal mol^{-1}, in excellent agreement with the 31.4 kcal mol^{-1} result of East and Allen [168].

SiF_n Radicals

The thermochemistry of the silicon fluoride radicals SiF_n is of considerable practical importance in chemical vapor deposition (CVD) and silicon etching processes. The experimental heat of formation of SiF_4 is accurately known [163,171], as apparently was the value for SiF_2. Experimental values of $\Delta H_f(SiF_3)$ differ by over 20 kcal mol^{-1}. Walsh's value of -239.0 ± 5.0 kcal mol^{-1} [172,173] was obtained from kinetic iodination studies, whereas Farber and Srivastara [174] find -259.3 ± 0.5 from effusion mass spectroscopy experiments. The ion–molecule reaction studies of Weber and Armentrout [175] give -258.0 ± 3.0 kcal mol^{-1}. For SiF_2, the -140.6 ± 0.3 kcal mol^{-1} value of Farber and Srivastara [174] was adopted by JANAF [163] and Walsh [172]. However, Weber and Armentrout [175] found a value of -142.4 ± 1.6 kcal mol^{-1} and recommend a value of -141 kcal mol^{-1}. The JANAF heat of formation [163] for SiF, -9.80 ± 3.0 kcal mol^{-1}, is based on the mass spectrometric measurements of Farber and Srivastara [174] and Ehlert and Margrave [176], and is also adopted by Walsh [172] and by Weber and Armentrout [175].

There have been three recent theoretical studies of the thermochemistry of SiF_n compounds. Ignacio and Schlegel [177] have used the isodesmic reactions

$$\frac{n}{4}\,SiF_4 + SiH_n \rightarrow \frac{n}{4}\,SiH_4 + SiF_n \tag{4.74}$$

$$SiH_{4-n}F_n + SiH_n \rightarrow SiH_4 + SiF_n \tag{4.75}$$

$$\frac{n}{4}\,SiF_4 + \frac{4-n}{n}\,Si \rightarrow SiF_n \tag{4.76}$$

$$SiH_{4-n}F_n + \frac{4-n}{4}\,Si \rightarrow \frac{4-n}{4}\,SiH_4 + SiF_n \tag{4.77}$$

to derive heats of formation of the SiF_n radicals from known heats of formation. The convergence of the calculated heats of reaction is studied in detail, at levels of theory from HF/6–31G(*d*) to MP4/6–311 + G(2*df*, 2*p*), and is an excellent example of the use of isodesmic reactions. Their final results, estimated to be accurate to ± 2 kcal mol^{-1}, are given in Table 4.4.

TABLE 4.4 Heats of formation of SiF_n compounds (kcal mol^{-1}) at 298 K

	BAC-MP4	ISO (IS)	ISO (G2)	G2	Experiment
SiF_4				−379.1	−386.2 ± 0.1 [171]
SiF_3	−237.4	−240.7	−238.1	−232.8	−239.0 ± 5.0 [173]
					−259.3 ± 0.5 [174]
SiF_2	−149.9	−153.0	−152.5	−149.0	−140.6 ± 0.3 [172]
					−141.0 [175]
SiF	−12.4	−14.2	−15.1	−13.4	−4.30 ± 3.0 [163]

Ho and Melius [53] have reported BAC-MP4 heats of formation for the SiF_n compounds, which are also given in Table 4.4. Their paper provides many details of the calculations; they are used to briefly illustrate the application of the BAC-MP4 methods to SiF_3. The MP4/6–31G* atomization energy of SiF_3 is 368.8 kcal mol^{-1}. This value includes PMP3 corrections for spin contamination and the vibrational zero-point energy of SiF_3 at the HF/6–31G* level of theory. Using the atomic heats of formation of silicon and fluorine, the MP4 heat of formation of SiF_3 is found to be −206.7 kcal mol^{-1} (see Table I of [53]). From the 1.5753 Å Si—F bond length (see Table II of [53]) and the BAC formulas (see above and [53]) the value of 9.99 kcal mol^{-1} is obtained for BAC (Si—F). Correcting for three Si—F bonds a BAC-MP4 heat of formation of − 236.7 is obtained for SiF_3 at 0 K. Using thermal corrections calculated from the theoretical geometry and vibrational frequencies, Ho and Melius [53] find a 298 K heat of formation of −237.4 kcal mol^{-1}

The highest levels of *ab initio* theory to be applied to the problem are found in the G1 and G2 studies of Michels and Hobbs [178]. They report heats of formation calculated from the isodesmic reaction (4.3) as well as those obtained from atomization energies. Both of these results, at the G2 level, are included in Table 4.4. The G2 value for $\Delta H_f(SiF_4)$ is 7.1 kcal mol^{-1} too high, a larger discrepancy than one might expect from the mean errors found for the G2 test set [77]. Michels and Hobbs [178] suggest this error is due to basis set deficiencies. It is also noted, as has been pointed out elsewhere [179], that there are very few fluorine containing compounds in the G2 test set and that the errors observed in the G2 atomization energies of such compounds tend to be larger than the G2 mean. It is interesting to compare G2 and 'isodesmic G2' results for the SiF_n heats of formation. For SiF, SiF_2, and SiF_3 they differ by 1.7, 3.5, and 5.3 kcal mol^{-1}, respectively. These differences are proportional to the number of Si—F bonds. If this average difference per bond, 1.72 kcal mol^{-1}, is used to extrapolate an 'isodesmic G2 heat of formation for SiF_4', −386.1 kcal mol^{-1} is obtained, in almost exact agreement with experiment. This argument suggests that the isodesmic G2 results of Michels and Hobbs should be significantly more accurate that the G2 atomization results.

Comparing the BAC-MP4, Ignacio and Schlegel [177], and isodesmic G2 results, they are found to be in agreement to within 2–3 kcal mol^{-1} on all the SiF_n species. The calculations clearly favor the −239.0 kcal mol^{-1} results of Dancaster and Walsh [173] for SiF_3, and suggest that the literature values for SiF_2 and SiF may be 10 kcal mol^{-1} or so too high. At present, it would appear that theory provides the most accurate values of the heats of formation of the SiF_n radicals.

SUMMARY

The examples discussed in this chapter demonstrate that there are a variety of methods that are currently available that can predict the energetics of radicals to the level of chemical accuracy. Thus, in cases where experimental measurement of the energetics limits the accuracy of the results, *ab initio* molecular orbital methods, presented in this chapter, can provide some assistance. The systems highlighted also demonstrate the consistency of results for a variety of methods. At this point the choice of method depends solely on the choice of molecules to be studied and available computational resources. For large radical systems, the computational cost could narrow the pool of methods.

Acknowledgment

J. S. F. would like to thank the John Simon Guggenheim Foundation for a Fellowship which provided the time to complete this manuscript.

REFERENCES

1. H. E. O'Neal and S. W. Benson (1973) Thermochemistry of Free Radicals, in J. K. Kochi (ed.), *Free Radicals*, Wiley, New York.
2. M. Szwarc and J. W. Taylor (1955) *J. Chem. Phys.* **23,** 2310.
3. J. A. Devore and H. E. O'Neal (1969) *J. Phys. Chem.* **73,** 2644.
4. J. T. Nijranen, D. Gutman and L. N. Krasnoperov (1992) *J. Phys. Chem.* **96,** 5881.
5. S. W. Benson (1960) *Foundation of Chemical Kinetics*, McGraw-Hill, New York.
6. S. Arrhenius (1889) *Z. Physik Chem.* **4,** 226.
7. J. A. Kerr (1966) *Chem. Rev.* **66,** 465.
8. J. A. Kerr and A. F. Trotman-Dickerson (1961) *Prog. Reaction Kinetics* **1,** 113.
9. E. L. Metcalfe (1963) *J. Chem. Soc.* 3560.
10. E. L. Metcalfe and A. F. Trotman-Dickerson (1962) *J. Chem. Soc.* 4620.
11. M. Szwarc (1950) *Chem. Rev.* **47,** 75.
12. D. F. McMillen and D. M. Golden (1982) *Annu. Rev. Phys. Chem.* **33,** 493.
13. D. M. Golden and S. W. Benson (1969) *Chem. Rev.* **69,** 125.
14. R. Walsh (1981) *Acc. Chem. Res.* **14,** 248.
15. J. J. Russell, K. A Sectula and D. Gutman (1988) *J. Am. Chem. Soc.* **110,** 3092.

16. G. B. KISTIAKOWSKY and E. R. VAN ARTSDALES (1944) *J. Chem. Phys.* **12,** 469.
17. J. M. NICOVICH, C. A. VAN DIJK, K. D. KREUTTER and P. H. WINE (1991) *J. Phys. Chem.* **95,** 9890.
18. J. M. NICOVICH, K. D. KREUTTER, C. A. VAN DIJK and P. H. WINE (1992) *J. Phys. Chem.* **96,** 2518.
19. P. W. SEAKING and M. J. PILLING (1991) *J. Phys. Chem.* **95,** 9874.
20. J. A. SECTULA, Y. FENG, D. GUTMAN, P. W. SEAKING and M. J. PILLING (1991) *J. Phys. Chem.* **95,** 1658.
21. I. SLAGLE and D. GUTMAN (1985) *J. Am. Chem. Soc.* **107,** 5342.
22. J. BERKOWITZ, W. A. CHUPKA, P. M. GUYON, J. H. HOLLOWAY and R. SPOHR (1971) *J. Chem. Phys.* **54,** 5165.
23. J. BERKOWITZ, G. B. ELLISON and D. GUTMAN (1994) *J. Phys. Chem.* **98,** 2744.
24. R. K. YOO, B. RUSCIC and J. BERKOWITZ (1992) *J. Chem. Phys.* **96,** 911.
25. S. W. BENSON (1978) *Chem. Rev.* **78,** 23.
26. D. GRILLER, J. M. KANABUS-KAMINSKA, and A. MACCOLL (1988) *J. Mol. Struct. (Theochem)* **163,** 125.
27. J. B. FORESMAN and A. FRISCH (1993) *Exploring Chemistry with Electronic Structure Methods,* Gaussian, Inc., Pittsburgh.
28. A. SZABO and N. S. OSTLUND (1989) *Modern Quantum Chemistry,* McGraw-Hill, New York.
29. W. J. HEHRE, L. RADOM, P. V. R. SCHLEYER and J. A. POPLE (1986) Ab initio Molecular Orbital Theory, Wiley, New York.
30. T. H. DUNNING and P. J. HAY (1977) *Methods of Electronic Structure Theory,* in H. F. Schaefer (ed.), Plenum Press, New York, pp. 1–27.
31. M. S. GORDON (1980) *Chem. Phys. Lett.* **76,** 163.
32. M. M. FRANCL, W. J. PIETRO, W. J. HEHRE, J. S. BINKLEY, M. S. GORDON, D. J. DEFREES and J. A. POPLE (1982) *J. Chem. Phys.* **77,** 3654.
33. R. KRISHNAN, J. S. BINKLEY, R. SEEGER and J. A. POPLE (1980) *J. Chem. Phys.* **72,** 650.
34. A. D. MCLEAN and G. S. CHANDLER (1980) *J. Chem. Phys.* **72,** 5639.
35. T. H. DUNNING JR (1989) *J. Chem. Phys.* **90,** 1007.
36. R. J. BARTLETT (1981) *Ann. Rev. Phys. Chem.* **32,** 59.
37. N. OLIPHANT and L. ADAMOWICZ (1991) *J. Chem. Phys.* **95,** 6645.
38. J. A. POPLE, M. HEAD-GORDON and K. RAGHAVACHARI (1987) *J. Chem. Phys.* **87,** 5968.
39. S. W. BENSON and J. H. BUSS (1958) *J. Chem. Phys.* **29,** 546.
40. S. W. BENSON, F. R. CRUICKSHANK, D. M. GOLDEN, G. R. HAUGEN, H. E. O'NEAL, A. S. RODGERS, R. SHAW and R. WALSH (1969) *Chem. Rev.* **69,** 269.
41. R. SHAW (1969) *J. Chem. Eng. Data* **14,** 461.
42. R. SHAW (1971) *J. Phys. Chem.* **75,** 4047.

43. H. K. Eigenmann, D. M. Golden and S. W. Benson (1973) *J. Phys. Chem.* **77,** 1687.
44. S. W. Benson (1976) *Thermochemical Kinetics* 2nd edn, John Wiley and Sons, New York.
45. M. Luria and S. W. Benson (1977) *J. Chem. Eng. Data* **22,** 90.
46. R. Shaw, D. M. Golden and S. W. Benson (1977) *J. Phys. Chem.* **81,** 1716.
47. S. E. Stein, D. M. Golden and S. W. Benson (1977) *J. Phys. Chem.* **81,** 314.
48. K. B. Wiberg (1984) *J. Comp. Chem.* **5,** 197.
49. M. R. Ibrahim and P.v.R. Schleyer (1985) *J. Comp. Chem.* **6,** 157.
50. K. B. Wiberg (1985) *J. Org. Chem.* **50,** 5285.
51. P. Ho, M. E. Coltrin, J. S. Binkley and C. F. Melius (1985) *J. Phys. Chem.* **89,** 4647.
52. P. Ho, M. E. Coltrin, J. S. Binkley and C. F. Melius (1986) *J. Phys. Chem.* **90,** 3399.
53. P. Ho and C. F. Melius (1990) *J. Phys. Chem.* **94,** 5120.
54. C. F. Melius (1990) Thermochemical Modeling: I. Application to Decomposition of Energetic Materials, in S. N. Bulusu (ed.), *Chemistry and Physics of Energetic Materials,* Kluwer, Dordrecht.
55. H. B. Schlegel (1986) *J. Chem. Phys.* **84,** 4530.
56. R. C. Bingham, M. J. S. Dewar and D. H. Lo (1975) *J. Am. Chem. Soc.* **97,** 1285.
57. M. J. S. Dewar and W. Thiel (1977) *J. Am. Chem. Soc.* **99,** 4899.
58. M. J. S. Dewar, E. G. Zoebisch, E. F. Healy and J. J. P. Stewart (1985) *J. Am. Chem. Soc.* **107,** 3902.
59. J. J. P. Stewart (1989) *J. Comp. Chem.* **10,** 209.
60. J. A. Pople and D. L. Beveridge (1970) *Approximate Molecular Orbital Theory,* McGraw-Hill, New York.
61. J. J. P. Stewart (1990) Semiempirical Molecular Orbital Methods, in K. B. Lipkowitz and D. B. Boyd (eds), *Rev. Comput. Chem. 1,* VCH, New York.
62. J. A. Pople, D. L. Beveridge and P. A. Dobosh (1967) *J. Chem. Phys.* **47,** 2026.
63. M. J. S. Dewar and W. Thiel (1977) *J. Am. Chem. Soc.* **99,** 4907.
64. J. A. Pople, D. P. Santry and G. A. Segal (1965) *J. Chem. Phys.* **43,** S129.
65. J. J. P. Stewart (1990) *J. Computer-Aided Mol. Design,* **4,** 1.
66. J. J. P. Stewart (1989) *J. Comp. Chem.* **10,** 221.
67. C. W. Bauschlicher, S. R. Langhoff and P.R. Taylor (1990) *Adv. Chem. Phys.* **77,** 203.
68. T. J. Lee and G. E. Scuseria (1995) Achieving Chemical Accuracy with Coupled-Cluster Theory, in S. R. Langhoff (ed.), *Quantum Mechanical Electronic Structure Calculations with Chemical Accuracy,* Kluwer Academic Publishers, Dordrecht, pp. 47–108.

69. C. W. Bauschlicher and S. R. Langhoff (1991) *Science* **254,** 394.
70. J. A. Pople, B. T. Luke, M. J. Frisch and J. S. Binkley (1985) *J. Phys. Chem.* **89,** 2198.
71. W. J. Hehre, R. Ditchfield, L. Radom and J. A. Pople (1970) *J. Am. Chem. Soc.* **92,** 4796.
72. L. Radom, W. J. Hehre and J. A. Pople (1971) *J. Am. Chem. Soc.* **93,** 289.
73. P. George, M. Trachtman, C. W. Bock and A. M. Brett (1976) *J. Chem. Soc., Perkin, II,* 1222.
74. P. George, M. Trachtman, A. M. Brett and C. W. Bock (1977) *J. Chem. Soc., Perkin II,* 1036.
75. J. A. Pople, M. Head-Gordon, D. J. Fox, K. Raghavachari and L. A. Curtiss (1989) *J. Chem. Phys.* **90,** 5622.
76. L. A. Curtiss, C. Jones, G. W. Trucks, K. Raghavachari and J. A. Pople (1990) *J. Chem. Phys.* **93,** 2537.
77. L. A. Curtiss, K. Raghavachari, G. W. Trucks and J. A. Pople (1991) *J. Chem. Phys.* **94,** 7221.
78. J. A. Pople and L. A. Curtiss (1987) *J. Phys. Chem.* **91,** 155.
79. J. A. Pople and L. A. Curtiss (1987) *J. Phys. Chem.* **91,** 3637.
80. L. A. Curtiss and J. A. Pople (1988) *J. Phys. Chem.* **92,** 894.
81. J. A. Pople, P.v.R. Schleyer, J. Kaneti and G. W. Spitznagel (1988) *Chem. Phys. Lett.* **145,** 359.
82. L. A. Curtiss, K. Raghavachari and J. A. Pople (1993) *J. Chem. Phys.* **98,** 1293.
83. L. A. Curtiss, J. E. Carpenter, K. Raghavachari and J. A. Pople (1992) *J. Chem. Phys.* **96,** 9030.
84. J. M. L. Martin (1992) *J. Chem. Phys.* **97,** 5012.
85. J. M. L. Martin (1994) *J. Chem. Phys.* **100,** 8186.
86. C. W. Bauschlicher, S. R. Langhoff and P. R. Taylor (1990) *Chem. Phys. Lett.* **171,** 42.
87. C. W. Bauschlicher and S. R. Langhoff (1990) *Chem. Phys. Lett.* **173,** 367.
88. C. W. Bauschlicher and S. R. Langhoff (1991) *Chem. Phys. Lett.* **177,** 133.
89. S. R. Langhoff, C. W. Bauschlicher and P. R. Taylor (1991) *Chem. Phys. Lett.* **180,** 88.
90. C. W. Bauschlicher and S. R. Langhoff (1992) *Chem. Phys. Lett.* **193,** 380.
91. C. W. Bauschlicher and H. Partridge (1993) *Chem. Phys. Lett.* **208,** 241.
92. C. W. Bauschlicher and H. Partridge (1994) *J. Phys. Chem.* **98,** 1826.
93. C. W. Bauschlicher (1994) *J. Phys. Chem.* **98,** 2564.
94. K. A. Peterson, R. A. Kendall and T. H. Dunning, Jr. (1993) *J. Chem. Phys.* **99,** 1930.
95. D. E. Woon and T. H. Dunning, Jr. (1993) *J. Chem. Phys.* **99,** 1914.

96. K. A. Peterson, R. A. Kendall and T. H. Dunning, Jr. (1993) *J. Chem. Phys.* **99,** 9790.

97. K. A. Peterson, D. E. Woon and T. H. Dunning, Jr. (1994) *J. Chem. Phys.* **100,** 7410.

98. M. R. Nyden and G. A. Petersson (1981) *J. Chem. Phys.* **75,** 1843.

99. G. A. Petersson and M. R. Nyden (1981) *J. Chem. Phys.* **75,** 3423.

100. G. A. Petersson and S. L. Licht (1981) *J. Chem. Phys.* **75,** 4556.

101. G. A. Petersson, A. K. Yee and A. Bennett (1985) *J. Chem. Phys.* **83,** 5105.

102. G. A. Petersson, A. Bennett, T. G. Tensfeldt, M. A. Al-Laham, W. A. Shirley and J. Mantzaris (1988) *J. Chem. Phys.* **89,** 2193.

103. G. A. Petersson and M. A. Al-Laham (1991) *J. Chem. Phys.* **94,** 6081.

104. G. A. Petersson, T. G. Tensfeldt and J.A. Montgomery, Jr. (1991) *J. Chem. Phys.* **94,** 6091.

105. J. A. Montgomery, Jr., J. W. Ochterski and G. A. Petersson (1994) *J. Chem. Phys.* **101,** 5900.

106. R. G. Parr and W. Yang (1989) *Density Functional Theory of Atoms and Molecules*, Oxford, New York.

107. P. C. Hohenberg and W. Kohn (1968) *Phys. Rev.* **A136,** 864.

108. T. Ziegler (1991) *Chem. Rev.* **91,** 651.

109. P. M. W. Gill, B. G. Johnson, J. A. Pople and M. J. Frisch (1992) *Chem. Phys. Lett.* **197,** 499.

110. B. G. Johnson, P. M. W. Gill and J. A. Pople (1993) *J. Chem. Phys.* **98,** 5612.

111. A. D. Becke (1992) *J. Chem. Phys.* **96,** 2155.

112. G. Fitzgerald and J. Andzelm (1991) *J. Phys. Chem.* **95,** 10531.

113. J. Andzelm, C. Sosa and R. A. Eades (1993) *J. Phys. Chem.* **97,** 4664.

114. A. M. Wodtke and Y. T. Lee (1985) *J. Phys. Chem.* **89,** 4744.

115. J. Segall, R. Lavi, Y. Wen and C. Wittig (1989) *J. Phys. Chem.* **93,** 7287.

116. P. G. Green, J. L. Kinsey and R. W. Field (1989) *J. Chem. Phys.* **91,** 5160.

117. K. M. Ervin, S. Gronert, S. E. Barlow, M. K. Gilles, A. G. Harrison, V. M. Bierbaum, C. H. DePuy, W. C. Lineberger and G. B. Ellison (1990) *J. Am. Chem. Soc.* **112,** 5750.

118. B. K. Janousek, J. I. Brauman and J. Simons (1979) *J. Phys. Chem.* **71,** 2057.

119. J. Segall, Y. Wen, R. Lavi, R. Singer and C. Wittig (1991) *J. Phys. Chem.* **95,** 8078.

120. B. Ruscic and J. Berkowitz (1990) *J. Chem. Phys.* **93,** 5586.

121. D. P. Baldwin, M. A. Buntine and D. W. Chandler (1990) *J. Chem. Phys.* **93,** 6578.

122. M. Hawley and M. A. Smith (1989) *J. Am. Chem. Soc.* **111,** 8293.

123. J. Berkowitz, C. A. Mayhew and B. Ruscic (1988) *J. Chem. Phys.* **88,** 7396.

124. J. A. Blush and P. Chen (1992) *J. Phys. Chem.* **96,** 4138.
125. S. S. Parmar and S. W. Benson (1988) *J. Phys. Chem.* **92,** 2652.
126. J. J. Russell, S. M. Senkan, J. A. Seetula and D. Gutman (1989) *J. Phys. Chem.* **93,** 5184.
127. M. B. McElroy, R. J. Salawitch and S. C. Wotsy (1986) *Nature* **321,** 759.
128. L. T. Molina and M. J. Molina (1987) *J. Phys. Chem.* **91,** 433.
129. J. A. Coxem and D. A. Ramsay (1976) *Can. J. Phys.* **54,** 1034.
130. A. V. Balaev, Z. K. Nikitine, L. I. Fedorara and V. Y. Rosoloouski (1980) *Izv. Akad. Nauk SSSR Sev. Khim* **9,** 1963.
131. R. L. DeKock and C. P. Jasperse (1983) *J. Fluor. Chem.* **22,** 575.
132. (a) J. S. Francisco and S. P. Sander (1994) *Chem. Phys. Lett.* **223,** 439.
132. (b) T. J. Lee (1994) *J. Phys. Chem.* **98,** 3697.
132. (c) J. S. Francisco and S. P. Sander (1995) *Chem. Phys. Lett.* **241,** 33.
133. A. J. Colussi and M. A. Grela (1993) *J. Phys. Chem.* **97,** 3775.
134. R. R. Friedl and S. P. Sander (1989) *J. Phys. Chem.* **93,** 4756.
135. S. W. Benson and J. H. Buss (1957) *J. Chem. Phys.* **27,** 1382.
136. H. S. Johnston, E. D. Morris and J. Van Don Bogaerde (1969) *J. Am. Chem. Soc.* **91,** 7712.
137. M. A. A. Clyne, D. J. McKenney and R. J. Watson (1975) *J. Chem. Soc. Faraday I* **71,** 322.
138. J. M. Nicovich, K. D. Krufter, C. J. Shackelford and P. H. Wine (1991) *Chem. Phys. Lett.* **179,** 367.
139. S. Baer, H. Hippler, R. Rahn, M. Sietke, N. Setzinger and J. Troe (1991) *J. Phys. Chem.* **95,** 6463.
140. R. Mauldin, J. B. Burkholder and A. R. Ravishankara (1992) *J. Phys. Chem.* **96,** 2582.
141. J. L. Gole (1980) *J. Phys. Chem.* **84,** 1333
142. A. Hinchlifte (1980) *J. Mol. Struct.* **64,** 117.
143. J. A. Jafri, B. H. Langsfield, III, C. W. Bauschlicher, Jr. and J. H. Phillips (1985) *J. Chem. Phys.* **83,** 1693.
144. K. A. Peterson and H-J. Werner (1992) *J. Chem. Phys.* **96,** 8948.
145. J. S. Francisco and S. P. Sander (1993) *J. Chem. Phys.* **99,** 2897.
146. T. Rathman and R. N. Schindler (1992) *Chem. Phys. Lett.* **190,** 539.
147. M. Szwarc and J. W. Taylor (1955) *J. Chem. Phys.* **23,** 2310.
148. E. O'Neal and J. W. Benson (1962) *J. Chem. Phys.* **36,** 2196; **37,** 540.
149. J. G. Calvert and J. T. Gruver (1958) *J. Am. Chem. Soc.* **80,** 1313.
150. J. A. Kerr and J. G. Calvert (1965) *J. Phys. Chem.* **69,** 1022.
151. E. Murad and M. G. Inghram (1964) *J. Chem. Phys.* **41,** 404
152. D. N. Golden, R. Walch and S. W. Benson (1965) *J. Am. Chem. Soc.* **87,** 4053.

153. R. Walch and S. W. Benson (1966) *J. Phys. Chem.* **70,** 3751.
154. K. J. Hole and M. F. R. Mulcahy (1969) *J. Phys. Chem.* **73,** 177.
155. J. A. Devore and H. E. O'Neal (1969) *J. Phys. Chem.* **73,** 2644.
156. M. A. Haney and L. Franklin (1969) *Trans. Faraday Soc.* **65,** 1794.
157. W. Tsan (1984) *Int. J. Chem. Kinet* **16,** 1543.
158. J. L. Holmes and F. P. Lossing (1984) *Int. J. Mass Spectrom. Ion Processes* **58,** 113.
159. M. R. Nimlos, J. A. Soderquist and G. B. Ellison (1989) *J. Am. Chem. Soc.* **111,** 7675.
160. J. T. Niiranen, D. Gutman and L. N. Krasnoperov (1992) *J. Phys. Chem.* **96,** 5881.
161. J. S. Francisco and N. J. Abersold (1991) *Chem. Phys. Lett* **187,** 354.
162. B. J. Smith, M. T. Nguyen, W. J. Bouma and L. Radom (1991) *J. Am. Chem.* **113,** 6452.
163. M. W. Chase, Jr., C. A. Davies, J. R. Downey, Jr., D. J. Frurip, R. A. McDonald and A. N. Syverud (1985) *J. Phys. Chem. Ref. Data* **14,** Suppl 1.
164. H. Okabe (1970) *J. Chem. Phys.* **53,** 3507.
165. T. A. Spiglanin, R. A. Perry and D. W. Chandler (1986) *J. Phys. Chem.* **90.** 6184.
166. X. Liu and R. D. Coombe (1989) *J. Phys. Chem.* **91,** 7543.
167. D. R. Cyr, R. E. Continetti, R. B. Metz, D. L. Osborn and D. M. Neumark (1992) *J. Chem. Phys.* **97,** 4937.
168. A. L. L. East and W. D. Allen (1993) *J. Chem. Phys.* **99,** 4638.
169. C. F. Melius, Cited in Ref. 168.
170. J. A. Montgomery, Jr., Unpublished calculations.
171. G. K. Johnson (1986) *J. Chem. Thermodyn.* **18,** 801.
172. R. Walsh (1983) *J. Chem. Soc., Faraday Trans. I* **79,** 2233.
173. A. M. Doncaster and R. Walsh (1978) *Int. J. Chem. Kinet.* **10,** 101.
174. M. Farber and R. D. Srivastara (1978) *J. Chem. Soc., Faraday Trans I,* **74,** 1089.
175. M. E. Weber and P. B. Armentrout (1988) *J. Chem. Phys.* **88,** 6898.
176. T. C. Ehlert and J. L. Margrave (1964) *J. Chem. Phys.* **41,** 1066.
177. E. W. Ignacio and H. B. Schlegel (1990) *J. Chem. Phys.* **92,** 5404.
178. H. H. Michels and R. H. Hobbs (1993) *Chem. Phys. Lett* **207,** 389.
179. J. A. Montgomery, Jr., H. H. Michels and J. S Francisco (1994) *Chem. Phys. Lett.* **220,** 391.

5

Photoacoustic Calorimetry of Radicals and Biradicals

JOSHUA L. GOODMAN

INTRODUCTION

Our understanding of chemical reaction mechanisms often depends on the knowledge of the energetics and kinetics of reactive intermediates. For example, the enthalpic criterion for distinguishing between concerted and non-concerted thermal rearrangements often involves a comparison between two enthalpies of activation, one experimental and the other hypothetical, attributable to a non-concerted model. Commonly, models involving two non-interacting free radicals (biradicals) are discussed and the knowledge of their enthalpies of formation details their role on the reaction pathway. Unfortunately, only estimates of these enthalpies are currently available using semi-empirical calculations, or Benson group additivities [1]. In addition, it is often quite important to understand the reactivity of radicals and biradicals. Modern techniques such as time resolved EPR and absorption spectroscopy are useful in this regard. However, they have some potential disadvantages in that they require the molecule to be paramagnetic or 'visible'.

In this chapter, the way in which time-resolved photoacoustic calorimetry (PAC) can be used to obtain both the energetics and kinetics of radicals and biradicals in solution. PAC measures the magnitude and temporal profile of volume changes in solution following the thermal deposition of energy. These time-resolved volume changes can be directly related to heats of reaction of reactive intermediates of interest. The principles of this photoacoustic technique will be discussed first and then how it has been successfully applied to examine radicals and biradicals formed in photoreactions.

BACKGROUND

Although the photoacoustic (PA) effect was initially discovered by Alexander Graham Bell in 1880, its application has dramatically increased in the past thirty years with the advent of modern instrumentation [2]. The PA effect has been greatly utilized in chemistry as a calorimetric and spectroscopic tool. The principles of the PA effect have been well established and reviewed by Tam and Patel [3]. Briefly, upon the absorption of a photon by a molecule, the energy may be converted into thermal energy, E_{th}, which will result in an increase in temperature along a cylinder defined by the path of the excitation beam through the sample. This rise in temperature, ΔT, increases the volume of the solution, ΔV_{th}, along the irradiated cylinder, generating an outwardly propagating pressure wave. If all of the photon energy is converted into thermal energy, the increase in E_{th} along the laser beam is expressed as

$$E_{th} = E_o (1 - 10^{-A}) \quad (5.1)$$

where A is the absorbance of the sample and E_o is the energy of the laser pulse. The increase in temperature is given by

$$\Delta T = E_{th} / \rho C_p V_o \quad (5.2)$$

where ρ is the density of the solution, C_p is its heat capacity, and V_o is the initial volume.

This temperature rise can be detected directly (laser calorimetry and optical calorimetry), or indirectly by measuring the change in either the refractive index (thermal lensing, beam deflection or refraction and thermal grating) or the volume (photo or optoacoustic methods). This review will focus primarily on photoacoustic methods because they have been the most widely used to obtain thermodynamic and kinetic information about reactive intermediates. Other calorimetric methods are discussed in detail in a recent review by Braslavsky and Heibel [4].

The volume change of the medium induced by the heat deposition is related to the temperature change by the thermal coefficient of expansion of the solution, β.

$$\Delta V_{th} = \beta V_o \Delta T \quad (5.3)$$

Consequently, the magnitude of the volume change can vary greatly depending on β. For example, the volume change in organic solvents is typically orders of magnitude greater than in water, which has a small and temperature dependent β.

The volume change is measured by an acoustic wave detector, typically a piezoelectric transducer (see Instrumentation, p. 156). This detector is sensitive to the magnitude and the temporal profile of the acoustic wave. As will be seen, the amplitude of the wave provides valuable enthalpic information about reactive intermediates, and the temporal profile can reveal the dynamics of these intermediates.

For chemical systems of interest, photolysis produces intermediates, such as radicals or biradicals, whose energetics relative to the reactants are unknown. The energetics of the intermediate can be established by comparison of the acoustic wave generated by the non-radiative decay to create the intermediate, producing thermal energy E_{th}, with that of a reference or calibration compound whose excited-state decay converts the entire photon energy into heat, $E_{th}(\text{ref})$. The ratio of acoustic wave amplitudes, α_{th}, represents the fraction of the photon energy that is converted into heat.

$$\alpha_{th} = E_{th} / E_{th}(\text{ref}) \tag{5.4}$$

The enthalpy of the intermediate relative to the reactant, ΔH, is given by

$$\Delta H = E_{h\nu}(1 - \alpha_{th}) \tag{5.5a}$$

where $E_{h\nu}$ is the photon energy.

The above discussion assumes unit quantum efficiency for the photochemical reaction. Consider the more general case where the excited state either undergoes a photochemical reaction with quantum yield Φ, or decays back to its ground state non-radiatively with quantum yield $(1 - \Phi)$. The enthalpy of reaction is expressed as

$$\Delta H = E_{h\nu}(1 - \alpha_{th})/\Phi \tag{5.5b}$$

Consequently, if the reaction enthalpy is unknown for a given process, the quantum yield must be determined from other measurements. Conversely, if the reaction enthalpy is known, then the quantum yield for the photochemical reaction can be measured. For example, the quantum yield for the photodissociation of di-*t*-butylperoxide and hydrogen peroxide in various solvents has been determined [5]. In addition, PAC has been used to obtain quantum yields for excited state processes [4], such as fluorescence [6], triplet state formation [7], and ion pair formation and separation [8]. In systems in which competitive reactions occur, care must be taken to account accurately for the partitioning. For example, if a reactive intermediate yields two products, then the measured heat of reaction is the sum of the two individual heats of reaction multiplied by their respective yields. Consequently, there are three unknowns; the partitioning and the individual heats of reaction. Two of them must be known to properly evaluate the third.

In most photoacoustic experiments, it has been generally assumed that the experimental volume change resulted only from the thermal deposition of energy to the solvent. In such cases, the ratio of the volume change of the sample to that of the reference compound, α_{th}, could be related to the enthalpy of reaction using equations (5.4) and (5.5). However, many reactions also undergo a reaction volume change, which will also contribute to the observed acoustic wave. Complexation, ionization, covalent bond formation or breakage, and conformational changes can all contribute to the photochemical reaction volume change [9]. The total acoustic wave, ΔV, may be considered to have been produced by two different types of volume changes

$$\Delta V = \Delta V_{th} + \Delta V_{rx} \tag{5.6}$$

where ΔV_{th} and ΔV_{rx} are the thermal and reaction volume changes, respectively. If the reaction volume change is positive due to volume expansion, the acoustic wave will be larger than anticipated if only thermal expansion for the system is considered. This would result in an erroneous value for ΔH.

Several methods have been developed to resolve these two volume contributions under suitable experimental conditions [10]. This allows both the enthalpy of reaction and the photochemical reaction volume to be determined simultaneously. These methods all involve changing the solvent properties, β, ρ, and C_p, and thereby effectively varying the thermal contribution in a systematic manner. PAC has been employed to investigate several chemical [10(d),11] and biological [10(a)–(c),12] reactions in both aqueous and non-aqueous media to determine reaction volumes. In organic solvents, however, the thermal volume change usually dominates due to the large expansion coefficient β, and so the reaction volume contribution is often ignored. For example, a reaction volume change of 20 ml/mol will introduce an error of about 2 kcal/mol in the calculated heat of reaction in organic solvents, but an error of 20 kcal/mol in water because of the significantly smaller expansion coefficient in water.

Time resolution of the enthalpy changes is often possible and depends on a number of experimental parameters, such as the characteristics of the transducer (oscillation frequency and relaxation time) and the acoustic transit time of the system, τ_a, which can be defined by $\tau_a = r_o/v_a$ where r_o is the radius of the irradiated sample, and v_a is the speed of sound in the liquid. The observed voltage response of the transducer, $V(t)$ is given by the convolution of the time-dependent heat source, $H(t)$ and the instrument response function, $T(t)$ [3,13].

$$V(t) = H(t) * T(t)$$

Under conditions of short laser pulses and where τ_a is less than the transducer response time, τ_o, three transducer voltage responses can be described depending on the time scale of the heat deposition, τ_h.

When $\tau_h \ll \tau_o$, $H(t)$ is a effectively a delta function, an 'instantaneous' heat deposition, and the resultant voltage response of the transducer is simply the instrument response function, $T(t)$. Although heat depositions cannot be time resolved in this regime, their magnitude and consequently the enthalpy of reaction can be evaluated relative to the calibration compound. Calibration compounds are employed under these experimental conditions because they deposit a known amount of heat rapidly, and effectively yield the instrument response function. When $\tau_h \gg \tau_o$, no detector response is observed, i.e. long-time events are transparent to the transducer and ignored. When $\tau_h \cong \tau_o$, the voltage response can be analyzed to obtain both enthalpic and kinetic information about the heat deposition. Several deconvolution methods have been described for this purpose [13,14]. In many chemical systems, several heat depositions can contribute to $H(t)$ and each can be separately evaluated.

To date, most PAC experiments have measured kinetic processes on the 10 μs–10 ns time scale. Although PAC could indeed be utilized for slower kinetic events, other techniques such as thermal lensing and thermal beam deflection have typically been used [4]. Although, in theory, the time resolution of PAC can be increased, it will be difficult to do experimentally because of the inherent transit acoustic time limitations. Although sample cell designs may alleviate this problem to some extent [15], other technical considerations limit this technique to >10 ns. However, the limitations of the transit acoustic time can be fully overcome by using the transient grating technique [16], in which the volume changes are effectively detected by changes in the refractive index of the system. This technique has been used to monitor kinetic events on the picosecond time scale [16]. The lifetime and energy of the 'phantom' or non-detected twisted singlet state of tetraphenylethylene in various solvents has been determined [17].

The application of PAC to a simple photochemical reaction can be illustrated. Acetone is readily excited to its singlet excited state which rapidly undergoes efficient intersystem crossing to its triplet state. The triplet state decays in solution primarily by radiationless decay. The PAC experimental waveforms obtained from the photoexcitation of acetone in air and argon saturated cyclohexane are shown in Figure 5.1. In addition, the waveform obtained from the calibration compound 2-hydroxybenzophenone is also shown.

The calibration compound undergoes no photochemistry. It deposits all of the photon energy rapidly, $\leq$10 ns, as heat. Its waveform, T-wave, serves as the instrument response function and is used to deconvolute the experimental waves. As is readily apparent, the experimental waves E-Wave(1) and E-Wave(2) are phase shifted relative to the T-Wave. This indicates that energy is deposited to solution within the time resolution of the transducer. Deconvolution of the E-waves indicates two heat depositions. The first, defined as $\leq$10 ns, corresponds to the rapid formation of the triplet state of acetone. The amplitude of the heat deposition can be used to calculate the triplet state energy. The experimental value of 76 $\pm$ 2 kcal/mol is in good agreement with the spectroscopic value of 78 kcal/mol [18]. The second decay corresponds to the decay of the triplet state in air-saturated, 75 ns, or argon-saturated, 450 ns, cyclohexane. The shorter lifetime for the triplet state in air-saturated solvent suggests that it is effectively quenched by oxygen.

There are several unique features about PAC. First, it and the related methods are the only experimental techniques currently available that

FIGURE 5.1 Experimental waveforms from excitation (266 nm) of 2-hydroxybenzophenone (T-Wave), and acetone in air-saturated (E-Wave(1)) and argon-saturated (E-Wave(2)) cyclohexane. (With permission from Joshua L. Goodman.)

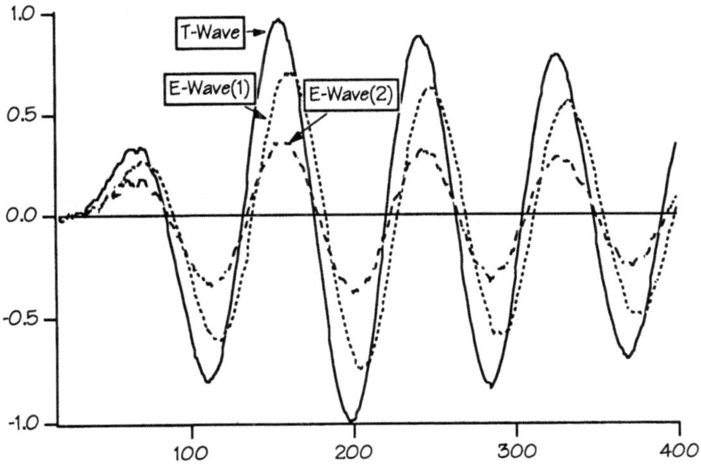

can measure the heats of reaction of radicals and biradicals on the μsec and faster timescale. This usually allows for an accurate determination of the heats of formation of these reactive intermediates. Second, PAC can monitor the reactions of transients which are optically transparent, i.e. do not have an UV-Vis optical absorbance. Hence, in addition to thermodynamics, PAC can also provide important kinetic information about these 'invisible' species.

INSTRUMENTATION

The experimental design for the photoacoustic experiment is relatively simple. The apparatus is quite similar to that employed for nanosecond absorption spectroscopy with the major difference being that a piezoelectric transducer is used to monitor the acoustic waves rather than a photomultiplier tube to analyze the incident light. A representative schematic for PAC is shown in Figure 5.2.

Nanosecond Nd:YAG, excimer, or nitrogen pulsed lasers are typically used as the light source. A fraction of the light beam is reflected to a photodiode that triggers the experimental data collection. The beam then impinges on the sample within the cell. The generated acoustic signal is detected by the piezoelectric transducer, amplified, digitized and stored in the computer. The acoustic waves are normalized for absorbance, which is determined *in situ* using the two energy meters and fluctuations in laser power. The computer analyzes the waveforms using a deconvolution program to obtain the requisite kinetic and thermodynamic information.

Several different cell designs have been used for PAC. In the cell design above, the beam passes through the cell wall at a right angle to that to which the transducer is attached. The time resolution in this cell design is

FIGURE 5.2 Schematic representation for photoacoustic calorimeter.

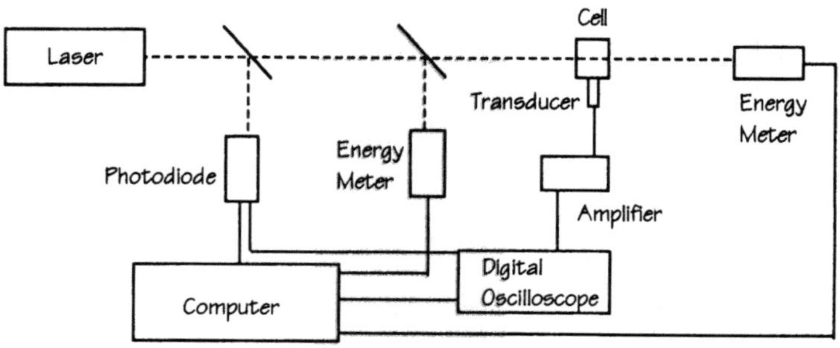

given by the beam diameter and the velocity of sound in the solvent, i.e. τ_a. An alternate front face design has been developed by Melton in which the beam passes through the cell wall parallel to that affixed to the transducer [15]. The time resolution in this cell is given simply by the cell thickness. Each design has certain advantages (and disadvantages) with regard to sensitivity, time resolution, sample preparation and ease of operation.

A variety of piezoelectric transducers have been employed for PAC. Ceramic transducers, usually lead zirconate titanate, are most commonly employed because of their sensitivity, time resolution and commercial availability. However, their acoustic response is often dominated by their own resonance, and so polymeric film detectors, such as polyvinylidene-difluoride, are often used. These piezoelectric materials are non-resonant, but not as sensitive as the ceramic detectors. Again, each detector has its own advantages (and disadvantages) [19].

The accuracy of the thermochemical data obtained by this technique has been examined in numerous systems. In general, the data compares well, ± 1 kcal/mol, with that obtained by other spectroscopic and calorimetric methods. The accuracy and reproducibility of the data is dependent on the magnitude and time scale of the heat depositions detected by PAC, which is associated with a given chemical process. Reactions which are highly exothermic are easy to detect, whereas ones that are not are difficult to detect. A thermoneutral reaction is invisible to PAC. Reactions that occur significantly slower than the response time of the transducer are not detected. If the reaction occurs either slightly slower or faster then the response times are difficult to resolve accurately. Clearly, the proper choice of the transducer is extremely important in order to resolve accurately a given chemical event.

Thermochemical data obtained from experiments in which only a fast heat deposition is detected are more reliable than those in which two or more heat depositions are observed. This is due in part to the necessity for deconvolution of the experimental data. With conventional transducers and low pulse energies, $<10\,\mu J$, the signal to noise ratio is typically >10, and so signal averaging is usually not necessary. Fluctuations in the measured heat depositions are usually small, $<2\%$ from experiment to experiment.

EXPERIMENTAL RESULTS

Radicals

Photoacoustic calorimetry can be used to obtain important kinetic and energetic information about radical intermediates. PAC studies have

$$(CH_3)_3CO{-}OC(CH_3)_3 \xrightarrow[\Delta H_1]{h\nu} 2\,(CH_3)_3CO\cdot$$

$$(CH_3)_3CO\cdot + XH \xrightarrow[\Delta H_2]{k_2} (CH_3)_3COH + X\cdot$$

SCHEME 5.1

typically involved the generation of the radicals of interest via an abstraction reaction by an initially formed reactive species, such as *t*-butoxy radical, or the triplet state of benzophenone. The kinetics and energetics of both the formation and subsequent reactions of the radical intermediate can then be monitored.

Photolysis of di-*t*-butylperoxide results in the efficient formation of *t*-butoxy radical, which can then be used to abstract hydrogen, X—H, from various substrates, presumably due to the formation of the strong O—H bond (104 kcal/mol).

The enthalpy of this abstraction reaction, ΔH_2, can be used to determine the bond dissociation energy of the X—H bond. In this manner, Peters measured the O—H bond energy in phenol, 84 kcal/mol, which is in good agreement with previous literature values [5(a)]. Griller [20(b)] and Peters [14(a)] also investigated the effect of para-substitution of phenols on the O—H bond dissociation energy. They found the values ranged from 87.2 kcal/mole for CF_3- to 78.1 kcal/mole for CH_3O^- substitution. The bond dissociation energies correlate well with both σ^+, and the rate constants for the hydrogen abstraction, k_2. They suggested that these observations support previous results that substituent effects reflect variations in bond energies rather than polar contributions to the transition state [20(b)].

Hydrogen abstraction has also been used by others to measure C–H bond dissociation energies in alkyl amines [21], ethers [22], and olefins [22]. Sn—H [22], Si—H [23], Ge—H [24] bond energies have also been determined for several substrates. Interestingly, substitution of silicon for carbon at the silicon center has a dramatic effect on lowering the Si—H bond dissociation energy. For example, the Si—H bond energy for $(CH_3Si)_3Si{-}H$ is 79.0 kcal/mol whereas it is 90.1 kcal/mol for $(CH_3CH_2)_3Si{-}H$ [23]. Hydrogen abstraction by *t*-butoxyl radical has also been used to measure metal hydride bond energies as that in the η^5-$C_5(CH_3)_5(P(CH_3)_3){-}IrH_2$ complex [25].

Photolysis of hydrogen peroxide yields the hydroxyl radical which has also been used to abstract hydrogen from various substrates [26]. Griller measured the C—H bond dissociation energies for several substrates in

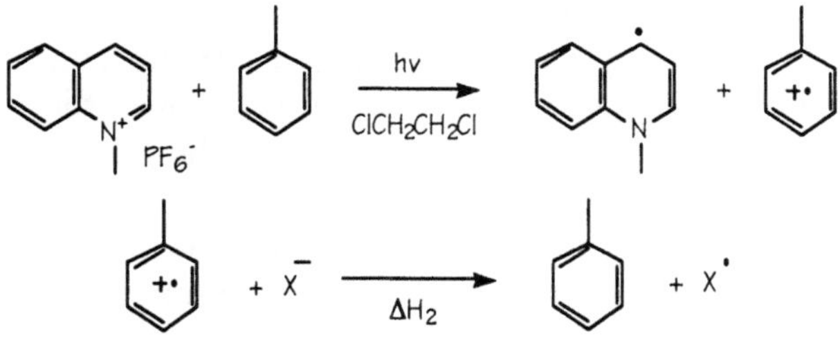

SCHEME 5.2

water. These workers found that the obtained heats of reaction in water were significantly less than in the gas phase. The difference was due to the change in the enthalpy of solvation that occurred when one equivalent of hydrogen peroxide was converted to two of water in the reaction.

Inorganic metal complexes can also be used to study atom transfer radical reactions. The rate and energetics of H, Cl, Br, and I abstraction from organic substrates by the excited state of the tetrakis (pyrophosphito) diplatiniate (II) complex were determined by PAC [27]. In principle, other excited state or $17e^-$ inorganic complexes could be employed to measure bond dissociation energies of substrates with readily abstractable atoms. Metal–metal bond dissociation energies can also be readily measured by PAC. For example, direct excitation of $Mn_2(CO)_{10}$ results in efficient homolytic cleavage of the Mn–Mn bond. The heat of this reaction yields the Mn—Mn bond energy, 38 kcal/mol [28]. Unfortunately, PAC has not been widely employed to investigate organometallic and inorganic photoreactions [29].

The redox potential of radicals can be determined by PAC. For example, consider the redox reactions in Scheme 5.2. Electron transfer to the excited state of *N*-methylquinolinium from toluene results in the efficient production of toluene cation radical, a powerful oxidant. This can oxidize substrates such as anions whose oxidation potentials are less than 2.3 V. The enthalpy of this reaction, ΔH_2, and the oxidation potential of toluene are used to measure the oxidation potential of the anion. The obtained value corresponds to the reduction potential of the radical, E_{ox} (anion) = E_{red} (radical).

Ion Radicals

The properties of ion radicals produced by photoinduced electron transfer have been investigated by PAC. Solvent and various salts were shown to have a pronounced effect on the ion pair energies in the photoreduction of benzophenone by 1,4-diazabicyclo[2.2.2]octane [30]. The addition of sodium and lithium perchlorate was found to lower the energy of the ion pair, presumably by the stabilizing interaction between the alkali metals and benzophenone anion radicals.

The photoreduction of quinones by triethylamine to form ion pairs was also investigated [31]. The results indicate that electron transfer followed by proton transfer permits the efficient accumulation of reactive but metastable products that persist long enough to be usable as reagents in other dark processes without the necessity for sacrificial reagents or complex catalysts. In a similar study, the energetics of proton transfer to benzophenone anion radical from aniline cation radical were examined [32].

PAC has been used to determine quantum yields for the formation of ion radicals following photoinduced electron transfer [8]. Knowledge of the ion radical pair energies from redox potentials enables the quantum yields to be determined using equation (5.56). The yields were verified using nanosecond absorption spectroscopy. Alternatively, if the quantum yields are known accurately, the energies can be obtained.

Biradicals

Biradicals (Carbenes). The first application of PAC to carbenes was by Peters who investigated the photolysis of diphenyldiazomethane to yield triplet diphenylcarbene [20(a),33]. The heat of reaction was found to be almost thermoneutral, which suggests that the heats of formation of diphenyldiazomethane and diphenylcarbene are similar. In addition, the rate and energetics of the reaction of the carbene with oxygen to form the carbonyl oxide were determined.

Several reactions of singlet halocarbenes have been investigated by PAC. The heats of reaction with alcohols can be measured and used to determine the heats of formation for the singlet carbenes [34]. For example, the heat of reaction of dimethoxycarbene, $C(OCH_3)_2$, with CH_3OH is -24.8 kcal/mol [34(a)]. Using the heats of formation of CH_3OH and the product, $CH(OCH_3)_3$, the calculated heat of formation of $C(OCH_3)_2$ is -54.7 kcal/mol. This is in good agreement with the AM1 value of -53 kcal/mol. However, the heats of formation of several carbenes has been found to be significantly less than the values obtained from semi-empirical calculations [34(a)]. Possible reasons for the discrepancy have been advanced, but not resolved.

With highly electrophilic carbenes, the reaction rates of carbenes with alcohols approach diffusion control. The alcohol acts as a nucleophile and

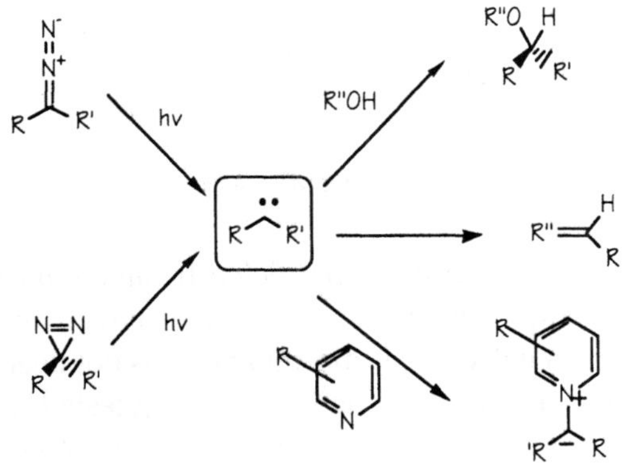

SCHEME 5.3

the reaction potentially occurs *via* an ylide intermediate. With nucleophilic carbenes the reaction rates are also fast and the alcohol acts as an electrophile. The alcohol protonates the carbene to form an ion pair, which subsequently collapses to product. The reaction rates with ambiphilic carbenes are often quite slow and the reaction may occur by a combination of pathways.

The heats and rates of reaction of carbenes with acetone [35], acetonitrile [34], and pyridine [36] to form ylides have also been measured and used to calculate the ylides' heats of formation. The heats of reaction of methylchloro- and phenylchlorocarbene with substituted pyridines were found to correlate well with the pK_a's of the pyridines. However, the correlation is not good for sterically demanding pyridines, presumably because of the large size of the carbene relative to the proton. In fact, some do not readily form ylides with the carbenes. The rates of ylide formation with the carbenes were also found to increase with the basicity of the pyridines.

The 1,2 hydrogen rearrangement of alkylhalocarbenes to vinyl halides has also been examined by PAC and several interesting observations were made. First, the heats of reaction were significantly larger than anticipated. These results together with chemical yield studies suggested that a second pathway for the formation of vinyl halides existed [37]. Further studies have shown that the diazirine precursor excited state can indeed partition between the product vinyl halide and the carbene, which can subsequently also form vinyl halide. This appears to be a general feature of diazirine photochemistry. Second, the obtained activation parameters for the rearrangement of several carbenes were found to be

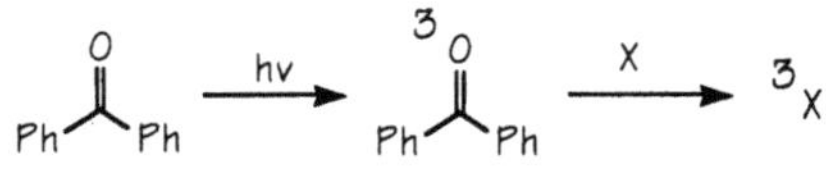

SCHEME 5.4

significantly different from those predicted by *ab initio* calculations [38]. In particular, the experimental energies are smaller and the entropies significantly more negative. This discrepancy has not been fully resolved, but quantum mechanical tunneling has been suggested to be involved in the rearrangement [39]. The reaction is thought to proceed classically at high temperatures as suggested by theory, but proceeds primarily by quantum mechanical tunneling at low temperatures.

1,2-Biradicals (alkene excited states). The energies of relaxed alkene triplet excited states, often referred to as '1,2-biradicals' can provide useful information about the excited state potential energy surface and its relationship to the ground state. Unfortunately, they are not readily obtained because triplet states typically undergo large geometrical changes upon formation. The rotation about the double bond usually results in a substantial lowering of its energy and rendering it difficult to measure spectroscopically. However, PAC can be used to determine relaxed triplet state energies and lifetimes. Furthermore, the energetic effect of substitution, strain and solvent can also be evaluated by PAC.

Several groups have investigated alkene triplet states [40,41]. Triplet sensitization of the alkenes typically using aromatic ketones produces the triplet state whose energy and lifetime can be determined.

A large number of alkenes have been examined and several generalizations have been advanced based on the results. First, the relaxed triplet state energies, E_T, are substantially lower than the vertical or unrelaxed triplet state energy when rotation about the C—C double bond is possible. For example, the E_T of 1-phenylcycloheptene was found to be similar to that of other acyclic styrene derivatives, and significantly lower than the vertical triplet energy. However, E_T of 1-phenylcyclohexene is 4 kcal higher than the E_T of its acyclic analog, and the E_T of 1-phenylcyclopentene is even higher and quite similar to the energy of its vertical triplet state (40,41). The data suggest that as the strain energy of twisting about the C—C double bond increases in the excited state, the E_T increases. The rotation angle for each of these triplet states is not experimentally known, but calculations suggest that the angle of the relaxed triplet state decreases with increasing strain.

Second, the E_T values for many unstrained alkenes are similar to those calculated for 1,2-biradicals using Benson group equivalent values. This supports the idea that unstrained relaxed triplet states are effectively 1,2-biradicals with a large twist angle (around 90 degrees) in which there is little interaction between the radical centers. In fact, the triplet energies are often quite similar to the transition state energies for alkene isomerization. This simple model has indeed been used to predict isomerization activation energies.

The *cis–trans* isomerization energies of several phenylcycloalkenes have been reported. The isomerization energies of 1-phenylcyclo-hexene, heptene, and octene, are 47.0, 29.0, and 13.3 kcal/mol respectively [40,41]. Clearly, the energy of the *trans* isomer relative to the *cis* increases as the ring size decreases. This effect has been attributed to the increase in the strain energy of the *trans* isomer as the ring size decreases.

The relaxed triplet energies of several *α,β*-unsaturated cyclic enones have been determined by PAC [19]. The effect of substitution and ring size was examined. Similar to alkene triplets, the energy of the relaxed triplet increased with increasing structural rigidity.

1,n-Biradicals. The triplet 1,3-biradicals cyclopentane-1,3-diyl and 2-isopropylidienecyclopentane-1,3-diyl were studied by PAC [42]. The triplet diyls were formed via photochemical decomposition of the appropriate azo precursors. The heats of formation of the diyls were determined using the PAC heats of reaction. For example, the ring closure of triplet cyclopentane-1,3-diyl to bicyclo[2.1.0]pentane is exothermic by 40.0 kcal/mol. Using this value and the experimental heat of formation of bicyclo[2.1.0]pentane, 37 kcal/mol [43], the heat of formation of triplet cyclopentane-1,3-diyl is 71.5 kcal/mol. This value is significantly higher than that obtained using Benson group equivalents, 66.4 kcal/mol [1]. This experimental heat of formation supports an upward revision of the Benson group equivalents for radicals [1(b),1(c),33]. In addition, PAC was used to measure the rates of ring closure and oxygen trapping of the diyls.

1,4-biradicals are reactive intermediates that are formed in a number of organic photoreactions. Two such reactions, the Norrish type II photocleavage, and the photodimerization of olefins have been studied using

SCHEME 5.5

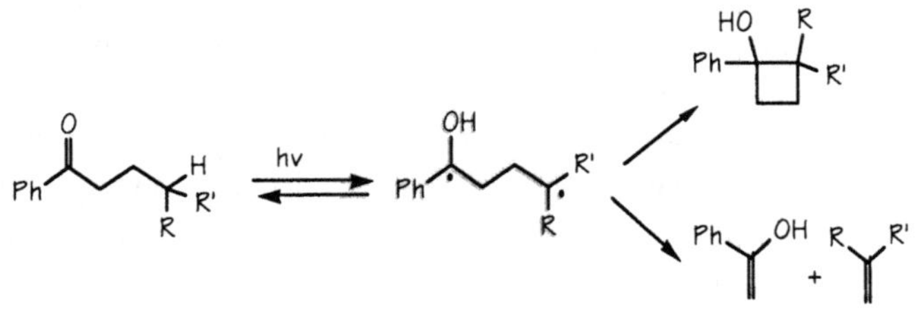

SCHEME 5.6

PAC. Aromatic ketones rapidly undergo intersystem crossing to their triplet state. If they possess a γ hydrogen, abstraction can occur to yield the 1,4-biradical. The biradical then typically decays to the ketone, cleavage to the olefin and enol, or cyclization to the cyclobutanol.

The heats of reaction to form the 1,4-biradical, and its subsequent decay to products have been measured by PAC [13(a)]. The effect of substitution at the γ carbon on the reaction energetics can be related to C—H bond dissociation energies and consequently radical stabilization energies [44].

The photodimerization of cyclopentenone is thought to involve the initial formation of a 1,4 biradical via ground state quenching of the triplet state. The heat of formation and lifetime of the biradical were determined. In both of these photoreactions, the heats of formation of the biradicals are in good agreement with the values calculated for such species using Benson group equivalents.

CONCLUSIONS

In this chapter, the application of photoacoustic calorimetry to determine the heats and rates of reaction of radicals and biradicals has been described. It can also be readily applied to study other reactive intermediates such as singlet and triplet states, strained rings, antiaromatics, ion pairs, metal and organometallic complexes, and odd electron species [4]. The technique is simple both in the experimental design and data analysis. It is an extremely versatile, highly sensitive analytical method that requires only that the reactive intermediate be generated rapidly (<10 μsec) under photochemical conditions and that it causes a volume change in solution. PAC can detect 'invisible' transients in that it does not require any given structural feature. Furthermore, PAC can serve as an excellent complementary technique to others such as absorption and emission spectroscopies.

Acknowledgment

Our work on the application of PAC to radicals and biradicals has been generously supported by the National Science Foundation, Alfred P. Sloan Foundation, and Petroleum Research Fund, to whom we are grateful.

REFERENCES

1. (a) Benson, S. W. (1970) *Thermochemical Kinetics*, 2nd edn, Wiley-Interscience: New York. (b) Doering, W. v. E. (1981) *Proc. Natl. Acad. Sci.* **78,** 5279. (c) Tsang, W. (1985) *J. Am. Chem. Soc.* **107,** 2872.
2. Bell, A. G. (1880) *Am. J. Sci.* **20,** 305.
3. (a) Patel, C. K. N. and Tam, A. C. (1981) *Rev. Mod. Phys.* **53,** 517. (b) Tam, A. C. (1986) *Rev. Mod. Phys.* **58,** 381.
4. Braslavsky, S. E. and Heibel, G. E. (1992) *Chem. Rev.* **92,** 1381.
5. (a) Peters, K. S. (1986) *Pure Appl. Chem.* **58,** 1263. (b) Burkey, T. J., Majewski, M. and Griller, D. (1986) *J. Am. Chem. Soc.* **108,** 2218.
6. (a) Cahen, D., Garty, H. and Becker, R. S. (1980) *J. Phys. Chem.* **84,** 3384. (b) Somoano, R. B. (1978) *Angew. Chem.* **90,** 250. (c) Gortz, W., Perkampus, H.-H. and Fresenius, Z. (1983) *Anal. Chem.* **316,** 180. (d) Kumar, D., Nauman, R. V., Mathers, T. L. and McGlynn, S. P. (1986) *J. Indian Chem. Soc.* **6,** 310. (e) Rothberg, L. J., Jedju, T. M. and Lawrence, K. J. (1987) *Quant. Spectrosc. Radiat. Transfer* **37,** 515. (f) Rudzki-Small, J., Hutchings, J. J. and Small, E. W. (1989) *SPE Fluorescence Detection III* **1054,** 26. (g) Rudzki-Small, J. and Larson, S. L. (1990) In *Time-Resolved Laser Spectroscopy in Biochemistry II* Lakowicz, J. R., ed. SPIE Proc. 1204: Bellingham, WA, p 126. (h) Arden, J., Deltau, G., Huth, V., Kringel, U., Peros, D. and Drexhage, K. H. (1990) *J. Luminesc.* **48/49,** 352. (i) Lesiecki, M. L. and Drake, J. M. (1982) *Appl. Optics* **21,** 557. (j) Bonch-Bruevich, A. M., Razumova, T. K. and Starobogatov, I. O. (1982) *J. Appl. Spectr.* **37,** 1413. (k) Negri, R. M., Zalts, A., San Roman, E. A., Aramendia, P. F and Braslavsky, S. E. (1991) *Photochem. Photobiol.* **53,** 317.
7. (a) Komorowski, S. J., Grabowski, Z. R. and Zielenkiewicz, W. J. (1985) *Photochem.* **30,** 141. (b) Boldridge, D. W., Scott, G. W. and Spiglanin, T. A. (1982) *J. Phys. Chem.* **86,** 1976. (c) Lynch, D. and Endicott, J. F. (1989) *Appl. Spectrosc.* **43,** 826. (d) Song, X. Q. and Endicott, J. F. (1991) *Inorg. Chem.* **30,** 2214. (d) Scaiano, J. C., Redmond, R. W., Mehta, B. and Arnason, J. T. (1990) *Photochem. Photobiol.* **52,** 655.
8. (a) Gould, I. R., Moser, J. E., Armitage, B., Farid, S., Goodman, J. L. and

HERMAN, M. S. (1989) *J. Am. Chem. Soc.* **111,** 1917. (b) DINNOCENZO, J. P., FARID, S., GOODMAN, J. L., GOULD, I. R. and TODD, W. P. (1991) *J. Am. Chem. Soc.,* **113,** 3601.

9. (a) ASANO, T. and LE NOBLE, W. J. (1978) *Chem Rev.* **78,** 407. (b) LE NOBLE, W. J. (1967) *Prog. Phys. Org. Chem.* **5,** 207. (c) LE NOBLE, W. J. (1980) *Angew. Chem., Int. Ed. Engl,* **19,** 841.
10. (a) CALLIS, J. B., PARSON, W. W and GOUTERMAN, M. (1972) *Biochim. Biophys. Acta* 267. (b) ORT, D. R. and PARSON, W. W. (1979) *Biophys. J.* **25,** 341, 355. (c) WESTRICK, J. A., GOODMAN, J. L. and PETERS, K. S. (1987) *Biochem.* **26,** 8313. (d) HERMAN, M. S. and GOODMAN, J. L. (1989) *J. Am. Chem. Soc.* **111,** 1849.
11. (a) GOODMAN, J. L. and HERMAN, M. S. (1989) *Chem. Phys. Lett.* **163,** 417. (b) HERMAN, M. S. and GOODMAN, J. L. (1989) *J. Am. Chem. Soc.* **111,** 9105. (c) HUNG, R. R. and GRABOWSKI, J. J. (1992) *J. Am. Chem. Soc.* **11,(4)** 351.
12. (a) ORT, D. R. and PARSON, W. W. (1978) *J. Biol. Chem.* **253,** 6158. (b) ARATA, H. and PARSON, W. W. (1981) *Bioph. Bioch. Acta* **636,** 70. (c) WESTRICK, J. A. and PETERS, K. S. (1990) *Bioph. Chem.* **37,** 73. (d) PETERS, K. S., WATSON, T. and MARR, K. (1991) *Annu. Rev. Biophys. Biophys. Chem.* **20,** 343. (e) WESTRICK, J. A., PETERS, K. S., ROPP, J. D. and SLIGAR, S. G. (1990) *Biochem.* **29,** 6741. (f) MARR, K. and PETERS, K. S. (1991) *Biochem.* **30,** 1254. (g) LARSON, S. L. and RUDZKI-SMALL, J. (1990) *J. Biophys.* **57,** 229.
13. (a) RUDZKI, J. E., GOODMAN, J. L. and PETERS, K. S. (1985) *J. Am. Chem. Soc.* **107,** 7849. (b) LAI, H. M. and YOUNG, K. (1982) *J. Acoust. Soc.* **72,** 2000. (c) HERITIER, J.-M. (1983) *Opt. Commun.* **44,** 267.
14. (a) PETERS, K. S. and SNYDER, G. J. (1988) *Science* **241,** 1053. (b) HEIHOFF, K. BRASLAVSKY, S. E. and SCHAFFNER, K. (1987) *Biochem.* **26,** 1422. (c) HEIHOFF, K. and BRASLAVSKY, S. E. (1986) *Chem. Phys. Lett.* **131,** 183. (d) BRASLAVSKY, S. E. and HEIHOFF, K. (1989) In *Handbook of Organic Photochemistry;* Scaiano, J. C., ed. CRC Press: Boca Raton, FL, Vol. 1, p 327. (e) RUDZKI-SMALL, J., LIBERTINI, L. J. and SMALL, E. W. *Biophys. Chem.* **42,** 29. (f) VIAPPIANI, C. and RUDZKI-SMALL, J. (1992) In *Time-Resolved Laser Spectroscopy in Biochemistry III* Lakowicz, J. R., ed. SPIE Proc. 1640: Bellingham, WA.
15. (a) MELTON, L. A., NI, T. and LU, Q. (1989) *Rev. Sci. Instr.* **60,** 3217. (b) ARNAUT, L. G., CALDWELL, R. A., ELBERT, J. E. and MELTON, L. A. (1992) *Rev. Sci. Instr.* **63,** 5381.
16. (a) MILLER, R. J. D., PIERRE, M., ROSE, T. S. and FAYER, M. D. (1984) *J. Phys. Chem.* **88,** 3021. (b) FAYER, M. D. (1982) *Annu. Rev. Phys. Chem.* **33,** 63. (c) MILLER, R. J. D. (1989) In *Time Resolved Spectroscopy,* Clark, R. J. H., Hester, R. E. ed., Wiley: New York, pp 1–54.
17. (a) MORAIS, J., MA, J. and ZIMMT. M. B. (1991) *J. Phys. Chem.* **95,** 3885. (b) ZIMMT, M. B. (1989) *Chem. Phys. Lett.* **160,** 564.
18. BIRKS, J. (1975) *Organic Molecular Photophysics,* Wiley: New York.
19. (a) SCHUSTER, D. I., HEIBEL, G. E., CALDWELL, R. A. and TANG, W. (1990) *Photochem. Photobiol.* **52,** 645. (b) CALDWELL, R. A., TANG, W., SCHUSTER, D. I. and HEIBEL, G. E. (1991) *Photochem. Photobiol.* **53,** 159.

20. (a) Peters, K. S. (1990) In *Kinetics and Spectroscopy of Carbenes and Biradicals;* Platz, M. S., ed., Plenum Press: New York, p. 37 (b) Mulder, P., Saastad. O. W. and Griller, D. (1988) *J. Am. Chem. Soc.* **110,** 4090.
21. Dinnocenzo, J. P. and Banach, T. E. (1989) *J. Am. Chem. Soc.* **111,** 8646.
22. (a) Griller, D. and Wayner, D. D. M. (1986) *Rev. Chem. Inst.* **7,** 31. (b) Clark, K. B., Culshaw, P. N., Griller, D., Lossing, F. P., Simoes, J. A. M. and Walton, J. C. (1991) *J. Org. Chem.* **56,** 5535.
23. Kanabus-Kaminska, J. M., Hawari, J. A., Griller, D. and Chatgilialoglu, C. (1987) *J. Am. Chem. Soc.* **109,** 5267.
24. Clark, K. B. and Griller, D. (1991) *Organometall.* **10,** 746.
25. Nolan, S. P., Hoff, C. D., Stoutland, P. O., Newman, L. J., Buchanan, J. M., Bergman, R. G., Yang, G. K. and Peters, K. S. (1987) *J. Am. Chem. Soc.* **109,** 3143.
26. Kanabus-Kaminska, J. M., Gilber, B. C. and Griller, D. (1989) *J. Am. Chem. Soc.* **111,** 3311.
27. Herman, M. S. and Goodman, J. L. (1991) *Inorg. Chem.* **30,** 1147.
28. Goodman, J. L., Peters, K. S. and Vaida, V. (1986) *Organometall.* **5,** 815.
29. (a) Yang, G. K., Peters, K. S. and Vaida, V. (1986) *Chem. Phys. Lett.* **125,** 566. (b) Yang, G. K., Vaida, V. and Peters, K. S. (1988) *Polyhedron* **7,** 1619. (c) Klassen, J. K., Selke, M., Sorensen, A. A. and Yang, G. K. (1990) *J. Am. Chem. Soc.* **112,** 1267. (d) Sorensen, A. A. and Yang, G. K. (1991) *J. Am. Chem. Soc.* **113,** 7061. (e) Klassen, J. K. and Yang, G. K. (1990) *Organometall.* **9,** 874. (f) Burkey, T. J. (1989) *Polyhedron* **8,** 2681. (g) Morse, J. M. Jr., Parker, G. H. and Burkey, T. J. (1989) *Organometall.* **8,** 2471. (h) Burkey, T. J. (1990) *J. Am. Chem. Soc.* **112,** 8329. (i) Nayak, S. K. and Burkey, T. J. (1991) *Organometall.* **10,** 3745.
30. Goodman, J. L. and Peters, K. S. (1986) *J. Am. Chem. Soc.* **108,** 1700.
31. Ci, X., da Silva, R. S., Goodman, J. L., Nicodem, D. E. and Whitten, D. G. (1988) *J. Am. Chem. Soc.* **110,** 8548.
32. Rothberg, L. J., Simon, J. D., Bernstein, M. and Peters, K. S. (1983) *J. Am. Chem. Soc.* **105,** 3464.
33. Simon, J. D. and Peters, K. S. (1983) *J. Am. Chem. Soc.* **105,** 5156.
34. (a) LaVilla, J. A. and Goodman, J. L. (1989) *J. Am. Chem. Soc.* **111,** 712. (b) Du, X.-M., Fan, H., Goodman, J. L., Kesselmayer, M. A., Krogh-Jespersen, K., LaVilla, J. A., Moss, R. A., Shen, S. and Sheridan, R. S. (1990) *J. Am. Chem. Soc.* **112,** 1920.
35. LaVilla, J. A. and Goodman, J. L. (1988) *Tet. Lett.* **29,** 2623.
36. LaVilla, J. A. and Goodman, J. L. (1990) *Tet. Lett.* **31,** 6287.
37. (a) LaVilla, J. A. and Goodman, J. L. (1990) *Tet. Lett.* **31,** 5109. (b) Modarelli, D. A., Morgan, S. and Platz, M. S. (1992) *J. Am. Chem. Soc.* **114,** 7034 and references therein.
38. LaVilla, J. A. and Goodman, J. L. (1989) *J. Am. Chem. Soc.* **111,** 6877.

39. (a) Dix, E. J., Herman, M. S. and Goodman, J. L. (1993) *J. Am. Chem. Soc.* **115,** 10424. (b) Dix, E. J. and Goodman, J. L. (1994) *Res. on Chem Interm.* 149.
40. (a) Goodman, J. L., Peters, K. S., Misawa, H. and Caldwell, R. A. (1986) *J. Am. Chem. Soc.* **108,** 6803. (b) LaVilla, J. A. and Goodman, J. L. (1987) *Chem. Phys. Lett.* **141,** 149.
41. (a) Ni, T., Caldwell, R. A. and Melton, L. A. (1989) *J. Am. Chem. Soc.* **111,** 457. (b) Brennan, C. M. and Caldwell, R. A. (1991) *Photochem. Photobiol.* **53,** 165. (c) Strickland, A. D. and Caldwell, R. A. (1993) *J. Phys. Chem.* **97,** 13394.
42. Herman, M. S. and Goodman, J. L. (1988) *J. Am. Chem. Soc.* **110,** 2681.
43. Chang, S., McNally, D., Tebrany, S., Hickey, M. and Bold, R. (1970) *J. Am. Chem. Soc.* **92,** 3109.
44. Caldwell, R. A., Goodman, J. L. and Peters, K. S., unpublished results.

6

A Short and Illustrated Guide to Metal–Alkyl Bonding Energetics

·J. A. MARTINHO SIMÕES and
M. E. MINAS DA PIEDADE

INTRODUCTION

This volume is called *Energetics of Organic Free Radicals*. Why then a (short) chapter on metal–alkyl bonding energetics? The subject could have a direct impact on the main topic of the book if very accurate metal–alkyl bond dissociation enthalpy data were available, which, together with the enthalpies of formation of the parent organometallic molecules and of the organometallic fragments (produced upon dissociation of the metal–alkyl bonds), would lead to the enthalpies of formation of the alkyl radicals. However, very few gas-phase metal–carbon bond dissociation enthalpies are accurately known for coordinatively saturated organometallic complexes and the data bank on the energetics of unsaturated (neutral or ionic) metal species is rather limited [1]. The present information on thermochemistry of organometallic compounds is thus almost irrelevant to the energetics of organic free radicals. In reality, things happen the other way around: organometallic thermochemists often *need* thermochemical data of organic free radicals to derive metal–ligand bond dissociation enthalpy values. In other words, the energetics of organometallic molecules may be regarded as an application example of the energetics of free radicals. This would probably be enough to justify this chapter, but there is another reason, which will become apparent throughout the text: in general, the enthalpies of metal–ligand bonds and the corresponding ligand–hydrogen bonds follow nearly parallel trends. This is not too surprising, as one would expect that the energetics of a given moiety (a radical) is reflected by the energetics of the whole molecule to which that moiety belongs.

The discussion is limited to metal–alkyl (and occasionally aryl) complexes, but this choice is somewhat arbitrary since many of the conclusions presented below could be drawn by using other families of groups bonded to the metal center. This paper is *not* a review on metal–alkyl bond dissociation enthalpies. It does not list and assess all the available data nor is there a comprehensive discussion of their trends. The experimental or theoretical methodologies that have been used to obtain that type of information will not be discussed. Rather, an attempt is made to provide an overview of the field of metal–alkyl bonding energetics, stressing some features that appear particularly important or interesting and disclosing some of its most feeble assumptions. A few selected examples will be discussed, usually one for each group of the periodic table, within a framework that was developed some years ago. Finally, recognizing that the discussion of thermochemical data will benefit from a deeper understanding of the nature of chemical bonds (and vice versa), a final section on this subject is included.

A note on nomenclature is appropriate. By far the largest amount of the thermochemical values available for organometallic compounds is either the enthalpy of formation of the crystalline complexes or the enthalpy of a given reaction in solution. Solvation and sublimation enthalpy data are rare, so many bond dissociation enthalpies are reported in solution. In these cases the symbol $DH^{\circ}_{\mathrm{sln}}$ will be used, whereas gas phase values will be indicated by DH°. An average bond dissociation enthalpy (i.e. the enthalpy required to cleave n identical metal–ligand bonds, divided by n) will be represented by $\overline{DH^{\circ}_{\mathrm{sln}}}$ or $\overline{DH^{\circ}}$. Unless stated otherwise, all the data presented refer to 298 K.

Selected enthalpies of formation of alkyl radicals and of their parent hydrocarbons are given in Table 6.1.

THE METHOD OF ANALYSIS

In 1987 the authors published a paper describing a very simple procedure for assessing the thermochemical data of several families of inorganic and organometallic complexes [2]. The method, which has since been applied to discuss the data for a variety of compounds and also to predict new values [3–15], consists of plotting the enthalpies of formation of a series of molecules of the same family, $\Delta_f H_m{}^{\circ}(ML_n)$, versus $\Delta_f H_m{}^{\circ}(LH)$, with ML_n and LH in either their standard reference states ('rs', their stable physical states at 298.15 K and 1 bar) or in the gas phase. It has been observed that

TABLE 6.1 Selected auxiliary data: enthalpies of formation of alkyl radicals and carbon–hydrogen bond dissociation enthalpies (Values in kJ/mol)

R^a	$\Delta_f H^o{}_m(\text{RH,g})^b$	$\Delta_f H^o{}_m(\text{R,g})^c$	$DH^o(\text{R–H})^{c,d}$
Me	−74.4 ± 0.4	147 ± 1	439 ± 1
Et	−83.8 ± 0.4	121 ± 2	423 ± 2
Pr	−104.7 ± 0.5	100 ± 2	423 ± 2
i-Pr	−104.7 ± 0.5	90 ± 2	413 ± 2
Bu	−125.6 ± 0.7	(79 ± 2)	(423 ± 2)
s-Bu	−125.6 ± 0.7	68 ± 2	412 ± 2
i-Bu	−134.2 ± 0.7	(61 ± 4)	(413 ± 4)
t-Bu	−134.2 ± 0.7	50 ± 2	402 ± 2
Pe	−146.9 ± 0.9	(58 ± 2)	(423 ± 2)
neo-Pe	−168.1 ± 0.8	32 ± 8	418 ± 8
He	−167.1 ± 0.8	(38 ± 2)	(423 ± 2)
Hp	−187.7 ± 1.3	(17 ± 2)	(423 ± 2)
Oc	−208.6 ± 1.4	(−4 ± 2)	(423 ± 2)
Bz	50.4 ± 0.6	204 ± 6	372 ± 6

[a] Pe = pentyl; He = hexyl; Hp = heptyl; Oc = octyl; Bz = benzyl.
[b] Data from reference 23.
[c] Values in parentheses are estimates.
[d] Values selected from data in the following references: Berkowitz, J., Ellison, G. B. and Gutman, D. (1994) *J. Phys. Chem.* **98,** 2744; Tsang, W., see chapter 2 in this volume and reference 1.

many of the above plots, which involve reliable thermochemical data, define excellent straight lines. This empirical linear relationship may be expressed as equation (6.1).

$$\Delta_f H^o_m(\text{ML}_n) = a[\Delta_f H^o_m(\text{LH})] + b \tag{6.1}$$

Scheme 6.1 is helpful in understanding the meaning of this correlation. $\Delta_r H^o(6.2)$ and $\Delta_r H^o(6.3)$ are the enthalpies of hypothetical reactions of a 'family' of compounds ML_n, reacting with a given molecule HX (e.g. HCl). The reactants and products are in the standard reference states and

SCHEME 6.1.

$$\text{ML}_n\,(\text{rs}) + n\text{HX}\,(\text{rs}) \xrightarrow{\Delta_r H^o\,(6.2)} \text{MX}_n\,(\text{rs}) + n\text{LH}\,(\text{rs}) \tag{6.2}$$

$$\downarrow \Delta^g_{rs}H^o_m(\text{ML}_n) \quad \downarrow n\Delta^g_{rs}H^o_m(\text{HX}) \qquad\qquad \downarrow \Delta^g_{rs}H^o_m(\text{MX}_n) \quad \downarrow n\Delta^g_{rs}H^o_m(\text{HX})$$

$$\text{ML}_n\,(\text{g}) + n\text{HX}\,(\text{g}) \xrightarrow{\Delta_r H^o\,(6.3)} \text{MX}_n\,(\text{g}) + n\text{LH}\,(\text{g}) \tag{6.3}$$

in the gas phase, respectively in reactions (6.2) and (6.3), and $\Delta_{rs}^{g}H^{o}$ are vaporization or sublimation enthalpies.

The observation of the linear equation (6.1) for products and reactants in their standard reference states implies that $\Delta_r H^o(6.2)$ is constant for the series of compounds ML_n. By using Scheme 6.1, this enthalpy of reaction can be expressed in terms of the bond dissociation enthalpies (equation 6.4).

$$\begin{aligned}\Delta_r H^o(6.2) = & \; n[\overline{DH^o}(M-L) - \overline{DH^o}(L-H)] \\ & + [\Delta_{rs}^{g}H_m^o(ML_n) - n\Delta_{rs}^{g}H_m^o(LH)] \\ & + [n\Delta_{rs}^{g}H_m^o(HX) - \Delta_{rs}^{g}H_m^o(MX_n)] \\ & + n[DH^o(H-X) - \overline{DH^o}(M-X)]\end{aligned} \quad (6.4)$$

The third and fourth bracketed terms in equation (6.4) are constant for a given X, implying that $[\overline{DH^o}(M-L) - \overline{DH^o}(L-H)] + [\Delta_{rs}^{g}H_m^o(ML_n) - n\Delta_{rs}^{g}H_m^o(LH)]$ is also constant for the series of compounds that obey the linear correlation. It is therefore very likely that $\overline{DH^o}(M-L)$ and $DH^o(L-H)$ follow nearly parallel trends. Obviously, if the linear correlation holds for reactants and products in the gas phase, then $\Delta_r H^o(6.3)$ is constant and so will be $\overline{DH^o}(M-L) - \overline{DH^o}(L-H)$.

In conclusion, the observation of a linear correlation between $\Delta_f H_m^o(ML_n)$ and $\Delta_f H_m^o(LH)$ means that the energetics of a series of molecules ML_n reflect the energetics of L in LH, so that a variation of the enthalpy of formation of LH is proportional to a change in the enthalpy of formation of ML_n. The resulting parallel trend between M—L and L—H bond dissociation enthalpies was also independently noted by Bercaw and co-workers from a series of results involving several families of transition metal complexes [16,17].

BOND ENTHALPY DATA

Group 1

Thermochemical data for lithium organometallic compounds have been the subject of a recent review [11] and some of the conclusions are summarized here. The experimental enthalpies of formation of several crystalline or liquid lithium alkyls or aryls have been reported, but their accuracy is questionable. This and also the lack of experimental values for the enthalpies of vaporization of most of those compounds hinders the calculation of reliable lithium–alkyl and –aryl bond dissociation enthalpies. Fortunately, theoretical calculations of enthalpies of formation of

TABLE 6.2 Lithium–alkyl bond dissociation enthalpies, derived from theoretically calculated enthalpies of formation [18] (Values in kJ/mol)

LiR	DH^{o}(Li—R)
LiMe	201
LiEt	171
LiPr	171
Li(*i*-Pr)	145
LiBu	171
Li(*s*-Bu)	149
Li(*t*-Bu)	120

gaseous lithium alkyls, by Leroy and co-workers [18], allow one to derive the bond enthalpy data shown in Table 6.2. As discussed in [11], these data look sensible.

A comparison of the values in Table 6.2 with the respective carbon–hydrogen bond dissociation enthalpies in the corresponding alkanes (Table 6.1) shows that DH^{o}(Li—Me) is 30 kJ/mol higher than DH^{o}(Li—Et), whereas DH^{o}(Me—H) − DH^{o}(Et—H) = 16 kJ/mol. In other words, methyl lithium is stabilized relative to the *n*-alkyl family. Opposite conclusions are drawn for secondary and tertiary lithium alkyls. These features can also be evidenced by a plot of the enthalpies of formation of the gaseous lithium alkyls against the enthalpies of formation of the corresponding alkanes [11]. The point for Me lies below the line defined by Et, Pr, and Bu, while the points for *i*-Pr, *s*-Bu, and *t*-Bu are above the same line.

As discussed in the above-mentioned review, the fact that the primary *n*-butyl lithium is more stable than either its secondary or tertiary isomer, in contrast to what is observed for the alkanes, may be related to the rather ionic character of the lithium–carbon bond. If the lithium alkyls are better described by $Li^{+}R^{-}$, it is expected that a series of these molecules will reflect the variation in the energetics of the anions R^{-}. Indeed, a plot of $\Delta_f H^{o}_{m}$(LiR,g) against $\Delta_f H^{o}_{m}$(R^{-},g), shown in Figure 6.1 [11] supports this idea. Note, however, that the slope is less than unity (about 0.5), indicating that the energetics of lithium alkyls are less sensitive to the nature of R than the energetics of the anions. The value for *t*-butyl lies clearly above the line, which suggests that the structures of the *t*-butyl moiety in Li(*t*-Bu) and in the ground state of the anion are substantially different. Incidentally, the general trend followed by the lithium alkyl compounds

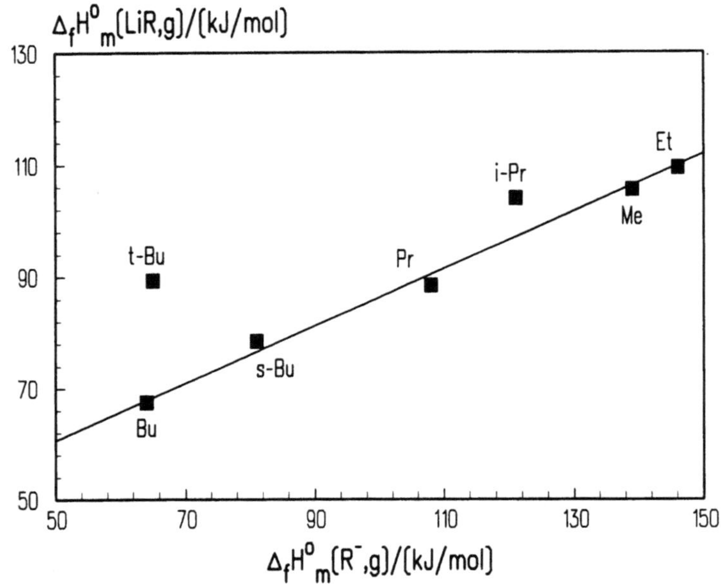

FIGURE 6.1 Quantum chemically derived standard enthalpies of formation of gaseous lithium alkyls vs. the standard enthalpies of formation of gaseous alkyl anions.

is similar to the one theoretically predicted for the scandium alkyl cations (see the last section of this chapter).

Group 2

The enthalpies of formation in ether of a variety of Grignard compounds, RMgBr, were determined by Holm [19,20]. The least squares line (correlation coefficient, $r = 0.998$) in Figure 6.2, where the data for R = alkyl vs. the gas phase enthalpies of formation of the corresponding alkanes are plotted, was drawn by using the values for the *n*-alkyls. As in the case of the lithium compounds mentioned above, it is noted that the point for Me falls below the line, whereas several non-linear alkyls are 'destabilized' relative to the *n*-alkyl family. The points for *i*-Bu and *neo*-Pe are fit by the linear correlation, indicating the absence of steric effects. Differential solvation enthalpies (recall that solution data for the Grignard compounds is being dealt with) may be in part responsible for these apparent discrepancies, but the good linear correlation in Figure 6.2 indicates that, in general, the importance of those effects is small, so that it may be concluded that Mg—R and R—H bond dissociation enthalpies follow parallel trends for the *n*-alkyls. It must stressed, however, that the slope of the line in Figure 6.2 is higher than one ($a = 1.202 \pm 0.038$), implying

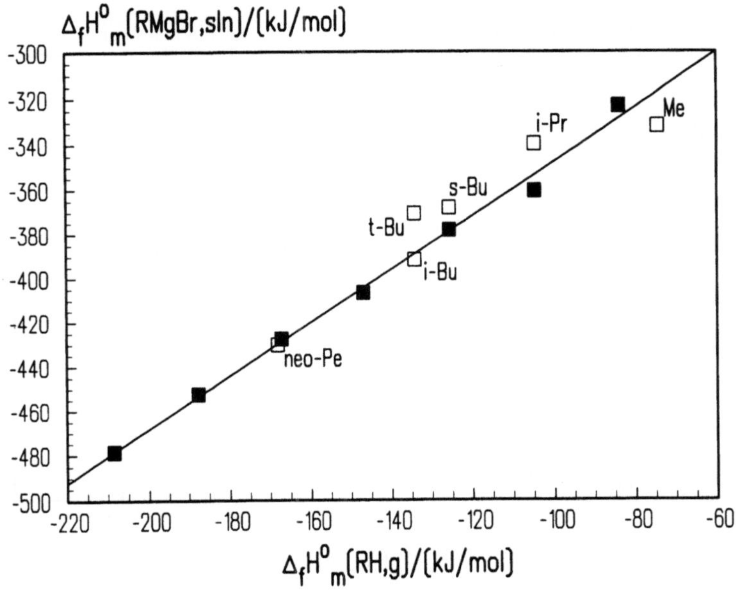

FIGURE 6.2 Standard enthalpies of formation of Grignard compounds RMgBr in ether vs. the standard enthalpies of formation of the gaseous alkanes RH. The darkened blocks represent the linear alkanes, from ethane to octane.

that the differences $DH^o_{sln}(Mg{-}R) - DH^o_{sln}(Mg{-}R')$ are slightly higher than $DH^o(R{-}H) - DH^o(R'{-}H)$. This is shown by equation 6.5, derived from the definitions of bond dissociation enthalpies and from equation 6.1.

$$DH^o_{sln}(Mg{-}R) - DH^o_{sln}(Mg{-}R')$$
$$= DH^o(R{-}H) - DH^o(R'{-}H) + (1 - a)[\Delta_f H^o_m(RH,g) - \Delta_f H^o_m(R'H,g)] \quad (6.5)$$

Group 3, Lanthanides, and Actinides

A major contribution to the actinide–alkyl bond dissociation enthalpies data bank has been provided by Marks and co-workers [1,21]. An example of this work is given in Figure 6.3, which shows a plot of *relative* values of the standard enthalpies of formation of several thorium complexes [22] vs. the standard enthalpies of formation of ligands RH in their standard reference states. In the absence of data for *n*-alkyls, the least squares line was drawn by considering all five points. Despite this choice, the correlation ($r = 0.997$) is satisfactory and the slope (0.998 ± 0.048) reflects the number of ligands R bonded to the metal atom. An interesting feature of the correlation is that it involves the enthalpies of formation of

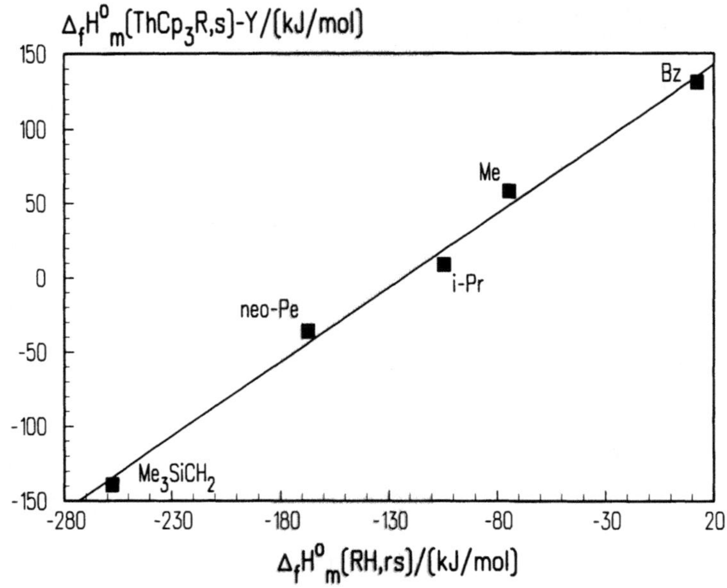

FIGURE 6.3 Relative standard enthalpies of formation of crystalline $ThCp_3R$ compounds (Cp = η^5-C_5H_5) vs. the standard enthalpies of formation of RH in their standard reference states (see text).

the *crystalline* complexes and, as remarked above, the enthalpies of formation of RH in their reference states, i.e. the liquid state for toluene [23] and tetramethylsilane [24], and the gaseous state for the remaining molecules (Table 6.1). The correlation would not be observed if only $\Delta_f H^o_m$(RH,g) had been used. This is possibly related to the fact that the sum $[DH^o(\text{Th}-\text{R}) - DH^o(\text{R}-\text{H})] + [\Delta^g_{rs}H^o_m(\text{ML}_n) - n\Delta^g_{rs}H^o_m(\text{LH})]$ is nearly constant for the ligands involved (see Scheme 6.1 and equation 6.4). Nevertheless, it is recognised that this less-than-ideal correlation (since it does not include only *n*-alkyls) apparently misses some information concerning, for instance, steric effects. Although the point for *neo*-pentyl lies above the line, indicating that the thorium complex is destabilized relative to other alkyls, the destabilization may be underestimated. Indeed, when the solution values for the Th–ligand bond dissociation enthalpies are considered [1], it is noted that their differences parallel the corresponding $DH^o(\text{R}-\text{H}) - DH^o(\text{R}'-\text{H})$, except the value for *neo*-pentyl: $DH^o_{sln}(\text{Th}-\text{CH}_2\text{SiMe}_3) - DH^o_{sln}$(Th–*neo*-Pe) is some 34 kJ/mol higher than $DH^o(\text{Me}_3\text{SiCH}_2-\text{H}) - DH^o$(*neo*-Pe—H). Finally, it is interesting to note that unlike the previous examples discussed, the thorium–methyl bond dissociation enthalpy is not high in comparison

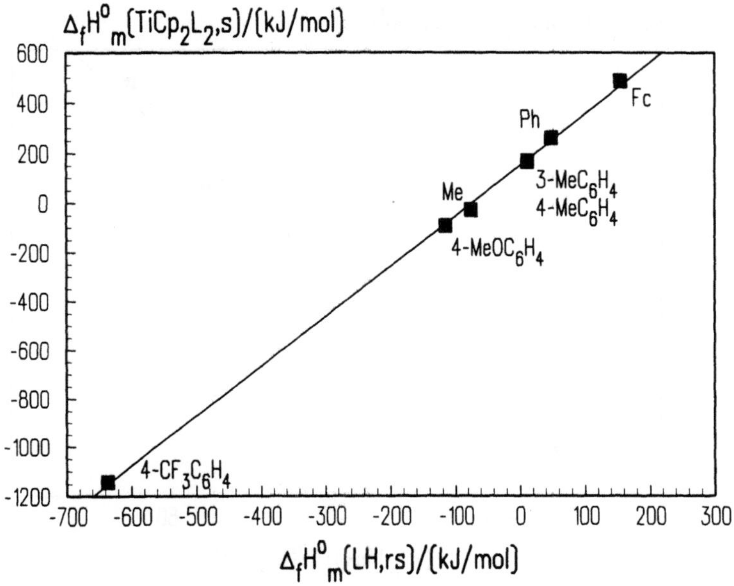

FIGURE 6.4 Standard enthalpies of formation of crystalline $TiCp_2L_2$ compounds ($Cp = \eta^5\text{-}C_5H_5$) vs. the standard enthalpies of formation of LH in their standard reference states.

with the data for other alkyls [1]. This feature is also apparent (albeit fortuitously) in Figure 6.3.

Group 4

Some of the available data for Group 4 metal–carbon bond dissociation enthalpies in coordinatively saturated compounds are subject to controversy [1]. At least in part, this may be caused by several assumptions required to derive those values, either in solution or in the gas phase. Probably less controversial are the results for the standard enthalpies of formation of the crystalline compounds, since these values are directly obtained from measured reaction enthalpies. The disadvantage of dealing with these numbers is, however, that only relative values (or trends) of bond enthalpies can be derived.

The example chosen to illustrate Group 6.4 data involves only one metal alkyl compound, $TiCp_2Me_2$, several metal–aryl complexes, $TiCp_2(C_6H_4R)_2$, and also the complex $TiCp_2Fc_2$ ($Cp = \eta^5\text{-}C_5H_5$, R = Me, OMe, $Fc = CpFeC_5H_4$) [1]. The plot in Figure 6.4 extends over a wide range of bond dissociation enthalpies and the least squares line ($r = 0.9996$; $a = 2.046 \pm 0.027$; $b = 149.9 \pm 6.8$) is consistent with the fact

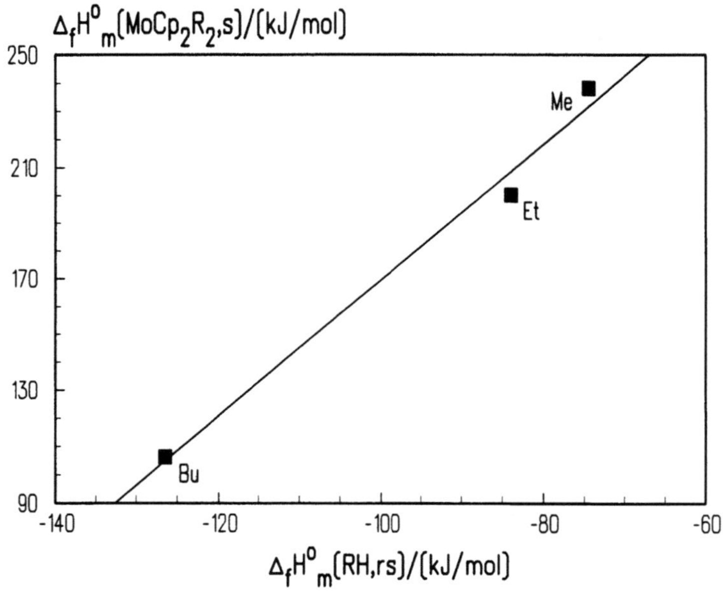

FIGURE 6.5 Standard enthalpies of formation of crystalline $MoCp_2R_2$ compounds ($Cp = \eta^5\text{-}C_5H_5$) vs. the standard enthalpies of formation of RH in their standard reference states.

that two ligands are bonded to the titanium atom. As remarked above for the thorium compounds, this implies that $[\overline{DH^o}(\text{Ti—L}) - \overline{DH^o}(\text{L—H})] + [\Delta^{g}_{rs}H^o_m(\text{ML}_n) - n\Delta^{g}_{rs}H^o_m(\text{LH})]$ is nearly constant for the ligands involved (see Scheme 6.1 and equation 6.4). It was somewhat surprising to us observing that the aryl, the ferrocenyl, and the methyl complexes could be all included in the same 'family'. For the Grignard compounds (Figure 6.2) the aryl data are not fit by the *n*-alkyl regression line.

Group 6

The metal–carbon bond enthalpy data for Group 5 complexes is rather scarce, but for Group 6 elements some information is available [1,9]. The enthalpies of formation of the crystalline complexes $MoCp_2R_2$ (R = Me, Et, Pr) are plotted against the enthalpies of formation of the gaseous alkanes in Figure 6.5. Although the correlation is not excellent ($r = 0.994$), it indicates that $[\overline{DH^o}(\text{Mo—R}) - DH^o(\text{R—H})] + [\Delta^{g}_{rs}H^o_m(\text{ML}_n) - n\Delta^{g}_{rs}H^o_m(\text{LH})]$ is constant within *c.* 10 kJ/mol. It is therefore likely that the molybdenum–alkyl and the alkyl–hydrogen bond dissociation enthalpies follow nearly parallel trends. This is indeed observed in the case of the complexes $Mo(Cp)(CO)_3R$ (R = Me, Et), for which solution values of Mo—Me and Mo—Et bond dissociation enthalpies have been reported by Hoff and co-workers [1,25]: the difference DH^o_{sln}(Mo—Me) − DH^o_{sln}(Mo—Et) matches DH^o(Me—H) − DH°(Et—H).

Group 7

The manganese complexes of the type $Mn(CO)_5L$ are one of the rare examples of coordinatively saturated organometallic complexes for which accurate gas phase bond dissociation enthalpies could be available without complicating assumptions. In fact, for L = Me, CF_3, Ph, and Bz, the experimental enthalpies of formation of the crystalline and the gaseous complexes are known [26]; for L = Me, CF_3, CH_2F, and CHF_2 the appearance energies of $Mn(CO)_5^+$ have been measured by photoionization mass spectrometry (PIMS) [27]. However, in order to derive DH^o (Mn—L) values from the former set of data, one needs the enthalpy of formation of $Mn(CO)_5$; in the case of the photoionization results, the adiabatic ionization energy of this species is required. Unfortunately, both quantities are rather uncertain. (They are, in fact, related: as $\Delta_f H^o_m[Mn(CO)_5^+]$ is known with good accuracy, it will be enough to measure one of them to derive the other). So, while for the time being one cannot list accurate manganese–carbon bond dissociation enthalpies for the above complexes, it is possible to derive data that rely on a given value of the enthalpy of formation of manganese pentacarbonyl radical. This has been done elsewhere [1]. Here, the usual method will be relied on to draw conclusions on the trend of DH^o(Mn—L) values. The enthalpies of formation of the gaseous Me, CF_3, Ph, and Bz complexes yield a rather good correlation when plotted against the enthalpies of formation of the gaseous hydrocarbons ($r = 0.9998$; $a = 1.008 \pm 0.016$; $b = -681.4 \pm 5.6$). This is illustrated in Figure 6.6 and implies (see Scheme 6.1 and equation 6.4) that DH^o(Mn—L) and DH^o(L—H) follow parallel trends (within *c.* ±10 kJ/mol, which is the typical uncertainty of the reported Mn—L bond dissociation enthalpies [1]). On this basis, it may be concluded that the PIMS result for DH^o(Mn—CF_3), which is some 20 kJ/mol lower than the one derived from the calorimetric experiments, is too low. It is also possible that the other two PIMS values are low limits: DH^o(Mn—Me) − DH^o(Mn—CH_2F) = 48 kJ/mol whereas DH^o(Me—H) − DH^o(CH_2F—H) = 15 kJ/mol; DH^o(Mn—Me) − DH^o(Mn—CH_2F) = 43 kJ/mol and DH^o(Me—H) − DH^o(CHF_2—H) = 9 kJ/mol [1]. Incidentally, the two latter manganese–fluoromethyl bond dissociation enthalpies are consistent with the respective carbon–hydrogen bond dissociation enthalpies:

$$DH^{\circ}(\mathrm{Mn{-}CH_2F}) - DH^{\circ}(\mathrm{Mn{-}CHF_2}) \approx DH^{\circ}(\mathrm{CH_2F{-}H}) - DH^{\circ}(\mathrm{CHF_2{-}H})\ [1].$$

Finally, it is noted that the correlation involving the reference states enthalpies of formation (Figure 6.6; $r = 0.9999$; $a = 1.047 \pm 0.011$; $b = -733.1 \pm 3.7$) is slightly better than the one obtained from the gas phase data.

Groups 8 and 9

Very few enthalpies of formation for Groups 8 and 9 organometallic complexes are available, but the information regarding metal–carbon bond dissociation enthalpies is not that scarce, in particular for cobalt, rhodium and iridium compounds [1]. The efforts to obtain $DH^{\circ}(\mathrm{Co{-}C})$ values have been fostered by the importance that the homolysis of

FIGURE 6.6 Standard enthalpies of formation of $Mn(CO)_5L$ compounds vs. the standard enthalpies of formation of RH. The darkened blocks and the open squares refer to reference states and gas phase data, respectively.

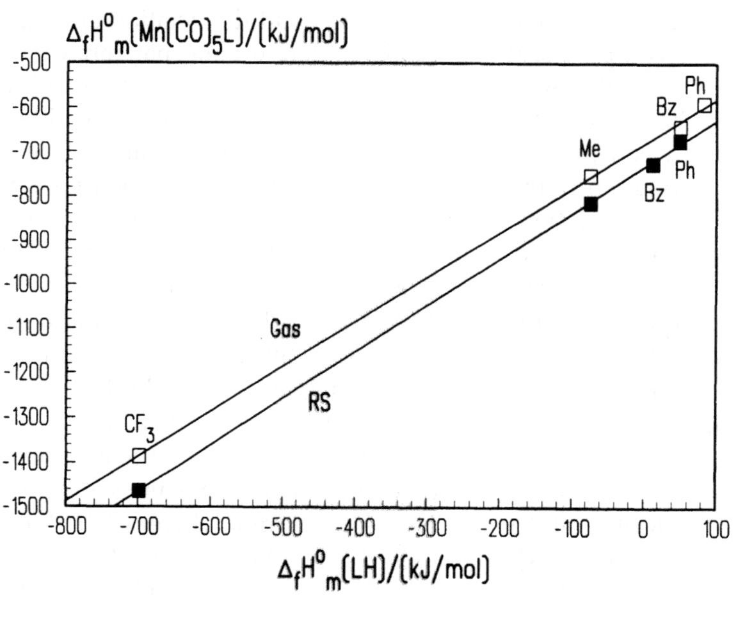

Co–adenosyl bond plays in the biochemistry of coenzyme B_{12} [28]. There is some controversy regarding those data, because they have been determined from kinetic methods in solution by estimating the activation barriers for radical recombination [1,29]. Such a discussion is outside the scope of the present chapter, and it is only wished to take a set of (hopefully internally consistent) DH^{o}_{sln}(Co—R) values [30] in cobalamines and plot them against the corresponding DH^{o}(R—H). This is made in Figure 6.7, by using a relative scale for the *y* axis: a value of DH^{o}_{sln}(Co—Me) = DH^{o}(Me—H) = 439 kJ/mol was arbitrarily assigned. The line in the figure corresponds to a perfect match between DH^{o}_{sln}(Co—R) – DH^{o}_{sln}(Co—R′) and DH^{o}(R—H) – DH^{o}(R′—H) and shows that ligands with more severe steric constraints (*neo*-pentyl and *i*-propyl) have weaker cobalt–carbon bonds (by *c.* 15 kJ/mol) than expected. Interestingly, the value for cobalt–benzyl bond dissociation enthalpy is some 40 kJ/mol higher than predicted.

A strong metal–benzyl bond is also observed in Figure 6.7 for the iridium complex Ir(Cl)(CO)(PMe_3)(I)Bz. Other data for this family of

FIGURE 6.7 Relative metal–carbon bond dissociation enthalpies (see text) in several cobalamin complexes, $B_{12}R$ (squares; solution data), and Ir(Cl)(CO)(PMe_3)(I)R compounds (circles; gas phase data) vs. R—H bond dissociation enthalpies. The minor divisions in the Y scale represent 10 kJ/mol.

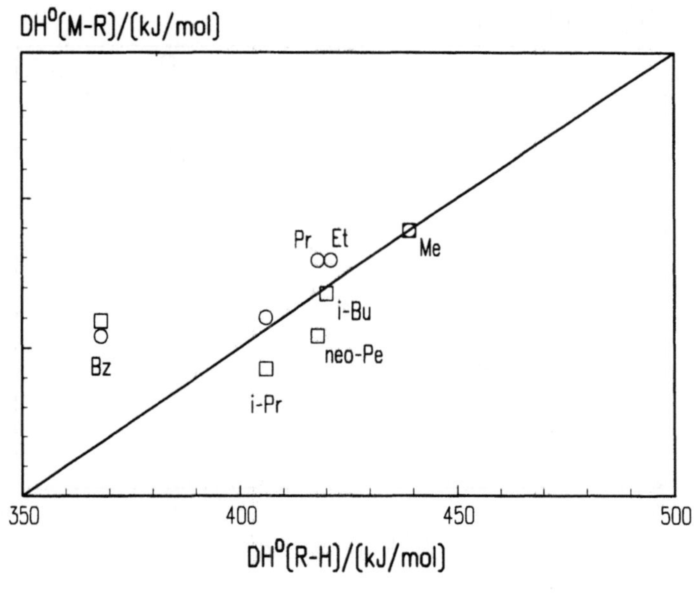

complexes, relative to DH^o(Ir—Me) = DH^o(Me—H) = 439 kJ/mol, is also presented in the figure and show that iridium–ethyl and –propyl bond dissociation enthalpies are slightly higher (by *c.* 10 kJ/mol) than predicted from the line of unit slope. These deviations are within the experimental uncertainties [1].

The metal–carbon and metal–hydrogen bonds in the iridium and rhodium complexes $M(Cp^*)(PMe_3)(H)R$ are sufficiently strong to ensure their thermodynamic stability, relative to the decomposition products, $M(Cp^*)(PMe_3)$ and RH ($Cp^* = \eta^5$-C_5Me_5). In other words, the oxidative additions of the hydrocarbons RH to the coordinatively unsaturated species $M(Cp^*)(PMe_3)$ are exothermic. Several iridium- and rhodium–carbon bond dissociation enthalpies [1] are represented in Figure 6.8, by using the same type of plot as in Figure 6.7. The data, which are relative to DH^o_{sln}(Ir—Me) and DH^o_{sln}(Rh—Pe), respectively, show that the metal–alkyl and alkyl–hydrogen bond dissociation enthalpies follow parallel trends. Note, however, how strong the metal–phenyl bond dissociation enthalpies are. This 'anomaly' is responsible for the thermodynamic selectivity of arene vs. alkane activation by those rhodium and iridium centers [31].

FIGURE 6.8 Relative metal–carbon bond dissociation enthalpies (see text) in several rhodium (squares) and iridium (circles) complexes, $M(Cp^*)(PMe_3)(H)R$, vs. R—H bond dissociation enthalpies. The minor division in the Y scale represent 10 kJ/mol.

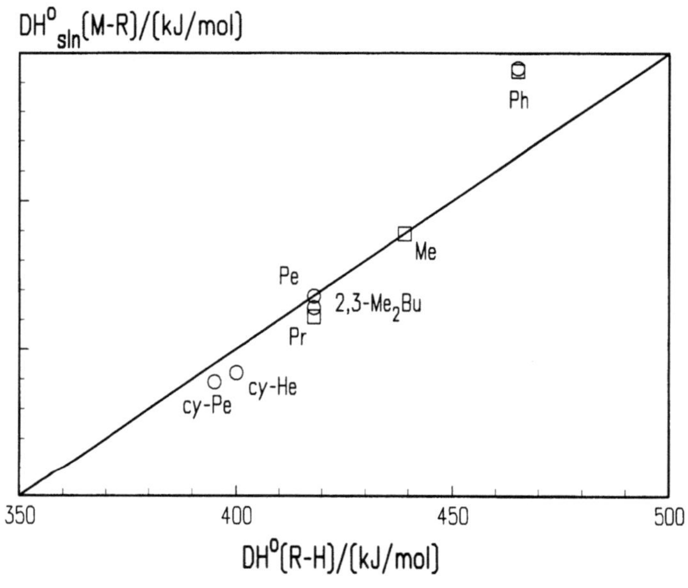

Groups 10–12

Thermochemical information on Group 10 organometallic complexes is much more abundant than for the copper, silver, gold triad [1]. However, in neither case could a set of data be found to apply this method of analysis. Things are somewhat better for Group 12 compounds, particularly for zinc and mercury. For instance, there are recent values for the enthalpies of formation of several homoleptic zinc alkyls, ZnR_2, which, together with older data are interesting to discuss [32,33]. A good linear correlation ($a = 2.162 \pm 0.065$; $b = 172.3 \pm 10.3$; $r = 0.9991$) involving $\Delta_f H^o_m(ZnR_2, 1)$ vs. $\Delta_f H^o_m(RH, rs)$ is observed in Figure 6.9. Neither the *neo*-Pe nor the Et values were included in that least squares analysis. However, while the destabilization of Zn(*neo*-Pe)$_2$ could be expected on steric grounds, there is no apparent reason to justify the deviation of $ZnEt_2$. It is therefore very likely that the reported enthalpy of formation of this compound is too high.

Similar conclusions can be drawn from a plot involving the enthalpies of formation of the gaseous zinc alkyls [32–36] and of the corresponding RH (23, 24), although this correlation is slightly worse ($a = 2.407 \pm 0.075$; $b = 236.4 \pm 11.1$; $r = 0.9990$) than the one shown in Figure 6.9. As usual, this implies that DH^o(Zn—R) and DH^o(R—H) follow parallel trends.

FIGURE 6.9 Standard enthalpies of formation of liquid ZnR_2 compounds vs. the standard enthalpies of formation of RH in their standard reference states.

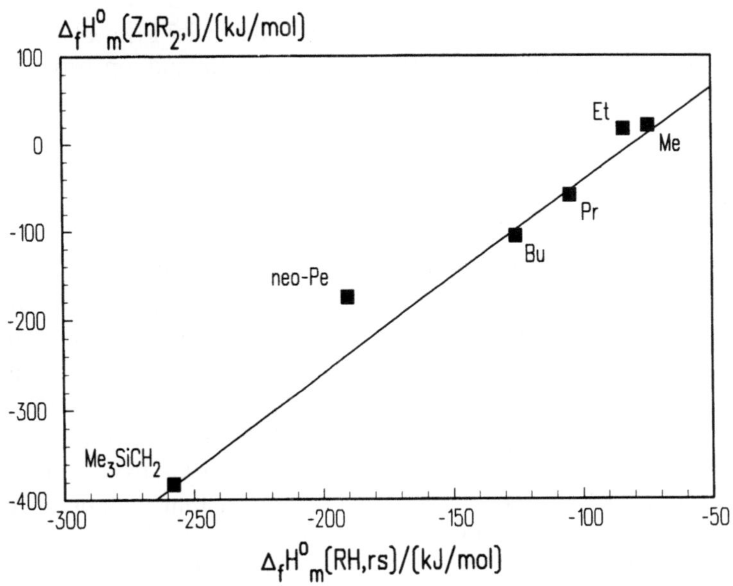

Dimethylzinc is one of the few organometallic compounds for which the Zn—Me stepwise bond dissociation enthalpies are known, albeit somewhat uncertain. Based on ion beam results, Armentrout and Beauchamp recommended DH°(Zn—Me) = 79 ± 13 kJ/mol in unsaturated species ZnMe [37]. On the other hand, Jackson, based on a RRKM analysis of pyrolysis data, arrived at DH°(MeZn—Me) = 266.5 ± 6.3 kJ/mol [38]. This value, together with $\Delta_f H^\circ_m$($ZnMe_2$, g) = 51.2 ± 1.2 kJ/mol [33,34], the enthalpy of formation of methyl radical (Table 6.1), and $\Delta_f H^\circ_m$(Zn, g) = 130.4 ± 0.2 kJ/mol [33], leads to $\Delta_f H^\circ_m$(ZnMe, g) = 170.7 ± 6.5 kJ/mol and DH°(Zn—Me) = 107 ± 6.6 kJ/mol, which is some 28 kJ/mol higher than the ion beam-derived result. In their review, McMillen and Golden recommended DH°(MeZn—Me) = 284.5 ± 16.7 kJ/mol [39] and, more recently, Smith and Patrick placed the same quantity in the range 276–298 kJ/mol [40]. These higher values for the *first* Zn—Me bond dissociation enthalpy in $ZnMe_2$ are of course more consistent with the recommendation of Armentrout and Beauchamp.

The accuracy of the calorimetry-based $\Delta_f H^\circ_m$($ZnMe_2$, g) = 51.2 ± 1.2 kJ/mol could be questioned, but Figure 6.9 suggests that the value is fairly reliable. Photoionization mass spectrometry results, which afforded the appearance energies of $ZnMe^+$ and Zn^+, and the ionization energy of $ZnMe_2{}^+$ [41] enable one to calculate $\Delta_f H^\circ_m$($ZnMe_2$, g) = 63.7 ± 2.4 kJ/mol. The disagreement is not large, but the calorimetric value is more consistent with the linear correlation.

The PIMS measurements were also used to derive the stepwise bond dissociation enthalpies in the cation $ZnMe_2{}^+$: DH°($MeZn^+$—Me) = 117.7 ± 2.7 kJ/mol and DH°(Zn^{+}—Me) = 280.8 ± 2.7 kJ/mol. The latter value is in fair agreement with that obtained from the ion beam experiments, 297 ± 13 kJ/mol [37].

Groups 13–15

There are two recent reviews on the thermochemistry of organometallic compounds of Groups 14 (germanium, tin, and lead) and 15 (arsenic, antimony, and bismuth) [12,14]. It is therefore preferable to illustrate metal–alkyl thermochemical data for an element of Group 13, for instance aluminum alkyls, which provide and example of what seems a fortuitous linear correlation of $\Delta_f H^\circ_m$(AlR_3, g) vs. $\Delta_f H^\circ_m$(RH, g). Most of the available enthalpies of formation of aluminum alkyls derive from static-bomb combustion calorimetry experiments, a technique that may be unreliable for studies involving aluminum compounds [42]. As a result, very large discrepancies are found in the above data [8]. Recently, a (hopefully) accurate value for the standard enthalpy of formation of triethylaluminum has been determined by reaction-solution calorimetry. This value was used to assess the literature data for the remaining alkyl compounds [8]. Although it was noted that the plot of $\Delta_f H^o_m(AlR_3, g)$ vs. $\Delta_f H^o_m(RH, g)$ for R = Me, Et, Pr, and Bu leads to an excellent correlation ($a = 3.819 \pm 0.073$; $b = 204.7 \pm 7.2$; $r = 0.9996$), a detailed discussion of the data afforded the conclusion that this is probably fortuitous. In addition, the slope obtained by the least squares fitting of those points is not around 3, as would be expected. Indeed, if only the two most reliable data points are considered (for Me and Et), the slope will be reduced to 3.3. With all these facts in mind, it has been stated that the present 'best' values for $\Delta_f H^o_m(AlR_3, g)$ are probably those estimated by drawing a line of slope 3 from $\Delta_f H^o_m(AlEt_3, g) = -112 \pm 5$ kJ/mol [8].

TRENDS OF METAL-ALKYL BOND DISSOCIATION ENERGIES: THEORY VERSUS EXPERIMENT

Theory and experiment go hand in hand in the study of the systematics of transition metal–alkyl bond dissociation energies. As referred to in the introduction, these theoretical and experimental studies are complementary because accurate experimental data can be used to assess the predictions of the theoretical methods, therefore contributing to a better understanding of the nature of the metal–alkyl bond. On the other hand, once the reliability of a given theoretical method to predict bond dissociation energies has been established, that method can be used to explain the experimentally observed trends along the periodic table and to identify experimental data that are likely to be wrong.

There are several recent theoretical studies of the systematics of transition metal–alkyl bond dissociation energies. The most extensive and accurate calculations were done using MCPF [43,44] and PCI-80 [45] *ab initio* methods and refer to MMe [43–45] and MMe^+ [43,44] species (M = first or second row metal). Some selected results from these studies, and also from similar studies by GVB [46] and density functional theory (DFT) methods [47], are compared in Figures 6.10–6.12 with the corresponding experimental bond dissociation enthalpies, at 298 K [37,48]. The theoretical data (DH^o) in Figures 6.10–6.13 also refer to 298 K, and were derived from the published dissociation energies DU^o_e and DU^o_0 (determined from the minimum of the potential energy curve of the dissociating molecule or the lowest energy level of the molecule at 0 K,

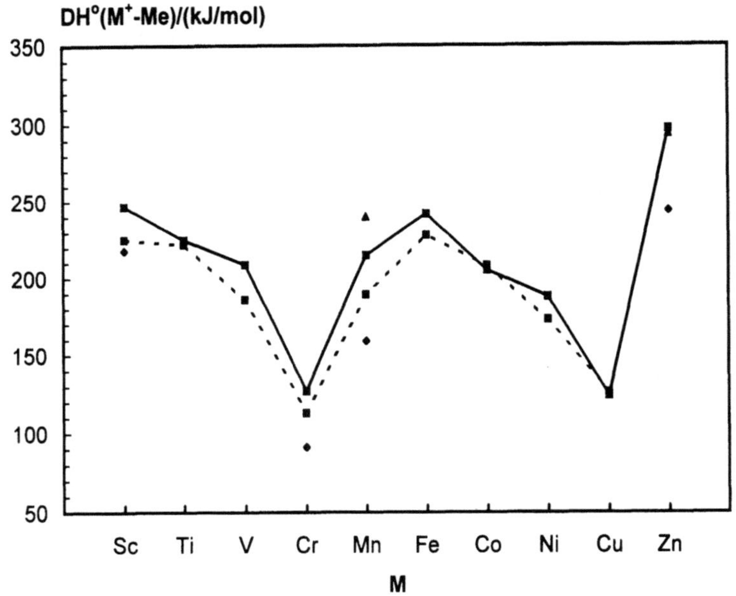

FIGURE 6.10 Comparison of theoretical and experimental bond dissociation enthalpies for first row MMe^+ species. ◆ GVB [46]; --■-- MCPF [44]; ▲ DFT [47]; —■— experimental [37,48].

respectively) by using [46]:

$$DH^\circ(\text{M}-\text{Me}) = DU_e^\circ(\text{M}-\text{Me}) - 7.6 \tag{6.6}$$

$$DH^\circ(\text{M}-\text{Me}) = DU_0^\circ(\text{M}-\text{Me}) + 8.8 \tag{6.7}$$

The above equations were applied to neutral and ionic species.

The accuracy of the theoretically predicted bond dissociation enthalpies depends on the method of calculation. For example, the results of high level *ab initio* methods such as MCPF and PCI–80, for M—Me and M^+—Me bond dissociation enthalpies, are expected to be accurate to < 20 kJ/mol [43–45], which is better than the uncertainty of several experimental values [37,48]. Discrepancies between the theoretical and the experimental values larger than 20 kJ/mol are likely to indicate that the latter may be in error. For example, the experimental $DH^\circ(M^+–Me)$ values for M=Ru, Rh, and Pd in Figure 6.11 are over 40 kJ/mol higher than the corresponding values predicted by MCPF calculations. This led Baushlicher *et al.* [43] to suggest that the energetics of these systems should be reinvestigated. The same is also true for various first row MMe neutral systems (Figure 6.12).

Despite the accuracy differences, in general the theoretical methods

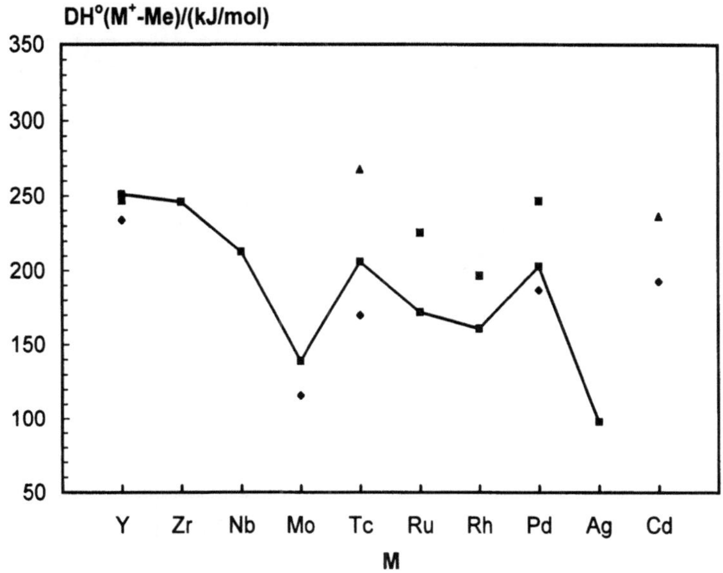

FIGURE 6.11 Comparison of theoretical and experimental bond dissociation enthalpies for second row MMe^+ species. ◆ GVB [46]; —■— MCPF [44]; ▲ DFT [47]; ■ experimental [37].

used to obtain the results in Figures 6.10–6.12 reproduce well the experimentally observed trends for $DH°(M{-}Me)$ and $DH°(M^+{-}Me)$ across the periodic table. These trends are mainly a function of three contributions [43–46,49]: (i) the energy, E^*, necessary to take a metal atom or ion in its ground state to an electron configuration suitable for bonding, if the ground state is not appropriate; (ii) the loss of exchange energy, E_{ex}, when the M—Me or M^+—Me bonds are formed; and (iii) the repulsion between the electrons on carbon and the metal *d* electrons, which increases with the increasing number of *d* electrons in the metal and thus contributes to a decrease of the metal–methyl bond dissociation enthalpy on going from left to right across a period. The contributions E^* and E_{ex} are dominant [43–45]. Table 6.3 shows the values of E^* and E_{ex}, and the corresponding promotion energy $E_p = E^* + E_{ex}$, for first row transition metal ions, obtained from the work of Carter and Goddard [49,50], on the assumption that the bonding state of the metal has a $4s^1d^{n-1}$ configuration.

As predicted by Hund's rule, an isolated metal atom or ion in the ground electronic state always adopts a configuration with the greatest number of unpaired electrons [49]. In this configuration, the stabilizing effect of the so-called exchange interactions, which contribute to the

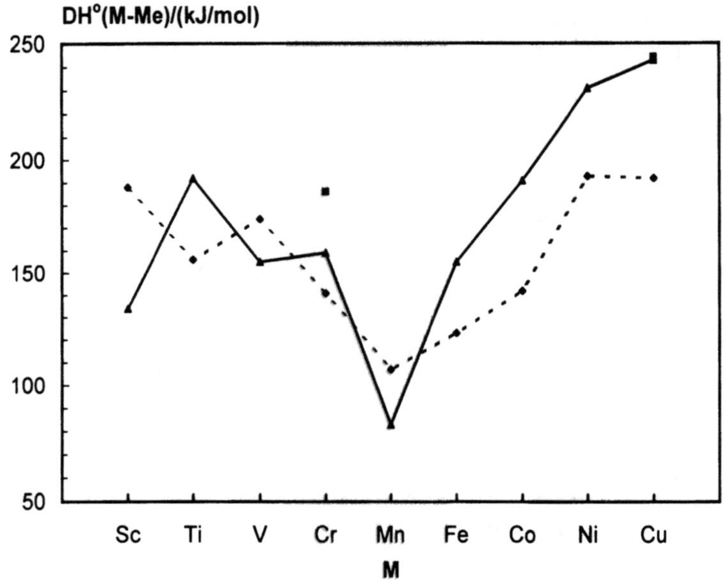

FIGURE 6.12 Comparison of theoretical and experimental bond dissociation enthalpies for first row MMe species. ---◆--- MCPF [44]; ■ DFT [47]; —▲— experimental [48].

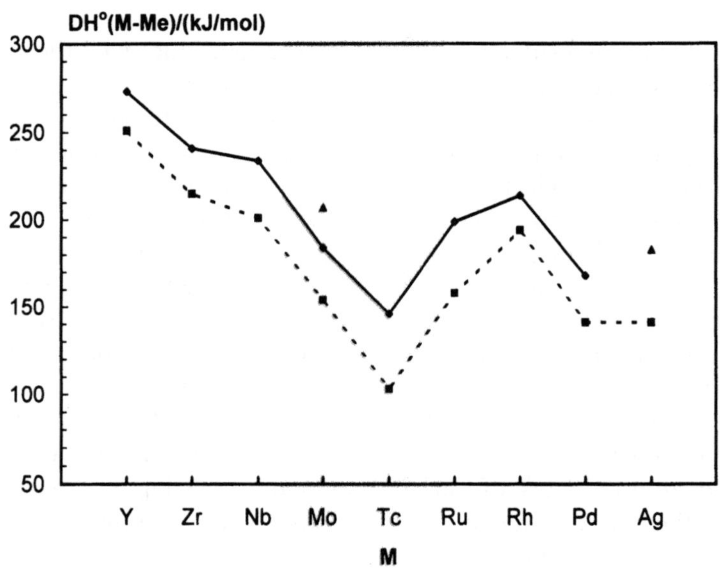

FIGURE 6.13 Theoretical bond dissociation enthalpies for second row MMe species. —◆— PCI-80 [45]; ---■--- MCFP [44]; ▲ DFT [47].

TABLE 6.3 Contributions to the promotion energy of first row transition metal ions assuming that the bonding state of the metal has a $4s^1d^{n-1}$ configuration (see text; data in kJ/mol) [49, 50].

Metal	Ground state	Bonding state	E^*	E_{ex}	E_p
Sc^+	s^1d^1	s^1d^1	0	15	15
Ti^+	s^1d^2	s^1d^2	0	27	27
V^+	d^4	s^1d^3	33	35	68
Cr^+	d^5	s^1d^4	147	42	189
Mn^+	s^1d^5	s^1d^5	0	50	50
Fe^+	s^1d^6	s^1d^6	0	42	42
Co^+	d^8	s^1d^7	41	30	71
Ni^+	d^9	s^1d^8	95	30	125

lowering of the electronic energy of the system, is maximum. The pairing of electrons upon covalent bond formation leads to a loss of exchange energy and, therefore, to a less stable (higher in energy) electronic state. This loss increases with the number of unpaired electrons initially present in the metal [49]. Thus, in the middle of a row, where the number of singly occupied *d* orbitals of the metal is largest, E_{ex} is also largest (Table 6.3) and the bond dissociation enthalpies go through a minimum. The influence of E^* on the observed bond dissociation enthalpy trends across the periodic table can be seen, for example, in the case of Fe^+, Co^+, and Ni^+. The observed decrease of $DH^o(M^+{-}Me)$ from the left to right along this series reflects the fact that E^* increases from Fe^+ to Ni^+ and is larger than E_{ex}, otherwise the inverse trend would be expected (see Table 6.3). The correlation between $DH^o(M^+{-}Me)$ and the E_p has been extensively discussed before [44,48]. This correlation is illustrated in Figure 6.4 for first row transition metal ions using the E_p values from Table 6.3. Similar correlations can also be derived for first row neutrals and second row neutral and ionic MMe species [44,48].

Theoretical and experimental information on bond dissociation enthalpies for binary species involving transition metal atoms or ions and alkyls larger than methyl is very limited [48,51]. Recent *ab initio* MCPF calculations of $DU_0^o(Sc^+{-}R)$ for R = Me, Et, Pr, *i*-Pr, and *t*-Bu [51], led to the conclusion that the $Sc^+{-}R$ bond strengths correlate with the electron affinities of the radicals, decreasing as methyl < primary < secondary < tertiary:

$$DU_0^o(Sc^+{-}Me) - DU_0^o(Sc^+{-}Et) = 20\ \text{kJ/mol},$$

$$DU_0^o(Sc^+{-}Et) - DU_0^o(Sc^+{-}i\text{-}Pr) = 14\ \text{kJ/mol},$$

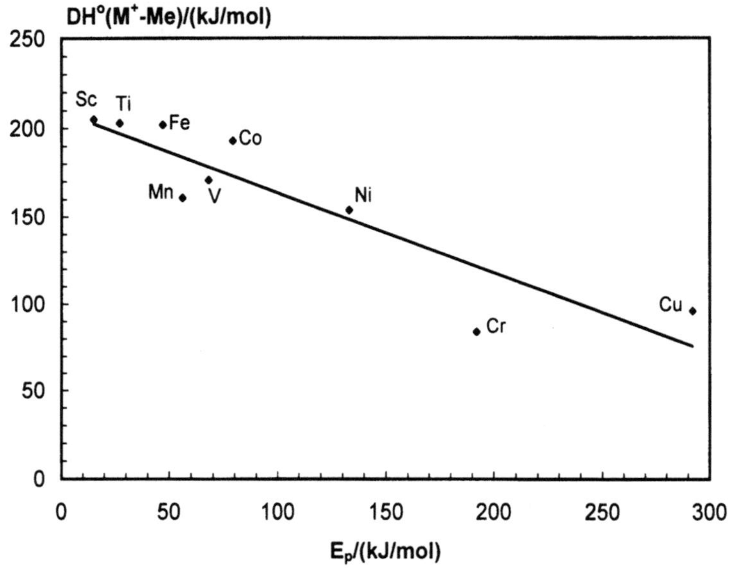

FIGURE 6.14 Plot of theoretically calculated $DH^0(M^+—Me)$ (data from ref. 44; see text) vs. E_p (Table 6.3), for first row transition metals.

and

$$DU_0^o(Sc^+—i\text{-}Pr) - DU_0^o(Sc^+—t\text{-}Bu) = 10\ kJ/mol.$$

The fact that $DU_0^o(Sc^+—Pr)$ is found to be larger than $DU_0^o(Sc^+—Et)$ by 22 kJ/mol has been attributed to the larger polarizability of propyl and to the ability of this ligand to 'solvate' the metal ion. The calculations indicate that in the most stable configuration for $Sc^+—Pr$ the β-methyl of the propyl ligand has a significant electrostatic interaction with the metal. Based on the obtained data and on values of alkyl–H bond dissociation enthalpies, the authors analysed the energetics of the σ-bond metathesis reactions [51],

$$Sc^+—Me + Pr—H \rightarrow Sc^+—Pr + Me—H \quad (6.8)$$

$$Sc^+—Me + i\text{–}Pr—H \rightarrow Sc^+—i\text{–}Pr + Me—H \quad (6.9)$$

and found that the activation of a primary C—H bond in reaction (6.8) is exothermic by 17 kJ/mol and the activation of a secondary C—H bond in reaction (6.9) is endothermic by 8 kJ/mol, despite the fact that $DH^o(Pr—H) - DH^o(i\text{–}Pr—H) = -10$ kJ/mol (Table 6.1). This is in good agreement with the observed preference (>90%) for attack at primary rather than secondary C-H bonds in reactions of Sc^+ with alkanes, studied by Fourier transform ion cyclotron resonance spectroscopy [52].

The trends observed for metal–alkyl bond dissociation enthalpies are not necessarily identical in the binary neutral or ionic MR (R = alkyl) species discussed above and in coordinatively saturated or unsaturated molecules containing more than one ligand, ML_nR_m. This is illustrated in Figure 6.15 for MMe^+ and $MMe_2{}^+$ species. The theoretical analysis of the relation between $DU^o_e(M{-}R)$ and $DU^o_e(ML_nR_{m-1}{-}R)$ indicates that the observed differences in the systematic variation of these values along a period are associated with the differences in promotion energies inherent to the formation of the M−R and the $ML_nR_{m-1}{-}R$ bonds [44,49]. A method to estimate bond dissociation energies in coordinatively saturated compounds from their corresponding values in coordinatively unsaturated species has been proposed [49].

Density functional theory calculations on MCp_2Me systems (M = Sc, Y, La, Mn, Tc, Re) indicate that the M−Me bond dissociation enthalpies increase on descending a group, due to a corresponding increase in the overlap of MCp_2 and Me bonding orbitals [53]. This is in good agreement with what is found experimentally for methyl and many other ligands in coordinatively saturated complexes [1]. In coordinatively unsaturated compounds of middle and late transition metals, however, an inverse trend may be observed. For example, MCPF *ab initio* calculations of

FIGURE 6.15 First and second DU^0_e for $MMe_2{}^+$ species of first row metals (data from ref. 44).

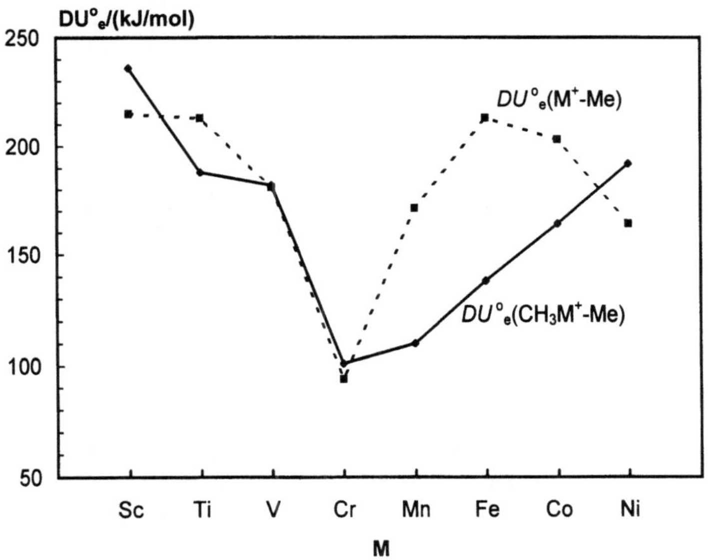

DU_e^o(M—Me) [44] in neutral and ionic MMe and MMe_2 species of first and second row metals indicate that

$$DU_e^o(\text{Mn—Me}) - DU_e^o(\text{Tc—Me}) = 5\ \text{kJ/mol},$$

$$DU_e^o(\text{Ni—Me}) - DU_e^o(\text{Pd—Me}) = 50\ \text{kJ/mol},$$

$$DU_e^o(\text{Fe}^+\text{—Me}) - DU_e^o(\text{Ru}^+\text{—Me}) = 45\ \text{kJ/mol},$$

$$DU_e^o(\text{Ni}^+\text{—Me}) - DU_e^o(\text{Pd}^+\text{—Me}) = 50\ \text{kJ/mol},$$

$$DU_e^o(\text{Cu}^+\text{—Me}) - DU_e^o(\text{Ag}^+\text{—Me}) = 28\ \text{kJ/mol},$$

$$DU_e^o(\text{MeNi—Me}) - DU_e^o(\text{MePd—Me}) = 41\ \text{kJ/mol},$$

and

$$DU_e^o(\text{MeFe}^+\text{—Me}) - DU_e^o(\text{MeRu}^+\text{—Me}) = 65\ \text{kJ/mol}.$$

The scarce experimental data available for MMe systems support the theoretical predictions:

$$DH^o(\text{Fe}^+\text{—Me}) > DH^o(\text{Ru}^+\text{—Me}),$$

$$DH^o(\text{Ni}^+\text{—Me}) > DH^o(\text{Pd}^+\text{—Me}),$$

and

$$DH^o(\text{Cu}^+|\text{Me}) > DH^o(\text{Ag}^+|\text{Me})\ [37,48].$$

Acknowledgements

This work was supported by Junta National de Investigação Científica, e Tecnológica, Portugal (Projects PBIC/C/CEN/1042/92 and STRDA/C/CEN 469/92).

REFERENCES

1. Martinho Simões, J. A. and Beauchamp, J. L. (1990) *Chem. Rev.* **90,** 629 and references cited therein.
2. Dias, A. R., Martinho Simões, J. A., Teixeira, C., Airoldi, C. and Chagas, A. P. (1987) *J. Organometal. Chem.* **335,** 71.

3. Dias, A. R., Martinho Simões, J. A., Teixeira, C., Airoldi, C. and Chagas, A. P. (1989) *J. Organometal. Chem.* **361,** 319.
4. Griller, D., Martinho Simões, J. A. and Wayner, D. D. M. (1991) In *Sulfur-Centered Reactive Intermediates in Chemistry and Biology,* Chatgilialoglu, C., Asmus, K.-D., eds, Plenum: New York.
5. Leal, J. P., Pires de Matos, A. and Martinho Simões, J. A. (1991) *J. Organometal. Chem.* **403,** 1.
6. Dias, A. R., Martinho Simões, J. A., Teixeira, C., Airoldi, C. and Chagas, A. P. (1991) *Polyhedron* **10,** 1433.
7. Martinho Simões, J. A. (1992) In *Energetics of Organometallic Species,* Martinho Simões, J. A., ed., NATO ASI Series C no. 367; Kluwer: Dordrecht.
8. Leal, J. P. and Martinho Simões, J. A. (1993) *Organometallics* **12,** 1442.
9. Diogo, H. P., Simoni, J. A., Minas da Piedade, M. E., Dias, A. R. and Martinho Simões, J. A. (1993) *J. Am. Chem. Soc.* **115,** 2764.
10. Dias, P. B., Minas da Piedade, M. E. and Martinho Simões, J. A. (1994) *Coord. Chem. Rev.* **135/136,** 737.
11. Liebman, J. F., Martinho Simões, J. A. and Slayden, S. W. (1995) In *Lithium Chemistry: Principles and Applications,* Schleyer, P. v R., Sapse, A.-M., ed., Wiley: New York.
12. Liebman, J. F., Martinho Simões, J. A. and Slayden, S. W. (1994) In *The Chemistry of Organic Arsenic, Antimony and Bismuth Compounds,* Patai, S., ed., Wiley: New York.
13. Leal, J. P. and Martinho Simões, J. A. (1993) *J. Organometal. Chem.* **460,** 131.
14. Liebman, J. F., Martinho Simões, J. A. and Slayden, S. W. (1995) In *The Chemistry of Organic Germanium, Tin and Lead Compounds,* Patai, S., ed., Wiley: New York.
15. Liebman, J. F., Martinho Simões, J. A. and Slayden, S. W. (1995) *Struct. Chem.* **6,** 65.
16. Bryndza, H. E., Fong, L. K., Paciello, R. A., Tam, W. and Bercaw, J. E. (1987) *J. Am. Chem. Soc.* **109,** 1444.
17. Bryndza, H. E., Dommaille, P. J., Tam, W., Fong, L. K., Paciello, R. A. and Bercaw, J. E. (1988) *Polyhedron* **7,** 1441.
18. Sana, M., Leroy, G. and Wilante, C. (1991) *Organometallics* **10,** 264.
19. Holm., T (1973) *J. Organometal. Chem.* **56,** 87.
20. Holm, T. (1981) *J. Chem. Soc., Perkin Trans. II* 465.
21. Giardello, M. A., King, W. A., Nolan, S. P., Porchia, M., Sishta, C. and Marks, T. J. (1992) In *Energetics of Organometallic Species,* Martinho Simões, J. A., ed., Kluwer: Dordrecht.
22. The relative values of the enthalpies of formation of the complexes were calculated from experimental enthalpies of reaction reported by Sonnenberger, D. C., Morss, L. R. and Marks, T. J. (1985) *Organometallics* **4,** 352.

23. PEDLEY, J. B., NAYLOR, R. D. and KIRBY, S. P. (1986) *Thermochemical Data of Organic Compounds*, Chapman and Hall, London.
24. STEELE, W. V. (1983) *J. Chem. Thermodyn.* **15,** 595.
25. NOLAN, S. P., LOPEZ DE LA VEGA, R., MUKERJEE, S. L. and HOFF, C. D. (1986) *Inorg. Chem.* **25,** 1160.
26. CONNOR, J. A., ZAFARANI-MOATTAR, M. T., BICKERTON, J., EL SAIED, N. I., SURADI, S., CARSON, R., AL-TAKHIN, G. and SKINNER, H. A. (1982) *Organometallics* **1,** 1166.
27. STEVENS, A. E. (1981) Ph.D. Thesis, California Institute of Technology.
28. DOWD, P. (1990) In *Selective Hydrocarbon Activation*, Davies, J. A., Watson, P. L., Liebman, J. F. and GREENBERG, A., ed., VCH: New York, Chapter 9.
29. WADDINGTON, M. D. and FINKE, R. G. (1993) *J. Am. Chem. Soc.* **115,** 4629.
30. The set of data was reported by SCHRAUZER, G. N. and GRATE, J. H. (1981) *J. Am. Chem. Soc.* **103,** 541. The value for R = Me, not included in the current chapter, was estimated by assuming a constant difference between $DH^{\circ}_{\mathrm{sln}}$(Co-*neo*-Pe) values in ethylene glycol [29] and in water.
31. JONES, W. D. and FEHER, F. J. (1989) *Acc. Chem Res.* **22,** 91.
32. GÜMRÜKÇÜOGLÜ, I. E., JEFFERY, J., LAPPERT, M. F., PEDLEY, J. B. and RAI, A. K. (1988) *J. Organometal. Chem.* **341,** 53.
33. PILCHER, G. and SKINNER, H. A. (1982) In *The Chemistry of the Metal–Carbon Bond*, Hartley, F. R., Patai, S., eds, Wiley: New York.
34. SOKOLOVSKII, A. E. and BAEV, A. K. (1982) *Zh. Obshch. Khim.* **54,** 103.
35. SOKOLOVSKII, A. E. and BAEV, A. K. (1984) *Vestsi Akad. Navuk BSSR, Ser. Khim. Navuk* **115.**
36. SOKOLOVSKII, A. E. and BAEV, A. K. (1984) *Russ. J. Phys. Chem.* **58,** 1635.
37. ARMENTROUT, P. B. and BEAUCHAMP, J. L. (1989) *Acc. Chem. Res.* **22,** 315.
38. JACKSON, R. L. (1989) *Chem. Phys. Lett.* **163,** 315.
39. MCMILLEN, D. F. and GOLDEN, D. M. (1982) *Ann. Rev. Phys. Chem.* **33,** 493.
40. SMITH, G. P. and PATRICK, R. (1983) *Int. J. Chem. Kinet.* **15,** 167.
41. DISTEFANO, G. and DIBELER, V. H. (1970) *Int. J. Mass Spectrom. Ion Phys.* **4,** 59
42. PILCHER, G. (1992) in *Energetics of Organometallic Species*, Martinho Simões, J. A., ed., Kluwer: Dordrecht.
43. BAUSCHLICHER, JR., C. W., LANGHOFF, S. R., PARTRIDGE, H. and BARNES, L. A. (1989) *J. Chem. Phys.* **91,** 2399.
44. ROSI, M., BAUSCHLICHER, JR., C. W., LANGHOFF, S. R. and PARTRIDGE, H. (1990) *J. Phys. Chem.* **94,** 8656.
45. SIEGBAHN, P. E. M. (1994) *J. Am. Chem. Soc.* **116,** 7722.
46. SCHILLING, J. B., GODDARD III, W. A. and BEAUCHAMP, J. L. (1987) *J. Am. Chem. Soc.* **109,** 5573. The bond dissociation *internal energies* from this paper were converted to bond dissociation *enthalpies*.
47. ZIEGLER, T., TSCHINKE, V. and BECKE, A. (1987) *J. Am. Chem. Soc.* **109,** 1351.

48. ARMENTROUT, P. B. and CLEMMER, D. E. (1992) In *Energetics of Organometallic Species*, Martinho Simões, J. A., ed., Kluwer: Dordrecht.
49. CARTER, E. A. and GODDARD III, W. A. (1988) *J. Phys. Chem.* **92,** 5679
50. In ref. [49] the authors use a different nomenclature from the one used here. The values E^* and E_p in the present paper correspond to the values E_p and E_{lost} included in Tables II and III of ref. [49] respectively.
51. PERRY, J. K. and GODDARD III, W. A. (1994) *J. Am. Chem. Soc.* **116,** 5013.
52. CRELLIN; K. C., GERIBALDI, S. and BEAUCHAMP, J. L., (1994) *Organometallics* **13,** 3733.
53. ZIEGLER, T., CHENG, W., BAERENDS, E. J. and RAVENEK, W. (1988) *Inorg. Chem.* **27,** 3458.

7

Resonance and 1,2-Rearrangement Enthalpies in Radicals: from Alkyl Radicals to Alkylcobalamins

ARTHUR GREENBERG and JOEL F. LIEBMAN

THE RESONANCE STABILIZATION OF RADICALS

Elsewhere in this volume thermodynamics and kinetics were interrelated, as were directly or indirectly determined enthalpies of formation and bond dissociation enthalpies. We consider here the interrelated concepts of the stability of radicals and of resonance energy, and immediately admit that there is room for considerably more discussion and dispute than space allows here. In the particular, no attempt will be made to ajudicate the different values of enthalpies of formation for each of the radicals of interest proffered in the various chapters of this volume. It seems wrong to do so, whether we wear the hats of chapter authors, or that of volume or series coeditors. Instead, for this chapter we will perhaps artificially and certainly effortlessly accept the recommendations found in but one of the chapters, namely that by Tsang of the Chemical Kinetics and Thermodynamics Division of the US National Institute of Standards and Technology. We trust the other contributors to this volume – and the reader alike – will not take umbrage or take flight at our choice, for we are fully confident that quantitatively little is changed, and qualitatively almost nothing is affected, by our choice.

When discussing the thermochemical stabilities of radicals, it should be quite apparent that it does not suffice to assert that the radical R′• is more stable than R″• *just* because the enthalpy of formation of the former is lower or more negative than the latter. After all, such logic is relatedly suspect for closed shell, and thus more conventional and generally isolable organic species. For example, it may justifiably be labelled a strawman to note that the enthalpy of formation of gaseous cyclopropane is *c*. 30 kJ mol^{-1} (7 kcal mol^{-1}) less than that of gaseous benzene, despite the strain in the former and resonance (and *a fortiori* aromaticity) of the latter. After all, we still want to say that allyl radical is 'somehow' more stable than *n*-propyl radical, even though the recommended enthalpy of formation of the former is some 71 kJ mol^{-1} (17 kcal mol^{-1}) higher than that of the latter. Despite interest in radicals, it is desirable to return briefly to closed shell species and so 'cap' the two radicals of interest. That is, allyl and propyl radicals are naturally related to the closed shell propene and propane, the two simplest species containing the allyl and n-propyl framework but now additionally containing another C—H bond. Indeed, these two closed shell hydrocarbons may be recognized as the two open shell radicals 'capped with monohydrogen'.

One natural interrelation among allyl radical, n-propyl radical, propene and propane, is comparison of the difference of the enthalpies of formation of propane and propene [1] with the earlier cited 71 kJ mol^{-1} (17 kcal mol^{-1}) difference for n-propyl and allyl radicals. The propene/propane enthalpy difference is almost 125 kJ mol^{-1} (30 kcal mol^{-1}) and so allyl radical is relatively stabilized over n-propyl radical by 125 − 71 = 54 kJ mol^{-1} (30 − 17 = 13 kcal mol^{-1}). Equivalently, two reactions may be considered:

$$CH_2CHCH_2\bullet + H_2 \rightarrow CH_3CH_2CH_2\bullet \quad (7.1)$$

$$CH_2CHCH_3 + H_2 \rightarrow CH_3CH_2CH_3 \quad (7.2)$$

These reactions are exothermic by 71 and 125 kJ mol^{-1} (17 and 30 kcal mol^{-1}) respectively. But what does the reactant H_2 have to do with anything? While the enthalpy of formation of H_2 is precisely 0.0 in any thermochemical set of units, this quantity is by definition, not by measurement. Taking the formal difference of reactions (7.1) and (7.2) above results in

$$CH_2CHCH_2\bullet + CH_2CHCH_3 \rightarrow CH_3CH_2CH_2\bullet + CH_3CH_2CH_3 \quad (7.3)$$

with an asociated exothermicity of 54 kJ mol^{-1} (13 kcal mol^{-1}). Reaction (7.3) may be readily recast in terms of the conventional reaction writing technique wherein all reactants and products appear with positive coefficients on opposite sides of the equation. The necessary, and particularly facile, transformation and algebraic manipulation results in

$$CH_2CHCH_2\bullet + CH_3CH_2CH_3 \rightarrow CH_3CH_2CH_2\bullet + CH_2CHCH_3 \quad (7.4)$$

with, of course, the same reaction enthalpy. We alternatively recognize that the C—H dissociation enthalpies of {3-}-propene and {1-}-propane differ by the same quantity as well, where the {3-}- and {1-}- convey it is the C_3—H and C_1—H bonds being discussed. Regardless of which reactions have been interrelated to derive this 54 kJ mol^{-1} (13 kcal mol^{-1}), we may now designate this quantity as the resonance energy of allyl radical.

The radical sites need not have been capped with monohydrogen but instead with a methyl group. As such, the last reaction written above would be recast as

$$CH_2CHCH_2\bullet + CH_3CH_2CH_2CH_3 \rightarrow CH_3CH_2CH_2\bullet + CH_2CHCH_2CH_3 \quad (7.5)$$

The new reaction is found to be exothermic by 55 kJ mol^{-1} (13 kcal mol^{-1}) and so is in concert with our earlier numerical finding. We may even ascribe this new value to the difference of the C_3—C_4 bond dissociation enthalpy of 1-butene and the C_1—C_2 bond dissociation enthalpy of *n*-butane. Accordingly we have another definition of the resonance energy of the allyl radical.

Are these two definitions conceptually indistinguishable? To assume so requires the difference of the enthalpies of formation of propene and 1-butene to be very nearly equal to the difference of the enthalpies of formation of propane and *n*-butane – which they are. Indeed, if the monohydrogen and methyl capping definitions of resonance energy are truly the same, then should not reaction (7.6) be thermoneutral for any R′ and R″ we choose and not just the above choice of R′ = CH_2CHCH_2 and R″ = $CH_3CH_2CH_2$.

$$R'H + R''CH_3 \rightarrow R'CH_3 + R''H \quad (7.6)$$

Yet this cannot be true when applied to arbitrary R′ and R″. For example, if R′ and R″ are the isomeric *n*- and isopropyl respectively, then the enthalpies of formation of their monohydrogen-capped species must be precisely identical since the species themselves are precisely identical: we cannot chemically distinguish {1-}- and {2-}-propane despite the notational neologism introduced above. By contrast, the methyl-capped *n*- and isopropyl radicals are distinct isomers. The difference of the enthalpies of formation of *n*- and iso-butane is almost 10 kJ mol^{-1} (3 kcal mol^{-1}). So, which capping definition do we prefer?

Intrinsically, we might think reaction (7.4) is to be preferred over reaction (7.5) because of the long known free radical halogenations of aliphatic hydrocarbons. Alternatively, C—C bond homolysis is seemingly much more common than C—H homolysis in hydrocarbons, as befits their relative bond strengths. This suggests reaction (7.5) is to be

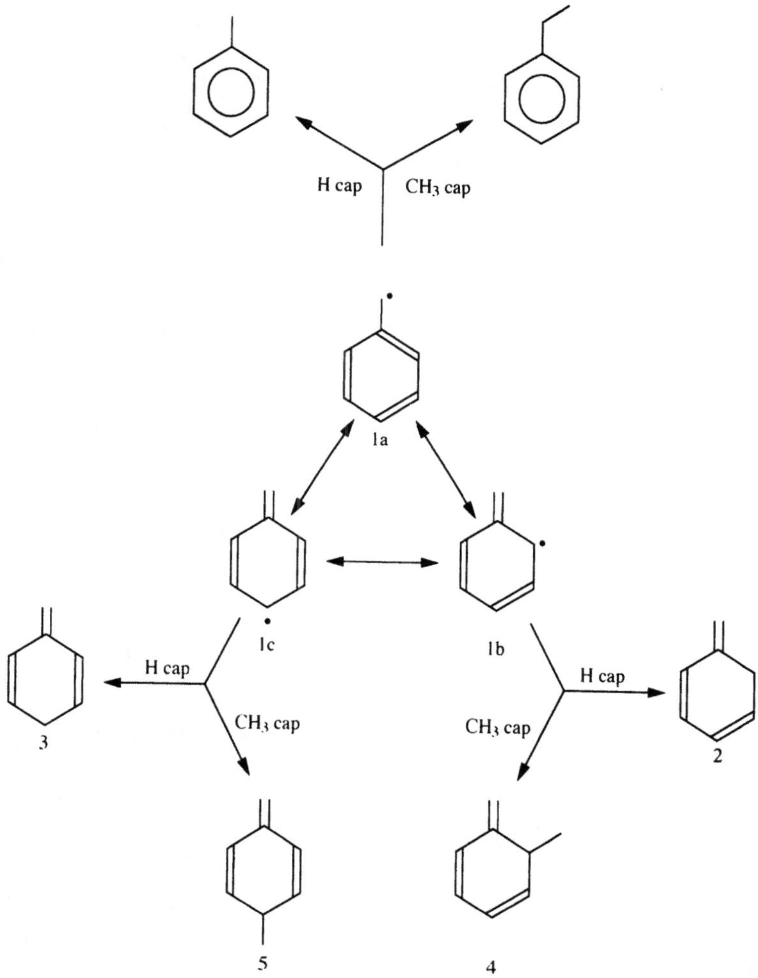

SCHEME 7.1.

preferred over (7.4). In fact, one opts for the use of reaction (7.4) because there are much more available thermochemical data for monohydrogenated than for corresponding methylated species.

For example, consider the benzyl radical (**1**) (Scheme 7.1). What is its resonance energy, or more precisely, what is its resonance energy beyond that of benzene itself? Analogous to reactions (7.4) and (7.5) are the new reactions (7.7) and (7.8):

$$C_7H_7\bullet + CH_3CH_2CH_3 \rightarrow CH_3CH_2CH_2\bullet + C_7H_8 \quad (7.7)$$

$$C_7H_7\bullet + CH_3CH_2CH_2CH_3 \rightarrow CH_3CH_2CH_2\bullet + C_7H_7CH_3 \quad (7.8)$$

These last two reactions are endothermic by the nearly identical values, 48 and 47 kJ mol^{-1} (both *c.* 11 kcal mol^{-1}) and so give the desired resonance energy quantity – if by C_7H_8 and $C_7H_7CH_3$ is meant toluene and ethylbenzene respectively. What else could be meant? In fact, the capping of benzyl radicals by monohydrogen could alternatively result in the isomeric 5-methylene-1,3-cyclohexadiene (**2**) and 3-methylene-1,4-cyclohexadiene (**3**), the so-called *o*- and *p*-isotoluenes, while the capping by methyl would result in the corresponding isomeric 6-methyl-5-methylene-1,3-cyclohexadiene (**4**) and 3-methyl-6-methylene-2,5-cyclohexadiene (**5**) derivatives, since they are all 'derived' from resonance structures (**1b**) and (**1c**). Reactions (7.4), (7.5), (7.7) and (7.8) all involve $-CH_2\bullet$ containing species, i.e. '"capping" resonance structure (**1a**)': to accommodate these alternative products, perhaps $>CH\bullet$ containing species should be considered. As such, perhaps the following reactions should have been considered, involving secondary C—H and $C-CH_3$ bonds for '"capping" resonance structures (**1b**) and (**1c**)'

$$C_7H_7\bullet + CH_3CH_2CH_3 \rightarrow CH_3CH\bullet CH_3 + C_7H_8 \qquad (7.9)$$

$$C_7H_7\bullet + CH_3CH(CH_3)CH_3 \rightarrow CH_3CH\bullet CH_3 + C_7H_7CH_3 \qquad (7.10)$$

Disappointingly, but perhaps not surprisingly, there are no thermochemical data for either of the desired methyl, methylenecyclohexadienes (**4**) and (**5**) but enthalpies of formation for both the methylenecyclohexadienes, (**2**) and (**3**) [2,3] are available.

Which reaction, (7.7) or (7.9), should be used to define the resonance energy of benzyl radical? The preference is for (7.7) because (a) simple, i.e. Hückel, molecular orbital logic places more radical character on the α-carbon of benzyl than on any other site and (b) the more stable product is formed upon capping the radical. It is fortunate that these definitions are consonant because references [2] and [3] report a *c.* 25 kJ mol^{-1} (6 kcal mol^{-1}) difference in the enthalpy of formation of 5-methylenecyclohexa-1,3-diene [4]! It can thus be concluded that the resonance stabilization of benzyl and allyl radicals are the same within 5 kJ mol^{-1} (1 kcal mol^{-1}). This result is surprising given the former radical is stabilized by a set of four, mostly inequivalent resonance structures and the latter radical by a set of two equivalent resonance structures. Returning to enthalpies of formation, it is seen that the difference of enthalpies of formation of benzyl and allyl radical is *c.* 36 kJ mol^{-1} (almost 9 kcal mol^{-1}), a value very close to the suggested nearly constant difference of closed shell, and unequivocally much more localized phenyl (C_6H_5-X) and vinyl ($CH_2=CH-X$) derivatives [5]. One is intrigued and admittedly somewhat confused by the near equality of the resonance energies of allyl and benzyl radical.

Let us now turn to propargyl radical. Analogous to reactions (7.4) and (7.5) are reactions (7.11) and (7.12)

$$C_3H_3\bullet + CH_3CH_2CH_3 \rightarrow CH_3CH_2CH_2\bullet + C_3H_4 \tag{7.11}$$

$$C_3H_3\bullet + CH_3CH_2CH_2CH_3 \rightarrow CH_3CH_2CH_2\bullet + C_3H_3CH_3 \tag{7.12}$$

These latter two reactions are endothermic by the nearly identical 49 and 48 kJ mol^{-1} (both *c.* 12 kcal mol^{-1}) – if by C_3H_4 and C_4H_6 is meant propyne and 1-butyne. In this case neither equivalence vs. inequivalence of resonance structures and their number, nor a nearly constant enthalpy of formation difference, offers an explanation for why the resonance energies of propargyl and allyl radicals are so similar.

Indeed, what else could be meant? The capping of propargyl radicals by monohydrogen and methyl could result in allene and 1,2-butadiene respectively. In fact, allene and propyne have comparable enthalpies of formation, the former more stable than the latter by somewhat less than 6 kJ mol^{-1} (something more than 1 kcal mol^{-1}). Relatedly, 1,2-butadiene and 1-butyne have comparable enthalpies of formation; the former is now less stable than the latter by almost 3 kJ mol^{-1} (1 kcal mol^{-1}). Perhaps one should have thus written reactions (7.13) and (7.14) involving olefinic C—H and C—CH_3 bonds

$$C_3H_3\bullet + C_2H_4 \rightarrow C_2H_3\bullet + C_3H_4 \tag{7.13}$$

$$C_3H_3\bullet + C_2H_3CH_3 \rightarrow C_2H_3\bullet + C_3H_3CH_3 \tag{7.14}$$

where the products are explicitly allene and 1,2-butadiene. This results in resonance energies for propargyl radical of 89 and 93 kJ mol^{-1} respectively (21 and 22 kcal mol^{-1}). While these values are comparable and so suggest monohydrogen and methyl capping results are conceptually equivalent, we are not comfortable with propargyl radical having a higher resonance energy than allyl radical: the latter radical has two equivalent resonance structures and the former radical has two inequivalent ones. Maybe we should thus content ourselves with propargyl and allyl radicals having nearly identical resonance energies.

Consider now the hydroxymethyl radical. Which monohydrogen capping reaction is preferable?

$$CH_2OH\bullet + CH_4 \rightarrow CH_3\bullet + CH_3OH \tag{7.15}$$

$$CH_2OH\bullet + CH_4 \rightarrow CH_3\bullet + CH_2OH_2 \tag{7.16}$$

We recognize $CH_2OH\bullet$ as a carbon-centered radical and CH_3OH is considerably more stable than its still unisolated isomer, the oxygen ylid CH_2OH_2. As such, reaction (7.15) is preferable, from which we derive a resonance energy for hydroxymethyl radical of 28 kJ mol^{-1} (7 kcal mol^{-1}). Yet, if the energetics of a substituted methyl radical are being investigated, perhaps a comparably substituted methyl radical and methane should be used as reference states. Equivalently, a primary C—H bond is being cleaved and formed instead of a methyl C—H bond. This suggests that reaction (7.15) should be replaced by

$$CH_2OH\bullet + CH_3CH_3 \rightarrow CH_3CH_2\bullet + CH_3OH \quad (7.17)$$

from which a resonance energy for hydroxymethyl radical of merely 10 kJ mol^{-1} (2 kcal mol^{-1}) is found.

Summarizing, a suggested definition for the resonance energy of a radical $R\bullet$ is as the endothermicity of the reaction of some localized radical $R^\bullet_{loc}$ and the monohydrogen-capped species RH and $R_{loc}H$

$$R\bullet + R_{loc}H \rightarrow RH + R^\bullet_{loc} \quad (7.18)$$

where the most stable $R_{loc}H$ and RH are chosen. In addition, R_{loc} is chosen to provide the most suitable mimic for the R by being primary, secondary, tertiary, or suitably unsaturated. But that assumes that the desired energetics data on the radicals – and/or their closed shell counterparts – are available!

ISOELECTRONIC ANALOGIES

Aminium and ammonium ions are conceptually derived from carbon-centered free radicals and their monohydrogen-capped derivatives by the formally simple isoelectronic replacement of the 'central' carbon atom by an atom of nitrogen.

$$RR'R''C\bullet \rightarrow RR'R''N^+ \quad (7.19)$$

$$RR'R''CH \rightarrow RR'R''NH^+ \quad (7.20)$$

Excluding the possibility of nuclear transformations, e.g. with suitably ^{14}C labelled species, reactions (7.19) and (7.20) are expertimentally unrealizable by synthetic chemists.

$$N^+ + RR'R''C\bullet \rightarrow C + RR'R''N^+ \quad (7.21)$$

$$N^+ + RR'R''CH \rightarrow C + RR'R''NH^+ \quad (7.22)$$

Relatedly, the 'exchange reactions' are likewise unrealizable by ion-molecule experimentalists.

However, aminium and ammonium ions may be conceptually derived from the corresponding amines by the loss of an electron and protonation respectively. As such, enthalpies of formation of aminium and ammonium ions are conceptually derivable from those of the amines by the well-defined measurable quantities of ionization potential and proton affinity [6]:

$$\mathrm{IP(RR'R''N)} = \Delta H_f(\mathrm{RR'R''N^+}) - \Delta H_f(\mathrm{RR'R''N}) \tag{7.23}$$

$$\mathrm{PA(RR'R''N)} = [\Delta H_f(\mathrm{RR'R''N}) + \Delta H_f(\mathrm{H^+})] - \Delta H_f(\mathrm{RR'R''NH^+}) \tag{7.24}$$

Indeed, conceptual derivation and experimental realization are both achievable assuming that protonation of the amine is on its nitrogen, and not some more basic site such as the β-carbon as found for some enamines or ring carbon as found for some anilines.

For the purposes at hand, the most relevant comparison among the carbon-centered radical, its hydrocarbon, and the corresponding aminium and ammonium ion, is the difference of the bond enthalpies $D(\mathrm{RR'R''C{-}H})$ and $D(\mathrm{RR'R''N^+{-}H})$. Through simple algebra, the application of definitions of bond enthalpies, ionization potentials and proton affinities, we find the desired difference quantity equals

$$\begin{aligned} '\Delta D' &\equiv D(\mathrm{RR'R''C{-}H}) - D(\mathrm{RR'R''N^+{-}H}) \\ &= [\Delta H_f(\mathrm{RR'R''C\bullet}) - \Delta H_f(\mathrm{RR'R''CH})] \\ &\quad - [\mathrm{IP(RR'R''N)} + \mathrm{PA(RR'R''N)} - \Delta H_f(\mathrm{H^+}] \end{aligned} \tag{7.25}$$

A particularly simple, seemingly unambiguous, and evocative series of species for bond enthalpy comparison are methane and its increasingly methylated derivatives, ethane, propane and isobutane, and the protonated forms of ammonia, and its increasingly methylated derivatives, methyl, dimethyl and trimethylamine. The difference quantity δD steadily increases upon methylation: for the non, mono, di and trimethyl case, it equals -83, -34, -1 and 7 kJ mol^{-1} (-20, -8, 0 and 2 kcal mol^{-1}) respectively [7] (see Table 7.1). It is important to note that methylation provides stabilization of each of the classes of species individually. For the carbon-centered radicals, sequential α-methylation decreases the enthalpy of formation by 28, 29 and 42 kJ mol^{-1} (6, 7 and 10 kcal mol^{-1}) and for the closed shell monohydrogen capped alkanes by but 10, 19 and 29 kJ mol^{-1} (2, 5 and 7 kcal mol^{-1}). Equivalently, the C—H bond energy of the alkanes decrease with increasing methylation although the effect is

TABLE 7.1 Bond enthalpy comparison of methane, protonated ammonia and their methylated derivatives (all values in kJ mol^{-1})

$\Delta H_f(RR'R''C\bullet)$	$\Delta H_f(RR'R''CH)$	RR'R''N	IP	PA	δD^a
CH_3 147	CH_4 −74.4	NH_3	980	854	−83
C_2H_5 119	C_2H_6 −83.8	CH_3NH_2	865	902	−34
$(CH_3)_2CH$ 90	$(CH_3)_2CH_2$ −104.7	$(CH_3)_2NH$	794	932	−2
$(CH_3)_3C$ 48	$(CH_3)_3CH$ −134.2	$(CH_3)_3N$	754	951	7

[a] $\delta D = [\Delta H_f(RR'R''C\bullet) - \Delta H_f(RR'R''CH)] - [IP(RR'R''N) + PA(RR'R''N) - \Delta H_f(H^+)]$

lessened with increasing substitution. Sequential α-methylation decreases the ionization potential of amines by 115, 71 and 40 kJ mol^{-1} (27, 17 and 10 kcal mol^{-1}).

Equivalently, the radical cations of amines are stabilized relative to their corresponding neutral by sequential methylation although the effect is ameliorated with increasing substitution. The proton affinity of the amines increases with sequential methylation by 42, 27 and 19 kJ mol^{-1} (10, 6 and 5 kcal mol^{-1}). Equivalently, the 'onium' ions of amines are stabiilized relative to their corresponding neutral although the effect becomes increasingly weakened here as well. α-Methyl groups stabilize unpaired electrons, or more properly, the interaction of occupied π-symmetry orbitals of methyl groups with half-filled orbitals: they stabilize positive charges wherever they may be in a molecule. That δD is not of one sign is frustrating: a simple but universal comparison that interrelates homolytic C—H bond enthalpies in hydrocarbons and ionization potentials and proton affinities of amines would have provided us with a major conceptual bridge between the energetics of gas phase ions and neutrals [8].

1,2-ALKYL AND HYDROGEN SHIFTS IN FREE RADICALS

This section considers the 1,2-rearrangement of radicals [9–13] (H or saturated alkyl migrating groups) from the standpoints of stabilities and resonance energies of transition states as well as thermodynamic driving forces. (The following sections consider related topics: Section 7.4 examines Vitamin B_{12}-induced rearrangements in light of the discussion in the present section and Section 7.5 very briefly looks at energetic relationships in 1,2-shifts in 1,3-diradicals.)

Radical rearrangements have been known for many years [9]. For

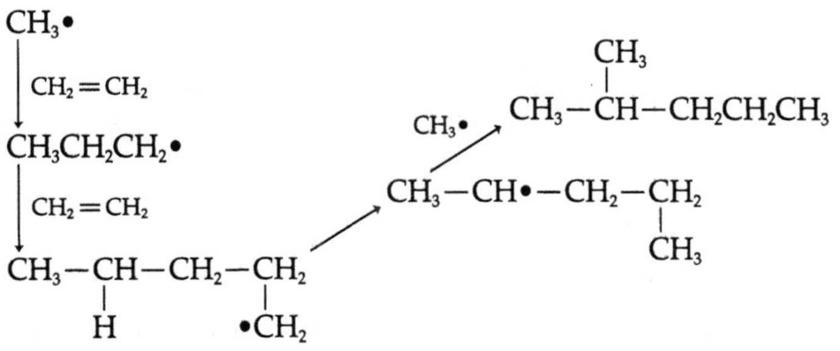

SCHEME 7.2.

example, the intramolecular 1,4-hydrogen shift depicted in Scheme 7.2 is also observed in longer *n*-alkyl radicals (which also undergo 1,5-H shifts), but such intramolecular hydrogen transfers are not observed for 1-butyl and 1-propyl radicals, which decompose differently [10]. The potential 1,3-shift in 1-propyl radical and higher species is also uncommon. Similarly, attack of isopropyl radical on acetylene forms the interesting homoallylic radical depicted in Scheme 7.3 upon intramolecular 1,4-H transfer and this species passes through a cyclopropylcarbinyl radical, which undergoes intramolecular elimination. There are, of course, many other reactions that compete with free radical rearrangements – dimerization, disproportionation, etc. The 1,4-, 1,5- and similar intramolecular hydrogen shifts have activation energies near 42 kJ mol^{-1} (10 kcal mol^{-1}), similar to the corresponding intermolecular reactions [9,10]. For example,

SCHEME 7.3.

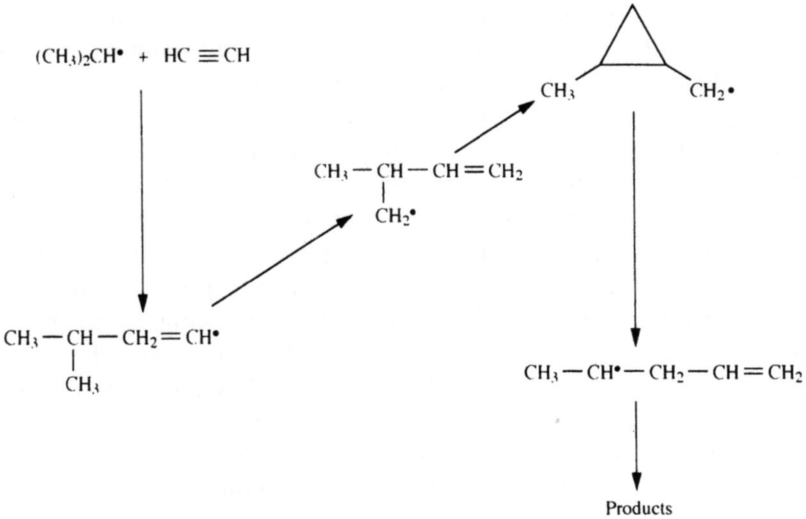

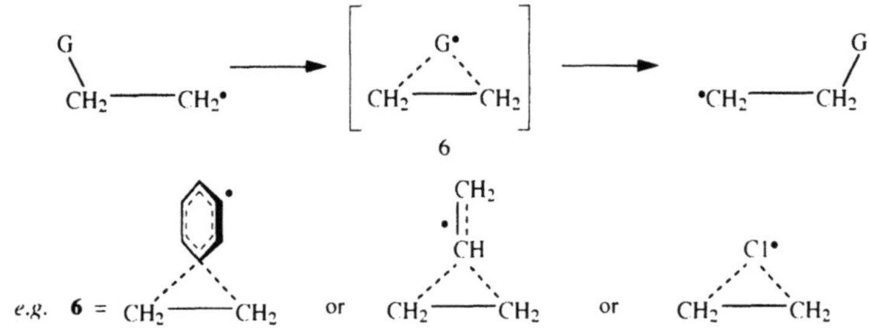

SCHEME 7.4 Migratory aptitude of group G [14]: $CH_2{=}CH > (CH_3)_3CO > C_6H_5 > (CH_3)_3C{-}C{=}C > CN$.

the activation energy for the transfer of a secondary hydrogen in propane to ethyl radical to form isopropyl radical and ethane is about 43 kJ mol^{-1} [10].

In contrast to the longer-range rearrangements, concerted intramolecular 1,2-hydrogen- or (saturated) alkyl-shifts are unknown for alkyl monoradicals [9–13]. 1,2-Shifts are known to occur in certain free radicals (Scheme 7.3) in which the migrating group is unsaturated (e.g. homoallylic, homobenzylic systems). Scheme 7.4 lists experimentally determined migratory aptitudes [14] for selected unsaturated groups. *α*-Chloro- and *α*-bromo-substituted species also undergo such rearrangements, usually explained in terms of relatively low-lying vacant *d* orbitals in the migrating groups.

What is the reason for the absence of 1,2-hydrogen or 1,2-methyl shifts in free radicals? This is an interesting question in light of the facile 1,2-rearrangements in the corresponding carbocations (carbenium ions). The generally accepted explanation of this dichotomy rests upon the idea that the transition state for the carbocation rearrangement has two electrons delocalized among three centers analogous to the aromatic cyclopropenium cation, while the corresponding radical transition state is 'isoelectronic' with the non-aromatic cyclopropenyl radical. One may attempt to quantify this difference by calculating the enthalpy change of the hypothetical isodesmic reaction depicted in equation (7.26) (unless otherwise noted, the $\Delta H_f°(g)$ values depicted in equation (7.26) and elsewhere are from the source cited in reference [6] of this chapter; free radical $\Delta H_f°(g)$ values are taken from Chapter 2 (Tsang, in this book) if available and otherwise from reference [6]). The approach taken in equation 7.26 subsumes that of the Dewar–Breslow definition of aromaticity. The very substantial exothermicity in equation (7.26) ($-$ 140 kJ mol^{-1} or $-$ 33 kcal mol^{-1}) is testimony primarily to the substantial resonance energy of the cyclopropenium cation, although independent evaluations of the resonance energies of all four species using the 'capping' models advanced in Section 7.1 would provide more specificity.

$$\text{Cyclopropenyl}\bullet \;+\; \text{Allylium}^{+} \;\rightarrow\; \text{Cyclopropenium}^{+} \;+\; \text{Allyl}\bullet \qquad (7.26)$$

$$(440\ \text{kJ mol}^{-1}) \quad (946\ \text{kJ mol}^{-1}) \quad (1075\ \text{kJ mol}^{-1}) \quad (171\ \text{kJ mol}^{-1})$$

$$\Delta H_r = -\,140\ \text{kJ mol}^{-1}$$

Perhaps more relevant is the enthalpy change in the transition from ethyl radical to symmetrically bridged transition state, depicted in equation (7.27). There is no experimental number for the

$$\overset{\bullet}{C}H_2\text{—}CH_2\text{—}H \longrightarrow CH_2\overset{H^{\bullet}}{\text{—}}CH_2 \qquad (7.27)$$

$$(119\ \text{kJ mol}^{-1}) \qquad (297\ \text{kJ mol}^{-1})$$

$$\Delta H_r = -201\ \text{kJ mol}^{-1}$$

bridged ethyl radical. However, the BAC-MP4 method of Melius [15] (based upon *ab initio* molecular orbital calculations at the MP4/6–31G** level of a Hartree–Fock 6–31G* optimized structure with zero point energy and thermal corrections to 298 K, bond length/bond energy and spin corrections) yields the enthalpy value for bridged ethyl radical employed in equation (7.27). The BAC-MP4 value for ethyl radical (118 kJ mol^{-1}) [15] is in excellent agreement with the experimental value (119 kJ mol^{-1}). Thus, it appears that the activation barrier for this rearrangement is about 178 kJ mol^{-1} (42–43 kcal mol^{-1}). This is also consistent with an experimental 35 kcal mol^{-1} barrier for the 1,2-H shift depicted in equation (7.28) [9,16].

$$CH_3CH_2CH_2^{\bullet} \rightarrow CH_3\text{—}CH\overset{H^{\bullet}}{\text{——}}CH_2 \rightarrow CH_3CH^{\bullet}CH_3 \qquad (7.28)$$

$$\Delta H_{\ddagger} = -35\ \text{kJ mol}^{-1}$$

In contrast to their radical cousins, formally primary carbenium ions such as ethylium *are* bridged [17] (i.e. the bridged structure is lower in energy than the classical localized structures in the gas phase). Thus, the published experimental $\Delta H_f°(g)$ for $C_2H_5^+$ (902 kJ mol^{-1}) is taken and used in equation (7.29) in order to make a comparison similar to that in equation (7.26). The result ($-$ 170 kJ mol^{-1} or $-$ 40 kcal mol^{-1}) is surprisingly similar to that from equation (7.26) and clearly indicates how unfavorable the 1,2-shift is in radicals relative to carbenium ions. It is important to realize that, while the H-bridged radical in equation (7.29) is a transition state (saddle point on the energy surface), the bridged ethylium cation is a ground state (minimum on the energy surface) [17]. Thus, their direct comparison in equation (7.29) is not strictly 'kosher' in the sense that the vibrational degrees of freedom and therefore zero-point-energy and thermal corrections are not really comparable. However, the discrepancies are small relative to the magnitude of the enthalpy change in equation (7.29). It is also worth noting that the aforementioned intramolecular 1,4-shift as well as intermolecular H-transfers (i.e. C---H---C, 2 electrons) are allyl-radical-like and presumably a (Dewar–Breslow-type) comparison can be made to the electronic energy of the cyclopropenyl-radical-like intramolecular 1,2-H-shift transition state, which also has strain as a destabilizing factor.

$$\underset{(297\ \mathrm{kJ\ mol^{-1}})}{\overset{\mathrm{H}^{\bullet}}{\mathrm{CH_2{-\!\!-}CH_2}}} + \underset{(946\ \mathrm{kJ\ mol^{-1}})}{\mathrm{Allyium^+}} \rightarrow \underset{(902\ \mathrm{kJ\ mol^{-1}})}{\overset{\mathrm{H}^{+}}{\mathrm{CH_2{-\!\!-}CH_2}}} + \underset{(171\ \mathrm{kJ\ mol^{-1}})}{\mathrm{Allyl^{\bullet}}} \qquad (7.29)$$

$$\Delta H_r = -170\ \mathrm{kJ\ mol^{-1}}$$

Is it possible, despite this high activation barrier, to induce a 1,2-H or 1,2-methyl shift if the reaction is strongly exothermic ('energy-driven')? Let us briefly explore some possibilities. In carbenium ion chemistry, the neopentyl to *t*-pentyl rearrangement (equation (7.30)); the value for the classical neopentyl cation is higher than that of the bridged ion which should be the true energy minimum [17]) has an extremely strong driving force (gas-phase data are shown, but the driving force in solution is similarly quite strong).

$$\underset{(>795\ \mathrm{kJ\ mol^{-1}})}{\mathrm{CH_3{-}\underset{CH_3}{\overset{CH_3}{C}}{-}CH_2^+}} \longrightarrow \underset{(661\ \mathrm{kJ\ mol^{-1}})}{\mathrm{CH_3{-}\overset{CH_3}{C_+}{-}CH_2{-}CH_3}};\quad \Delta H_r = -134\ \mathrm{kJ\ mol^{-1}} \qquad (7.30)$$

The origin of this transformation is the great stability of 3° relative to 1° carbocations. Indeed, if neopentyl is best described as bridged, then the classical structure depicted in equation (7.30) is higher in energy and reaction (7.30) is even more exothermic. In contrast, the corresponding rearrangement of neopentyl radical (equation (7.31)) is barely exothermic – the relatively mild stabilization achieved in transforming a 1° radical to a 3° radical is mitigated by decreased branching in the rearranged radical. The lack of a strong thermodynamic driving force in simple free radical rearrangements has been noted elsewhere [11].

$$CH_3-C(CH_3)_2-CH_2^{\bullet} \longrightarrow CH_3-\dot{C}(CH_3)-CH_2-CH_3; \quad \Delta H_r = -6\ \text{kJ mol}^{-1} \qquad (7.31)$$

(33 kJ mol^{-1}) (27 kJ mol^{-1})

If transformation of a 1° radical to a 3° radical does not supply a significant driving force, then what about the incorporation of benzylic stabilization (see Section 7.1)? Examining equation (7.32), it is noted that the rearrangement of the 'neophyl' radical is significantly exothermic. How does the system behave?

$$C_6H_5-C(CH_3)_2-CH_2^{\bullet} \longrightarrow C_6H_5-C^{\bullet}(CH_3)-CH_2-CH_3; \quad \Delta H_r = -68\ \text{kJ mol}^{-1} \qquad (7.32)$$

(183 kJ mol^{-1}) [18] (115 kJ mol^{-1}) [19]

Here, the experimental results are depicted in Scheme 7.5. It is fascinating that the 1,2-methyl shift does not occur in the neophyl radical despite the fact that this pathway has 46 kJ mol^{-1} more driving force than the 1,2-phenyl shift which does, however, occur (see Scheme 7.5). In Scheme

SCHEME 7.5.

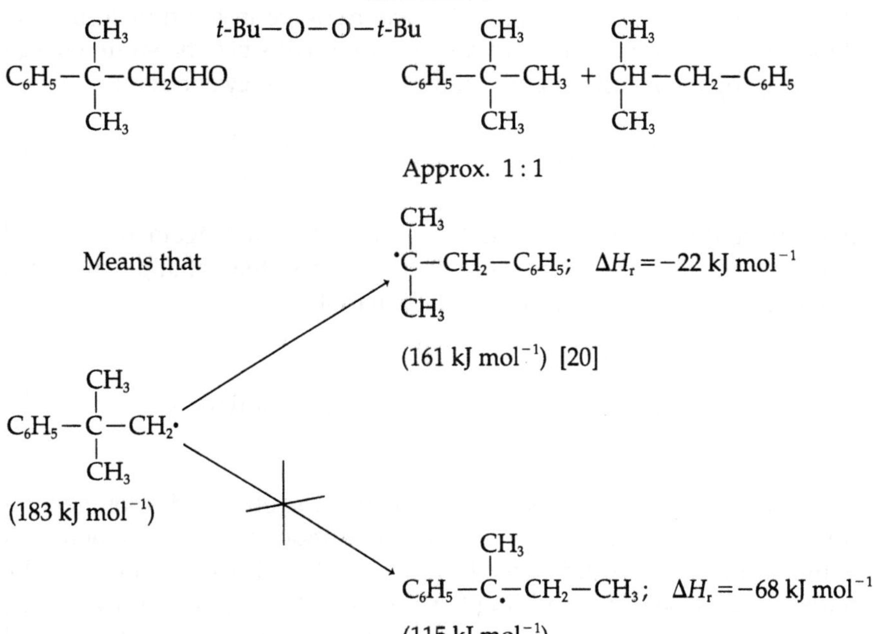

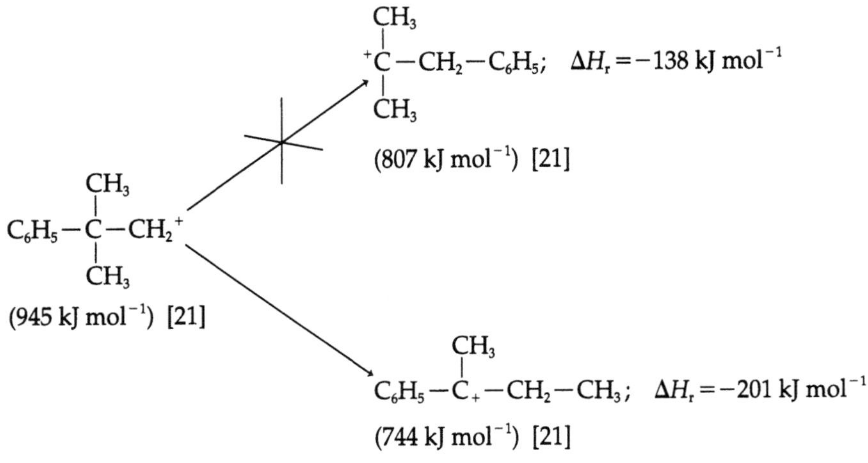

SCHEME 7.6.

7.6 one observes that the energetically favored 1,2-methyl shift does occur from neophyl cation. The differences in the enthalpies of the radical (Scheme 7.5) and the carbocation rearrangement (Scheme 7.6) pathways are not very different (-46 kJ mol^{-1} and -63 kJ mol^{-1} respectively), but the differences in kinetics are decisive.

What is the next logical molecular manipulation? If there can be a significant release in strain upon 1,2-alkyl shift, perhaps this factor can drive the reaction. Interestingly, the driving force in the transformation depicted in equation (7.33) will not be very great since the strain energy in cyclopropane is only slightly higher than that in cyclobutane.

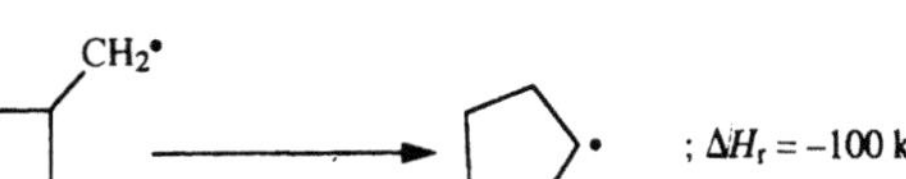

(7.33)

In contrast, the cyclobutylcarbinyl to cyclopentyl rearrangement depicted in equation (7.34) should be more promising and the enthalpy change is calculated to be -100 kJ mol^{-1} (-24 kcal mol^{-1}).

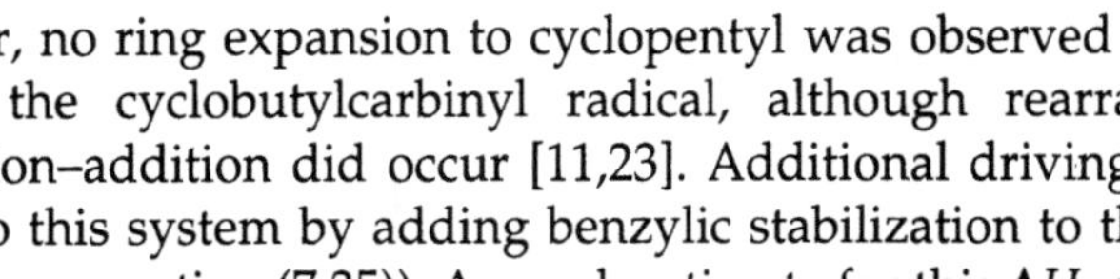

(7.34)

However, no ring expansion to cyclopentyl was observed upon generation of the cyclobutylcarbinyl radical, although rearrangement *via* elimination–addition did occur [11,23]. Additional driving force can be built into this system by adding benzylic stabilization to the rearranged radical (see equation (7.35)). A rough estimate for this ΔH_r can be derived

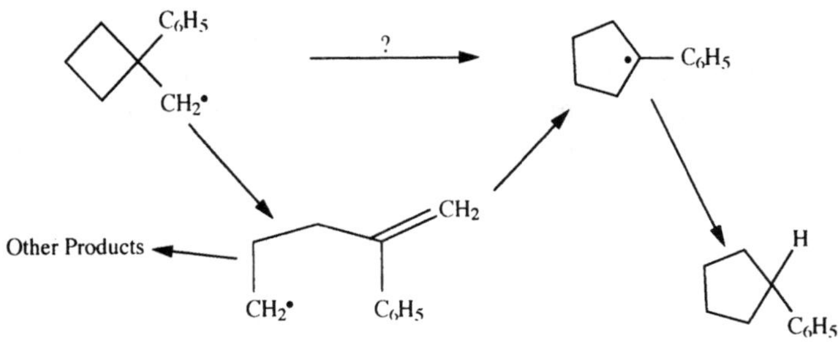

SCHEME 7.7.

from comparison of the exothermicity of the neopentyl rearrangement (−6 kJ/mol; equation (7.31)) and the methyl shift in Scheme 7.5 ($\Delta H_r = -68$ kJ mol^{-1}) and addition of the resulting increase in exothermicity (−62 kJ mol^{-1}) to that in equation (7.34) (−100 kJ mol^{-1}) in order to obtain −162 kJ mol^{-1} (−39 kcal mol^{-1}).

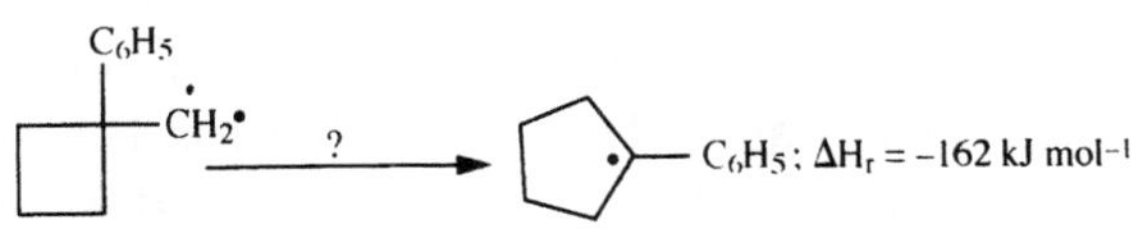

(7.35)

Although ring expansion of the 1-phenylcyclobutylcarbinyl radical does occur, it is thought to be a consequence of elimination–addition (Scheme 7.7) rather than a 1,2-alkyl shift [11,24] For the sake of comparison with equation (7.34), the corresponding carbenium ion rearrangement (equation (7.36)) is exothermic by 163 kJ mol^{-1} (39 kcal mol^{-1}).

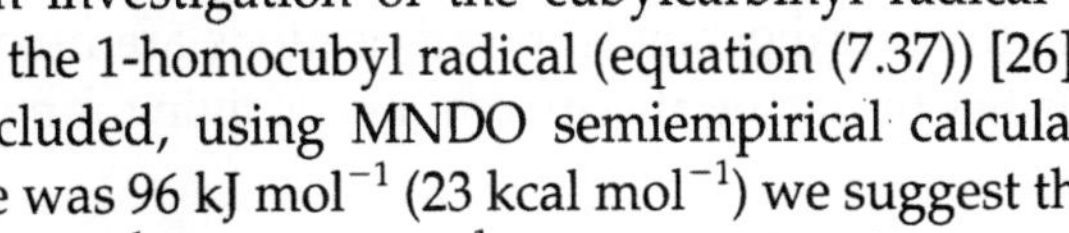

(964 kJ mol^{-1})[25] (801 kJ mol^{-1})

(7.36)

Release of some of the enormous strain energy in the cubane system prompted an investigation of the cubylcarbinyl radical which did *not* rearrange to the 1-homocubyl radical (equation (7.37)) [26]. Although the authors concluded, using MNDO semiempirical calculations, that the driving force was 96 kJ mol^{-1} (23 kcal mol^{-1}) we suggest that it is actually about 202 kJ mol^{-1} (48 kcal mol^{-1}) (equation (7.37)).

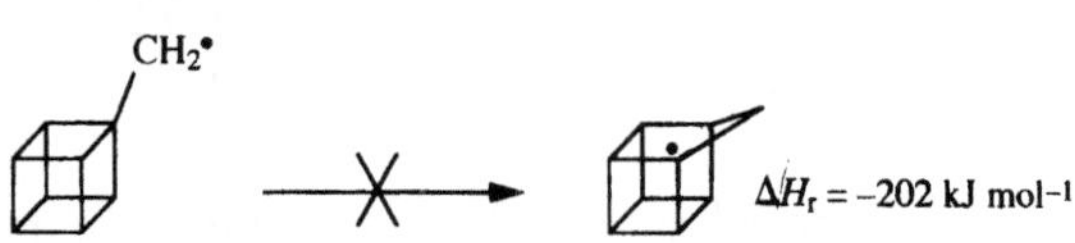

(789.5 kJ mol^{-1})[27] (587.4 kJ mol^{-1})[28]

(7.37)

For comparison's sake, the corresponding rearrangement of the cubylcarbinyl to 1-homocubyl cation is depicted in equation (7.38). Here, the exothermicity of the transformation of a 3° carbocation to a 1° carbocation with release of some of the strain in the cubane system is mitigated by 80.3 kJ mol^{-1} (19.2 kcal mol^{-1})[29] of extra strain in the appreciably nonplanar 1-homocubyl carbocation.

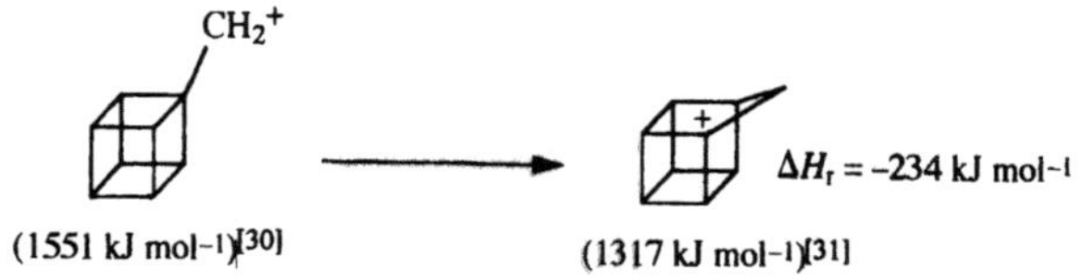

(7.38)

Although the carbenium ion rearrangement depicted in equation (7.38) is only 33 kJ mol^{-1} (8 kcal mol^{-1}) more exothermic than the nonexistent free radical rearrangement (equation (7.37)), it occurs extremely rapidly [26].

Since the rates of rearrangements (and other reactions) are determined by intrinsic rate factors, embodied in transition state structures, modified by thermodynamic driving forces, it is useful to reexamine these rearrangements in terms of Marcus Theory [32]. Marcus Theory conceptualizes observed activation energies in terms of intrinsic barriers and thermodynamic driving forces. Our earlier discussion indicates that a reasonable intrinsic barrier for 1,2-H shift in ethyl radical is about 180 kJ mol^{-1} (*c.* 40 kcal mol^{-1}), slightly lower in 1-propyl radical, with corresponding 1,2-methyl shifts slightly higher. Thus, if assuming an intrinsic barrier ($\Delta E_o^{\ddagger}$ in equation (7.39)) of 180 kJ mol^{-1} and combining it with

$$\Delta E^{\ddagger} = \Delta E_o^{\ddagger}[1 + \Delta E/(4\,\Delta E_o^{\ddagger})]^2 \qquad (7.39)$$

the enthalpy of reaction in equation (7.38) (roughly $-$ 230 kJ mol^{-1} or $-$55 kcal mol^{-1}), we may estimate an energy of activation of about 85 kJ mol^{-1} (20 kcal mol^{-1}) for the rearrangement of the cubylcarbinyl radical. These relationships are depicted in Figure 7.1. In contrast, the intrinsic barriers to rearrangements in carbocations are much lower and will lowered further by strong thermodynamic driving forces.

ADENOSYLCOBALAMIN CATALYSIS OF 1,2-SHIFTS IN RADICALS

Vitamin B_{12} is properly termed cyanocobalamin, while the related active coenzymes are either adenosylcobalamin or methylcobalamin [33–35].

Vitamin B_{12} and the related coenzymes have furnished the 'Mount Everest' or 'Moonshot' for crystallographers, synthetic organic chemists and organometallic chemists for about 50 years in the sense that the studies have defined the limits of chemical methodology more than once during this time period. Adenosylcobalamin was the first naturally occurring compound known to have a metal–carbon bond [33]. This weak Co—C bond (*c.* 120–170 kJ mol^{-1} or 30–40 kcal mol^{-1}) is the origin of the relative ease of formation of free radicals, particularly in the

FIGURE 7.1 Schematic depiction of the approximate 'intrinsic' energy of activation for typical degenerate 1,2-alkyl shifts in simple free radicals and comparison with the approximate energy of activation calculated (equation (7.39)) using the same 'intrinsic' energy of activation in concert with the calculated exothermicity of the cubylcarbinyl to 1-homocubyl radical rearrangement.

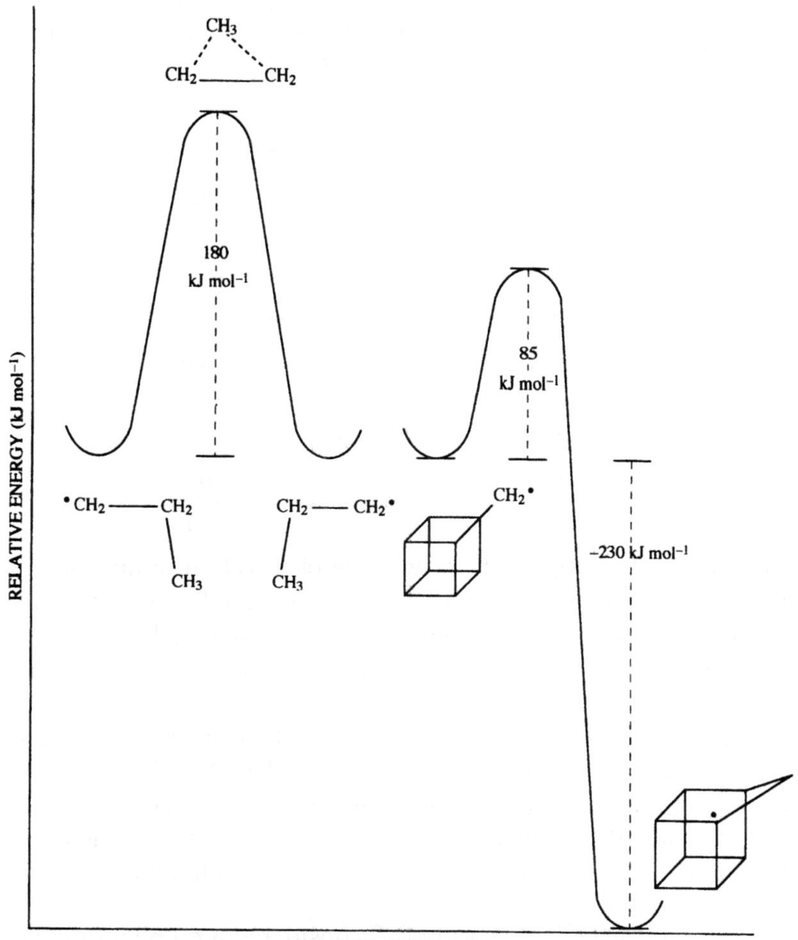

presence of enzyme and substrates that are known to increase the rates of bond breakage by at least 10 [11].

Adenosylcobalamin and methylcobalamin are coenzymes involved in the catalysis of a family of 1,2-rearrangements of the type depicted in equation (7.40).

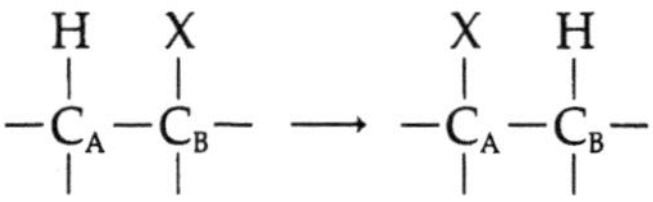

(7.40)

We will focus upon three examples of carbon-skeleton rearrangements, all catalyzed by adenosylcobalamin [33–35]. These are (1) the methylmalonyl-CoA mutase reaction (equation (7.41)); (2) the *α*-methyleneglutarate mutase reaction (equation (7.42)), and (3) the glutamate mutase reaction (equation (7.43)). Each of these is thought to proceed by a radical pathway typified in Scheme 7.8 (adapted from references [34] and [35]).

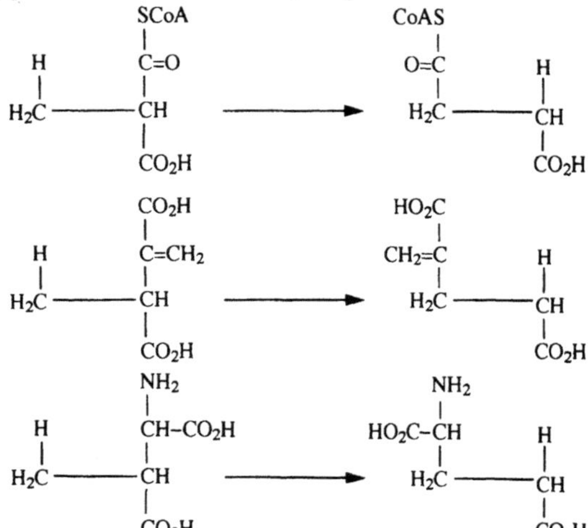

(7.41)

(7.42)

(7.43)

In this regard, it is important to note the observation of an ESR signal formed only upon addition of the substrate methylmalonyl-CoA to methylmalonyl-CoA mutase coenzyme B_{12} holoenzyme [36]. This is the first evidence for a radical presence in a coenzyme B_{12} dependent carbon skeleton rearrangement reaction [36].

In analyzing these reactions it is helpful to consider a framework involving three mechanistic extremes (Scheme 7.9) [37]. These include (a) concerted migration in which the symmetrically bridged species is a transition state; (b) stepwise migration in which the symmetrically bridged species is a true intermediate; and (c) elimination–addition involving an olefinic intermediate.

An early MNDO SCF-MO study [37] considered a number of possible

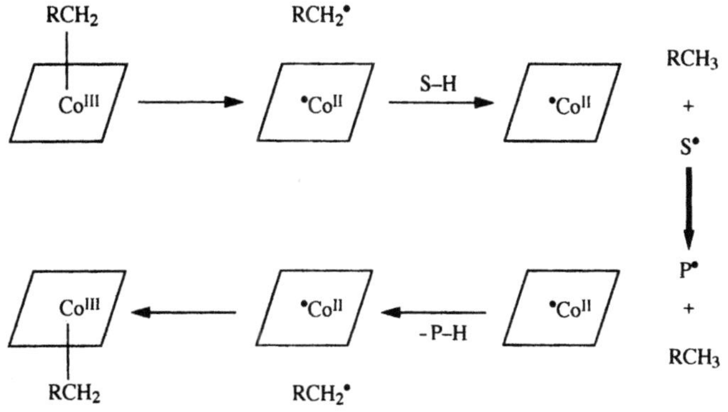

SCHEME 7.8.

group migrations in the context of Scheme 7.9. The enthalpy of activation for 1,2-H shift *via* concerted mechanism (a) in Scheme 7.9 was calculated to be 219 kJ mol^{-1} (52.3 kcal mol^{-1}), which is about 40 kJ mol^{-1} (10 kcal mol^{-1}) higher than Melius' BAC-MP4 value (equation 7.27). The enthalpy of activation for elimination–addition was calculated using MNDO to be slightly higher [37]. The concerted 1,2-methyl shift was calculated using MNDO to have an enthalpy of activation of 245 kJ mol^{-1} (58.6 kcal mol^{-1}), some 23–24 kcal mol^{-1} higher than the barrier for elimination–addition [37]. Presumably, the BAC-MP4 value would again lower this somewhat, but it is clear that concerted 1,2-shifts involving simple saturated alkyl groups have very high activation barriers.

In considering the three examples of B_{12}-catalyzed rearrangements chosen here, we first note that the methylmalonyl-SCoA rearrangement can, in principle, be regarded simply as a homoallylic rearrangement (Scheme 7.4) involving an intermediate such as (7). Thus, facile rearrangement would not be remarkable, consistent with the well-known rapidity of such rearrangements [38]. Similarly, the rapid rearrangement involved in the α-methyleneglutarate rearrangement (equation (7.42)) is another homoallylic case. For comparison, the MNDO value for the enthalpy of activation for concerted migration of COSH (a model for malonyl SCoA) is 100 kJ mol^{-1} (23 kcal mol^{-1}) lower than the value for 1,2-methyl shift [37]. The MNDO technique predicts that concerted migration of COSH is competitive with elimination–addition [37].

The most intriguing of the three reactions depicted here is the glutamate mutase reaction (equation 7.43)). The transition state for concerted 1,2-rearrangement is calculated by MNDO to be very high since the migrating carbon is saturated [37]. However, the activation barrier for elimination–addition is calculated by MNDO to be quite low [37].

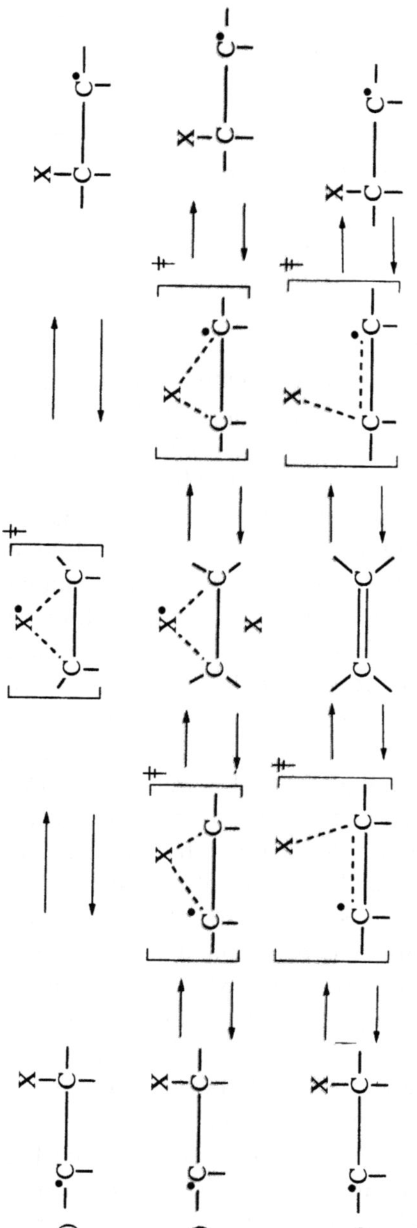

SCHEME 7.9.

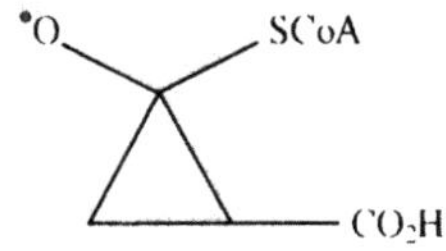

(7)

Dowd [35] has beautifully summarized a great deal of the physical organic chemistry employed in the exploration of vitamin B_{12} chemistry. He points out some problems in free-radical mechanisms. For example, in the free radical mechanism proposed for methylmalonate rearrangement, the second H-transfer occurs from 5′-deoxyadenosine (**8**) to succinyl-CoA radical (**9**) (equation (7.44)). Dowd estimates that, unlike most other such rearrangements, this one is significantly endothermic (by 7–8 kcal mol^{-1} or 30–35 kJ mol^{-1}). He used the activation barrier (15 kcal mol^{-1}) for the thermoneutral reaction between $CH_3\bullet$ and CH_4 to estimate a kinetic barrier of roughly 22–23 kcal mol^{-1} (*c.* 100 kJ mol^{-1}) [35], a very formidable barrier indeed, even if the next step (Co—C bond formation) is exothermic by over 120 kJ mol^{-1}.

Another interesting point was raised by Dowd in connection with his synthesis of putative alkyl-cobalamin intermediates potentially related to these rearrangements. Thus, he employed the methylmalonate homologue (**10**). In the presence of tri n-butyltin hydride, the true free radical cyclizes to a cyclohexylcarbinyl radical (Scheme 7.10) at a rate some 30 000 times more rapidly than the known isolated methylmalonyl-to-succinyl free radical rearrangement. Scheme 7.10 shows no products derived from succinyl in the malonate free radical pathway. In contrast, when substrate (**10**) is reacted with 'super-reduced' vitamin B_{12} [vitamin B_{12s} – a species containing Co(I); note that reduced vitamin B_{12} or vitamin B_{12r} contains Co(II) – see Scheme 7.8, while vitamin B_{12} and its derived coenzymes contain Co(III)] to produce the alkyl-cobalamin intermediate (**11**), succinate (**12**) is the major product (Scheme 1.10) [35]. These experimental findings have been cited in support of the view that succinate (**12**) is *not* a product of free radical rearrangement [35].

To conclude this section, note the very intriguing suggestion by Rétey that it may be necessary to reexamine our fundamental views of enzyme catalysis [39,40]. The usual paradigm is to consider enzymes to function

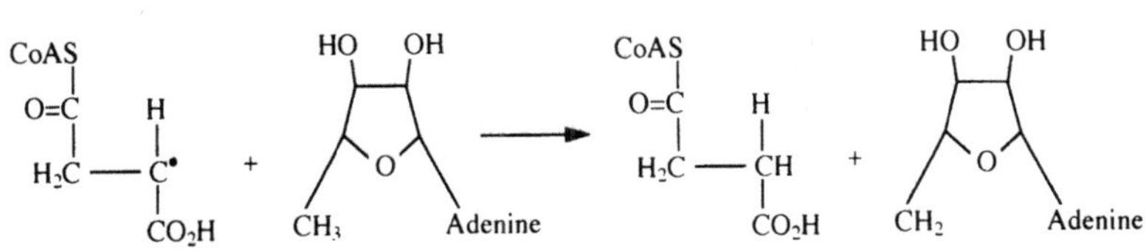

(9) (8) (7.44)

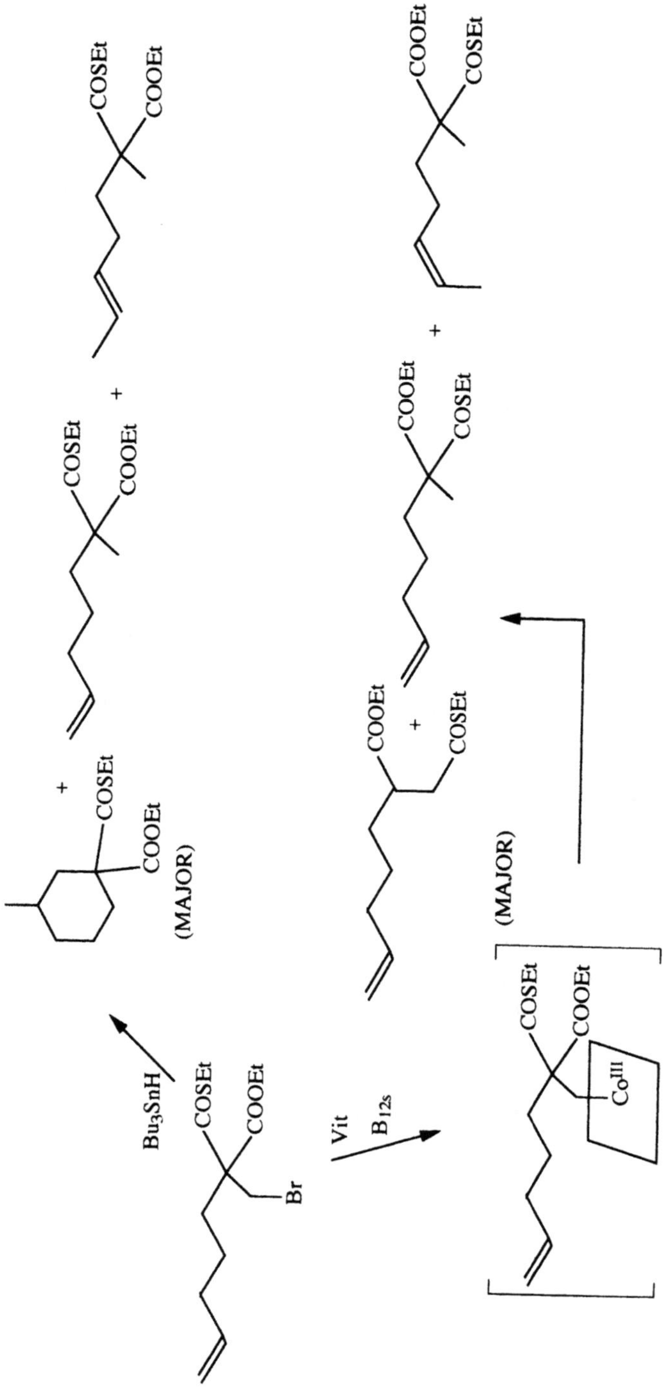

SCHEME 7.10.

through enormous acceleration of a given pathway. (The reader is reminded of the observation by Zhao *et al.* of ESR signals associated with the vitamin B_{12}-induced rearrangement of methylmalonyl-CoA [36] as well as related ESR work [41].) Rétey notes that in promoting 1,2-rearrangements in radicals, the enzyme may function through 'negative catalysis' [39] in the sense of inhibiting competing reactions (e.g. fragmentation or dimerization, which might be slow in a sterically encumbered environment) so that 1,2-rearrangement occurs as if by default. While this stimulating idea appears to be plausible, it is well to recognize that some of the rearrangements are, as noted previously, potentially homoallylic in character and, thus, intrinsically rapid. On the other hand, the sole rearrangement involving saturated carbon (the glutamate mutase reaction) should be exceedingly slow based upon the kinetic and thermodynamic arguments advanced in this chapter. If a free radical pathway *is* followed here, it might be explicable in terms of stabilizing association of the rearrangement transition state with the coenzyme, enzyme or both rather than 'passive victory over other decomposition pathways *via* default'. Recognizing the intrinsic high activation barrier for unassisted rearrangement of glutamate radical, Dowd has described a potential alternative pathway for glutamate rearrangement involving possible enzymatic conversion of the amino group into Schiff's bases and even ketones prior to putative homoallylic rearrangement and regeneration of the amino group [35].

In attempting to assess the mechanistic alternatives, it would be wonderful to better understand the nature of the vitamin B_{12} coenzyme binding sites in the various enzymes that use them as cofactors and also to 'see' the nature of the species in the various states of substrate, intermediate and product binding. In this regard, we note the recently published X-ray crystallographic structure of the 27-kilodalton methylcobalamin-containing fragment of methionine synthase from *Escherichia coli* [42]. Three points will be summarized here which relate to that study: (a) the surprising finding that the dimethylimidazole ligand, which is known to be attached to cobalt in free methylcobalamin, is displaced by a histidine residue on the protein to a hydrophobic pocket of the protein; (b) that the methyl group on the cofactor is extremely hindered, thus suggesting dramatic conformational changes during active chemistry; and (c) that the sequence containing the critical histidine residue is conserved in the adenosylcobalamin-dependent enzymes methylmalonyl-CoA mutase and glutamate mutase suggesting that these may share structural and mechanistic commonality with methionine synthase [42]. We also take note of the synthesis of molecular mimics designed to probe the structure of the post-homolysis intermediate formed through homolytic cleavage of the Co—C bond in adenosylcobalamin [43]. The study suggests a distance of 10 Å between the cobalt and 5′-deoxyadenosyl carbon paramagnetic centers following homolytic cleavage in the methylmalonyl-CoA-coenzyme B_{12} complex [43].

APPARENT 1,2-SHIFTS IN 1,3-DIRADICALS

One may very briefly note here that 1,3-diradicals *appear* to undergo 1,2-shifts of hydrogen. An example is depicted in equation (7.45), in which triplet CHT has attacked 2-butene [44].

$$\bullet CHT-\overset{\overset{CH_3}{|}}{C}H-CH\bullet-CH_3 \rightarrow CH_3-CHT-CH{=}CH-CH_3 \qquad (7.45)$$

The very strong thermodynamic driving force derives from the formation of an olefinic linkage although the nature of the conversion to the singlet is not obvious. Indeed, the simplest version of this is the apparent rearrangement of 1,3-propanediyl (trimethylene) to propene as a critical step in the thermal isomerization of cyclopropane to propene (equation (7.46)) [13]. This rearrangement presumably focuses on singlet species.

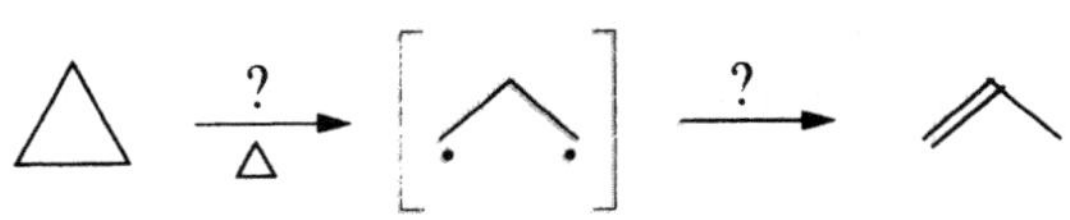

(7.46)

Benson [45] first estimated the enthalpy of formation of trimethylene at 67 kcal mol^{-1} or *c.* 280 kJ mol^{-1} (227 kJ mol^{-1} higher than cyclopropane and, thus, 260 kJ mol^{-1} higher than propene). This still placed trimethylene (of unknown electronic state) some 10 kcal mol^{-1} (42 kJ mol^{-1}) below the isomerization transition state. A wealth of elegant, challenging and subtle chemistry has been employed to investigate this deceptively simple looking system. We note a recently published, high-level *ab initio* study of the singlet trimethylene reaction surface [46]. At present, we simply note that the SD-CI calculations in that study indicated that the various singlet trimethylene minima are all around 250 kJ mol^{-1} higher in energy than cyclopropane (and therefore 285 kJ mol^{-1} higher than propene) [46]. While this value is 80 kJ mol^{-1} more exothermic than the cubylcarbinyl to 1-homocubyl radical rearrangement (equation (7.37)), the nature of the putative transition state (if indeed a diradical is formed *en route* between cyclopropane and propene) should be very different than that depicted in equation (7.27). A significant stabilizing interaction should occur between the two spins as they pair to form an olefinic linkage even as the transition state is early due to great exothermicity.

We close this chapter by noting the recent experimental observation of trimethylene and its associated lifetime on the order of 100 femtoseconds [47]. It is hard to imagine a discrete H-shift occurring on such a time scale.

Acknowledgements

We are pleased to acknowledge very helpful conversations and correspondence with Edward F. Hunter, Joseph J. Gajewski and Janos Rétey.

REFERENCES AND EXPLANATORY NOTES

1. Implicitly, all enthalpies of formation of closed shell, neutral organic compounds in this chapter refer to the gas phase and are taken from:
PEDLEY, J. B., NAYLOR, R. D. and KIRBY, S. P. (1986) *Thermochemical Data of Organic Compounds*, 2nd Edn., Chapman & Hall, London.
2. The enthalpies of formation of both 5-methylene-1,3-cyclohexadiene and 3-methylene-1,4-cyclohexadiene are found in:
(a) BARTMESS, J. E. (1982) *J. Am. Chem. Soc.*, **104**, 335.
(b) BARTMESS, J. E. and GRIFFITH, S. S. (1990) *J. Am. Chem. Soc.*, **112**, 2931.
3. Only the enthalpy of formation of 5-methylene-1,3-cyclohexadiene is found in:
BALLY, T., HASSELMANN, D. and LOOSEN, K. (1985) *Helv. Chim. Acta*, **68**, 345.
4. Admittedly, our preference is for the results in refs. [2] because reference 2(b) presents MNDO and molecular mechanical calculations that reproduce their purported enthalpy of formation difference.
5. (a) LIEBMAN, J. F. (1986) In *Molecular Structure and Energetics: Studies of Organic Molecules* (Vol. 3) Liebman, J. F. and Greenberg, A., Eds. VCH: Deerfield Beach, pp. 267–328.
(b) GEORGE, P., BOCK, C. W., and TRACHTMAN, M. (1987) in *Molecular Structure and Energetics: Biophysical Aspects* (Vol. 4); Liebman, J. F. and Greenberg, A., Eds. VCH: Deerfield Beach, pp. 163–187.
(c) LUO, Y.-R. and HOLMES, J. L. (1992) *J. Phys. Chem.*, **96**, 568.
6. For some qualitative relations and extensive numerical values of ionization potentials and proton affinities, the reader is referred to 'the GIANT tables':
LIAS, S. G., BARTMESS, J. E., LIEBMAN, J. F., HOLMES, J. L., LEVIN, R. D. and MALLARD, W. G. (1988) Gas-Phase Ion and Neutral Thermochemistry, *J. Phys. Chem. Ref. Data*, **17**, Supplement 1.
7. For this analysis, we accepted the recommendation of revised proton affinity values suggested by:
HUNTER, E. F., LIAS, S. G. and AUSLOOS, P. in a manuscript in preparation (also for *J. Phys. Chem. Ref. Data*) that will supplant the proton affinity tables used to generate numerous enthalpies of formation in ref. [6].
8. In fact, we have been 'too rigorously honest': there is the desired conceptual bridge between homolytic X-H bond enthalpies in neutral molecules and homolytic (X′) + -H

bond enthalpies in their isoelectronic cationic counterparts. For a given pair of neutral molecule and related cation, the latter homolytic bond enthalpy is generally larger; see:

STANTON, R. E. (1963) *J. Chem. Phys.*, **39**, 2368 and LIEBMAN J. F., in *Molecular Structure and Energetics: Biophysical Aspects* (Vol. 4), Liebman, J. F. and Greenberg, A., Eds. VCH: Deerfield Beach, pp. 49–70.

9. KERR, J. A. and LLOYD, A. C. (1968) *Quart. Rev. (London)*, **22,** 549.
10. KERR, J. A. (1973) in *Free Radicals*, Kochi, J. K. (ed), Vol. 1, Wiley, New York, pp. 1–36.
11. WILT, J. W. (1973) in *Free Radicals*, Kochi, J. K. (ed), Vol. 1, Wiley, New York, pp. 333–501.
12. BECKWITH, A. L. J. and INGOLD, K. U. (1980) in *Rearrangements in Ground and Excited States*, Vol. 1, de Mayo, P. (ed), Academic Press, New York, pp. 161–310.
13. GAJEWSKI, J. J. (1981) *Hydrocarbon Thermal Isomerizations*, Academic Press, New York, pp. 35–41.
14. LINDSAY, D. A., LUSZTYK, J. and INGOLD, K. U. (1984) *J. Am. Chem. Soc.*, **106,** 7087.
15. MELIUS, C. F., Personal Communication.
16. PACANSKY, J., WALTMAN, R. J. and BARNES, L. A. (1993) *J. Phys. Chem.*, **97,** 10694.
17. HEHRE, W. J., RADOM, L., SCHLEYER, P. v. R and POPLE, J. A. (1986) *Ab Initio Molecular Orbital Theory*, Wiley, New York, pp. 383–385.
18. Estimated $\Delta H_f[C_6H_5C(CH_3)_2\bullet] = \Delta H_f[C_6H_5C(CH_3)_3] + \Delta H_f[(CH_3)_2CHCH_2\bullet] - \Delta H_f[CH(CH_3)_3]$.
19. Estimated $\Delta H_f[C_6H_5C(CH_3)C_2H_5)\bullet] = \Delta H_f[C_6H_5C(CH_3)_2\bullet] + \Delta H_f[C_2H_5C(CH_3)_2\bullet] - \Delta H_f[(CH_3)_3C\bullet]$
($\Delta H_f[(CH_3)_3C\bullet]$ is obtained from W. Tsang, Chapter 2, this book.)
20. Estimated $\Delta H_f[C_6H_5CH_2C(CH_3)_2\bullet] = \Delta H_f[(CH_3)_3C\bullet] + \Delta H_f[C_6H_5CH_2CH(CH_3)_2] - \Delta H_f[(CH_3)_3CH]$
($\Delta H_f[(CH_3)_3C\bullet]$ is obtained from W. Tsang, Chapter 2, this book.)
21. All values estimated using the methods outlined in Refs [18–20] and data from Ref. [6].
22. ΔH_f[cyclobutylcarbinyl] = ΔH_f[methylcyclobutane] + $\Delta H_f[(CH_3)_2CHCH_2\bullet] - \Delta H_f[(CH_3)_3CH]$.
23. KAPLAN, L. (1968) *J. Org. Chem.*, **33,** 2531.
24. WILT, J. W., MARAVETZ, L. L. and ZAWADZKI, J. F. (1966) *J. Org. Chem.*, **31,** 3108.
25. Estimated in a manner analogous to that in Ref. [22].
26. EATON, P. E. and YIP, Y. C. (1991) *J. Am. Chem. Soc.*, **113,** 7692.
27. ΔH_f[cubylcarbinyl] = ΔH_f[cubane] + $\Delta H_f[(CH_3)_3CCH_2\bullet]$ $\Delta H_f[(CH_3)_3CH]$
28. From the experimental strain energy of 490 kJ mol^{-1} (117 kcal mol^{-1}) for homocubane (see BECKHAUS, H. D., RÜCHARDT, C. and SMISEK, M. (1984) *Thermochim. Acta*, **79,** 149)

added to the sum of Benson Group Increments (see BENSON, S. W. (1970) *Thermochemical Kinetics*, 2nd Edn., Wiley, New York).

29. DELLA, E. W., HEAD, N. J., JANOWSKI, W. K. and SCHIESSER, C. H. (1993) *J. Org. Chem.*, **58,** 7876.
30. This is calculated analogously to the cubylcarbinyl radical (see ref. [27]).
31. This is calculated analogously to the 1-homocubyl radical (see ref. [28] with the addition of 80.3 kJ mol^{-1} (19.2 kcal mol^{-1}) for extra strain in the carbocation (see ref. [29]).
32. SHAIK, S., SCHLEGEL, H. B. and WOLFE, S. (1992) *Theoretical Aspects of Physical Organic Chemistry: The S_N^2 Mechanism*, Wiley, New York, pp. 33–37.
33. ROSSI, M. and GLUSKER, J. P. (1988) in *Environmental Influences and Recognition in Enzyme Chemistry*, Liebman, J. F. and Greenberg, A. (eds.), VCH Pub., New York, pp. 1–58.
34. ABELES, R. H., FREY, P. A. and JENCKS, W. P. (1992) *Biochemistry*, Jones and Bartlett Pub., Boston, pp. 554–559.
35. DOWD, P. (1990) in *Selective Hydrocarbon Activation: Principles and Progress*, Davies, J. A., Watson, P. L., Liebman, J. F. and Greenberg, A. (eds.), VCH Pub., New York, pp. 265–303.
36. ZHAO, Y., SUCH, P. and RÉTEY, J. (1992) *Angew. Chem. Int. Ed. Engl.*, **31,** 215.
37. RUSSELL, J. J., RZEPA, H. S. and WIDDOWSON, D. A. (1983) *J. Chem. Soc. Chem. Commun.* 625.
38. AFFIO, A., GRILLER, D., INGOLD, K. U., BECKWITH, A. L. J. and SERELIS, A. K. (1980) *J. Am. Chem. Soc.*, **102,** 1734.
39. RÉTEY, J. (1990) *Angew. Chem. Int. Ed. Engl.*, **29,** 355.
40. POPPE, L., HULL, W. E. and RÉTEY, J. (1993) *Helv. Chim. Acta*, **76,** 2367.
41. ZHAO, Y., ABEND, A., KUNZ, M., SUCH, P. and RÉTEY, J. (1994) *Eur. J. Biochem.*, **225,** 891.
42. DRENNAN, C. L., HUANG, S., DRUMMOND, J. T., MATHEWS, R. G. and LUDWIG, M. L. (1994) *Science*, **266,** 1669. (See also Perspective by STUBBE, J. (1994) *Science*, **266,** 1663.)
43. POPPE, L. and RÉTEY, J. (1995) *Archiv. Biochem. Biophys.*, **316,** 541.
44. MCKNIGHT, E. S. and ROWLAND, F. S. (1966) *J. Am. Chem. Soc.* **88,,** 3179.
45. BENSON, S. W. (1976) *Thermochemical Kinetics*, 2nd Edn., Wiley, New York, pp. 117–128.
46. GETTY, S. J., DAVIDSON, E. R. and BORDEN, W. T. (1992) *J. Am. Chem. Soc.*, **114,** 2085.
47. PEDERSON, S., HEREK, J. L. and ZEWAIL, A. H. (1994) *Science*, **266,** 1359. (See also Perspective by BERSON, J. A. (1994) *Science*, **266,** 1338).

8

Solvent Effects in the Reactions of Neutral Free Radicals*

James M. Tanko and N. Kamrudin Suleman

INTRODUCTION

It is a commonly held misconception that reactions of neutral free radicals are insensitive to the effect of solvent. Indeed, while it is true that in general the rate and selectivity of radical reactions are only nominally influenced by solvent in comparison to reactions of polar species (i.e., nucleophiles and electrophiles), there are several examples in the free radical literature where solvent profoundly affects the outcome of a process. Moreover, because of the extraordinary reactivity of many free radicals, solvent can influence the outcome of radical reactions in ways not typically observed in reactions of nucleophiles and electrophiles.

In general, reaction rate will vary with solvent any time a change of solvent affects the Gibbs energy of the reactant(s) and transition state differently (Figure 8.1). Consequently, any discourse on solvent effects must begin with a discussion of solute–solvent interactions and an interpretation of solvent effects on the basis of appropriate linear free energy relationships and solvent parameters.

As is the case for polar processes, solvent polarity may be a factor affecting the reactivity of a free radical. However, for neutral free radicals, other solvent properties may prove more important in determining the outcome of a reaction: Through its *internal pressure*, solvent may impede (or promote) a process where the volumes of the reactants and

* Dedicated to Prof. Glen A. Russell on the occasion of his seventieth birthday.

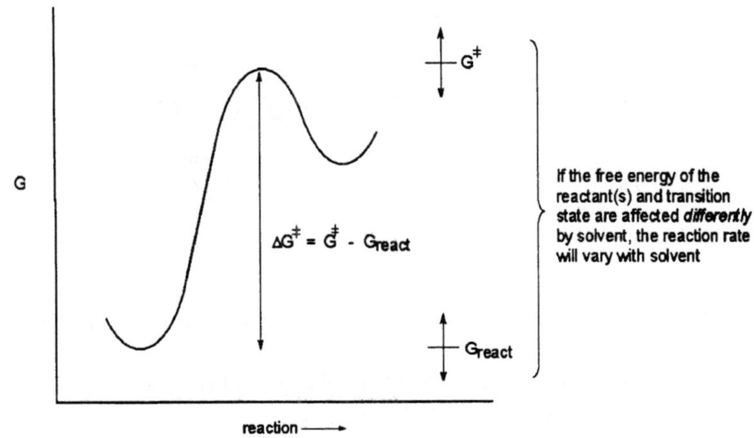

FIGURE 8.1 Solvent effects on reaction rate.

the transition state are different. The radical may chemically interact with the solvent so as to generate a new chemical species (radicaloid). Through its viscosity, the solvent may hinder the approach of two species participating in a bimolecular reaction (i.e., *diffusion control*), or perhaps hinder the separation of two reactive species resulting in *cage-effects*. The theory and mathematical basis for each of these effects is briefly presented in the following sections.

Solvent Polarity Effects

Solvent polarity will influence the rate of a chemical reaction if the dipole moment of the reactants and transition state are different. Although solvent polarity effects are much more common in polar processes, there have been several examples where polarity also has been shown to influence outcome of reactions involving neutral free radicals.

In terms of a quantitative measure of solvent polarity, by far the most common is dielectric constant (ε). For two oppositely charged particles in a medium, the potential energy of attraction between the particles (U) is attenuated by the dielectric constant of the medium according to equation (8.1), where ε_o is the permittivity of vacuum, q^+ and q^- are the charges of the particles, and r is the distance separating them.

$$U = -\left(\frac{1}{4\pi\varepsilon_o}\right)\frac{q^+q^-}{\varepsilon r} \tag{8.1}$$

This attenuation arises because the dipoles of the solvent molecules align so as to shield the oppositely charged particles from each other.

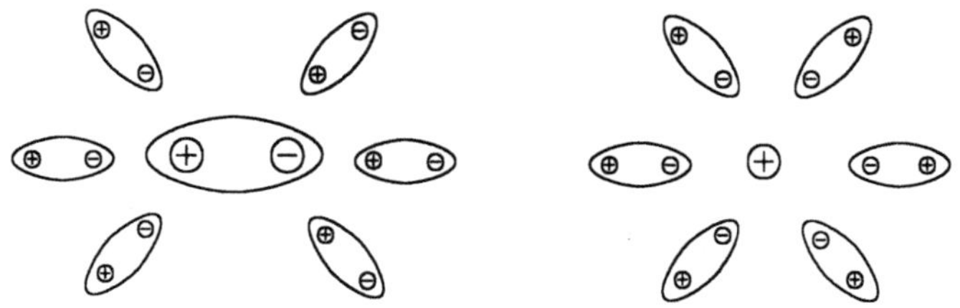

FIGURE 8.2 Solvation of (a) a polar molecule and (b) an ion in a dipolar solvent.

Presumably ε provides a measure of a solvent's ability to stabilize either a polar molecule or ion (Figure 8.2).

The theory pertaining to the influence of dielectric constant on reaction rate has been developed by Kirkwood [1,2]. The Gibbs energy of solvation of a polar molecule in a polar solvent is expressed by equation (8.2) where N is Avogadro's number, μ and r refer to the dipole moment and radius, respectively, of the dissolved species, and ε is the solvent dielectric constant. The dielectric function $(\varepsilon - 1)/(2\varepsilon + 1)$ is generally referred to as the Kirkwood function.

$$\Delta G = \frac{-N\mu^2}{4\pi\varepsilon_o r^3} \frac{(\varepsilon - 1)}{(2\varepsilon + 1)} \tag{8.2}$$

Accordingly, for the reaction A + B → transition state → products, the variation of rate constant with solvent is expressed by equation (8.3) where the μ and r terms refer to the dipole moments and radii of the reacting species (A and B) and the transition state (ts). A linear correlation between $\ln(k)$ and the Kirkwood function $(\varepsilon - 1)/(2\varepsilon + 1)$ is anticipated if solvent polarity is important [1,2]. (Similarly, the correlation between $\ln(k)$ and $1/\varepsilon$ is linear because $(\varepsilon - 1)/(2\varepsilon + 1)$ is linearly related to $1/\varepsilon$ for most solvents [3].)

$$\ln(k) = \ln(k_o) - \frac{1}{4\pi\varepsilon_o} \frac{N}{RT} \frac{(\varepsilon - 1)}{(2\varepsilon + 1)} \left\{ \frac{\mu_A^2}{r_A^3} + \frac{\mu_B^2}{r_B^3} - \frac{\mu_{ts}^2}{r_{ts}^3} \right\} \tag{8.3}$$

The major drawback of this treatment is that because dielectric constant is a bulk property, it neglects specific solvent–solute interactions that might be occurring on a molecular level (e.g. hydrogen bonding). A 'molecular level' solvent polarity scale, based upon the solvent-dependent UV/VIS absorption of (**1**) is the Dimroth E_T scale, where $E_T = hc/\lambda_{max}$. (E_T ranges from 30.9 kcal/mol in hexane to 63.1 kcal/mol in water) [4]. Other solvent polarity scales common for polar reactions such as Winstein's Y-scale [5] are generally not used to correlate radical reactions, mostly for practical reasons: Insolubility of neutral radical precursors in protic solvents and the incompatibility of alcohol solvents (which are potential hydrogen atom donors) with many free radicals.

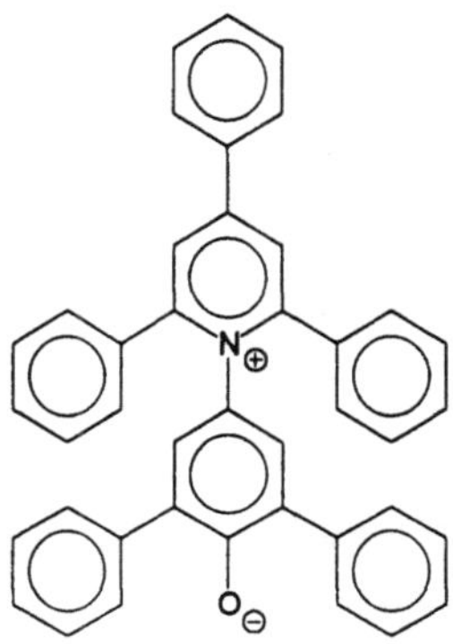

(1)

The unique spectroscopic properties of free radicals themselves can also be used to evaluate solvent polarity at a molecular level. Nitroxides (**2**) are stable and persistent free radicals, readily characterized by ESR spectroscopy. Because of the electronegativity difference between nitrogen and oxygen, the N—O bond in these radicals is polarized, as illustrated by resonance form (**3**). The relative contribution of dipolar structure (**3**) to the resonance hybrid increases with increasing solvent polarity, resulting in an increased spin-density at nitrogen that is reflected in the nitrogen hyperfine coupling constant a_N [6]. Consequently, several studies have examined the relation between the rate of a radical reaction in a given solvent and the value of a_N in the same solvent in an attempt to assess the effect of solvent polarity on a molecular level (see below).

(**2**) (**3**)

Solvent (Internal) Pressure Effects

Internal pressure (P_i) is the 'pressure' exerted by a solvent on a solute, and can be qualitatively (and quantitatively) understood on the basis of the *Cavity Model.* Imagine a pure liquid, held together by van der Waals interactions as the sole intermolecular force of attraction (Figure 8.3(a)). In order to accomodate a solute S, the solvent molecules must be 'pulled

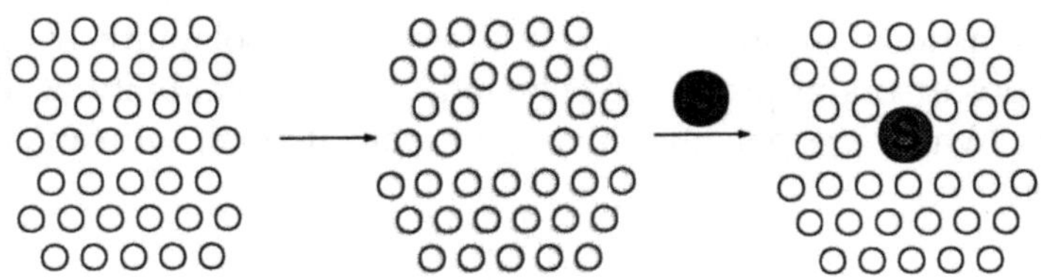

FIGURE 8.3 The cavity model. (a) Pure liquid solvent; (b) cavity (hole) created in solvent; (c) solute in cavity.

apart' in order to create a hole (cavity) for the solute (Figure 8.3(b)). Once placed in the cavity, the solvent 'snaps back' surrounding and exerting pressure on the solute (Figure 8.3(c)).

Formally, the *internal pressure* of a liquid is expressed as the first derivative of energy with respect to volume ($P_1 = (\delta E/\delta V)_T$ [7]. A related quantity is the *cohesive energy density* (ced) which is related to the heat of vaporization and molar volume of the liquid as shown in equation (8.4) [8]. The cohesive energy density is related to the *Hildebrand solubility parameter* (δ): $\delta = \text{ced}^{1/2}$.

$$\text{ced} = \delta^2 = \frac{\Delta E_{\text{vap}}}{V_{\text{m}}} = \frac{\Delta H_{\text{vap}} - RT}{V_{\text{m}}} \tag{8.4}$$

A greater number of δ values are available [8] for solvents in comparison to a rather limited list of available P_is [9]. In the absense of strong intermolecular interactions (e.g. hydrogen bonding), δ^2 and P_i are approximately equal.

Internal pressure is expected to affect reaction rate if the *volume* of the reactants and the transition state are different. For the reaction A + B → transition state → product, the mathematical relationship between the rate constant and δ is presented in equation (8.5) where δs refer to the Hildebrand parameters of the reactants (A and B), solvent (S), and transition state (ts), and Vs refer to the molar volumes of A, B, and ts [2, 10–13].

$$\ln(k) = \ln(k_o) + 1/(RT)\,[V_A(\delta_A - \delta_S)^2 + V_B(\delta_B - \delta_S)^2 - V_{ts}(\delta_{ts} - \delta_S)^2] \tag{8.5}$$

Assuming V_A, V_B, V_{ts}, δ_A, and δ_B do not vary with solvent, differentiation of equation (8.5) with respect to δ_S^2 yields equation (8.6), where $\Delta V_{\text{act}} = V_{ts} - V_A - V_B$ is defined as the *volume of activation*. It is noteworthy that equation (8.6), which describes the effect of *internal pressure* on reaction rate, is similar in form to that which describes the effect of *external pressure* (equation (8.7)) [14]. In order to test for a dependance on internal pressure, ln(k) is typically plotted against δ_S^2 or P_i in accordance with equation (8.6).

$$\frac{\delta \ln(k)}{\delta(\delta_S^2)} = \frac{-\Delta V_{act}}{RT} - \frac{(\delta_A V_A + \delta_B V_B - \delta_{ts} V_{ts})}{RT\delta_S} \tag{8.6}$$

$$\frac{\delta \ln(k)}{\delta P} = \frac{-\Delta V_{act}}{RT} \tag{8.7}$$

Several important assumptions are associated with this model of which the most important is that this theory applies to regular solutions (i.e. there is a random distribution of solute and solvent molecules). Specific orientation-dependent interactions (ion–dipole, dipole–dipole, hydrogen bonding) are not accounted for. Moreover, the assumption that V_A, V_B, and V_{ts} are solvent-independent is dubious in reactions involving charged or highly polar species. As a consequence, this model is best applied to reactions of non-polar species in relatively non-polar solvents.

Solvent Viscosity Effects

The potential effect of solvent viscosity (η) on reaction rate/selectivity does not arise because of differential solvation of the reactant(s) and transition state per se. Rather, viscosity may exert an effect by hindering the mobility of solutes in solution. In radical chemistry, the effect of this hindered mobility on reactivity/selectivity is manifested in two ways.

The rates of bimolecular reaction between two extremely reactive species in solution may be limited by the rate at which the two species diffuse together. Consider the bimolecular reaction of A and B, which diffuse together forming a *diffusive* caged-pair $(A/B)_{cage}$ from which A and B subsequently react to form product P (Scheme 8.1). The rate constants k_{diff} and k_{-diff} refer to diffusion of A and B in and out of the cage, respectively; k_o is the intrinsic rate constant of the reaction.

A simple steady-state treatment reveals that the observed rate constant for the reaction (k_{obs}) is expressed by equation (8.8).

SCHEME 8.1

$$k_{obs} = \frac{k_{diff}\, k_o}{k_{diff} + k_o} \tag{8.8}$$

If A and B are so reactive that every encounter leads to reaction ($k_o > k_{-diff}$), then $k_{obs} \approx k_{diff}$. Such reactions are referred to as *diffusion-controlled* because the reaction rate constant is governed solely by diffusion.

The Smoluchowski equation (equation (8.9)) shows that the diffusion-controlled rate constant (k_{diff}) dependent on the diffusion coefficient (D) of the reacting species and the reaction distance (R_{AB}) .

$$k_{diff} = \frac{4\pi\, N\,(D_A + D_B)\, R_{AB}}{1000}\ (M^{-1}\ s^{-1}) \tag{8.9}$$

The Stokes–Einstein equation reveals that the diffusion coefficient is inversely related to viscosity: $D = kT/6\pi r\eta$, where k is the Boltzmann constant, T is temperature in Kelvin, and r is the radius of the solute [3]. Thus to a first approximation k_{diff} is inversely related to solvent viscosity (η).

The other way solvent viscosity can alter the outcome of a chemical process is by hindering the diffusion of two reactive species away from each other. Again consider two reactive species A and B, but instead of having to diffuse together in order to react, imagine that they are generated simultaneously in a solvent cage as a *geminate* caged-pair from a common precursor molecule M (Scheme 8.2). Increased solvent viscosity should favor production of product P ($k_o > k_{-diff}$), while different products arising from individual reactions of free A and free B are formed at lower viscosities ($k_{-diff} > k_o$).

Disclaimer

The aim of this chapter is to provide the reader with several examples where each of these types of phenomena (polarity, internal pressure, viscosity, and direct reaction/complex formation) have been shown to

SCHEME 8.2

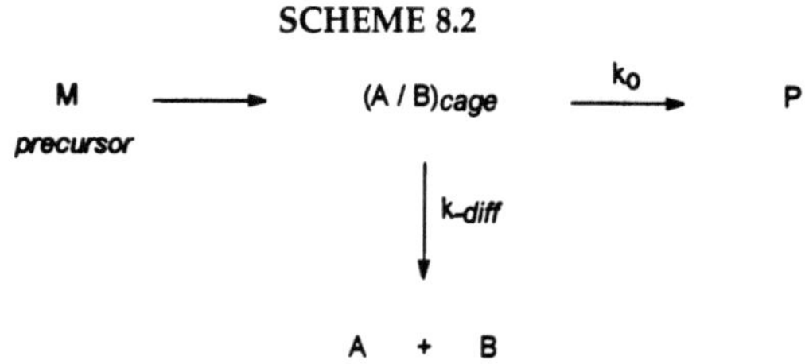

influence the rate and selectivity of reactions involving neutral free radicals. This overview is intended to be selective rather than exhaustive. The topics discussed in this chapter have been selected because they provide reasonably unambiguous illustrations of a particular phenomenon or effect.

SOLVENT POLARITY EFFECTS

β-Scission Reactions of Alkoxyl Radicals

t-Butyl hypochlorite (*t*-BuOCl) is a common reagent for the free radical chlorination of alkanes. The reaction proceeds via a chain mechanism (Scheme 8.3) and is initiated either photochemically or thermally using a free radical initiator [15].

The first propagation step in Scheme 8.3 involves the abstraction of a hydrogen atom by the *t*-butoxyl radical from the alkane substrate. However, *t*-butoxyl radical also undergoes a competing unimolecular bond cleavage to yield acetone and a methyl radical (equation (8.10)). This interfering reaction is referred to as β-scission because the C—C bond that ruptures is located β to the radical center. The occurrence of this competing β-scission process results in diminished yields of the desired RCl. (In addition to acetone, CH_3Cl is produced as a by-product.)

$$CH_3-\underset{CH_3}{\overset{CH_3}{\underset{|}{\overset{|}{C_\beta}}}}-O_\alpha\bullet \xrightarrow{k\beta} \underset{CH_3}{\overset{CH_3}{>}}C{=}O + CH_3\bullet \qquad (8.10)$$

In the early 1960s, Walling and Wagner [16] studied the effect of

SCHEME 8.3

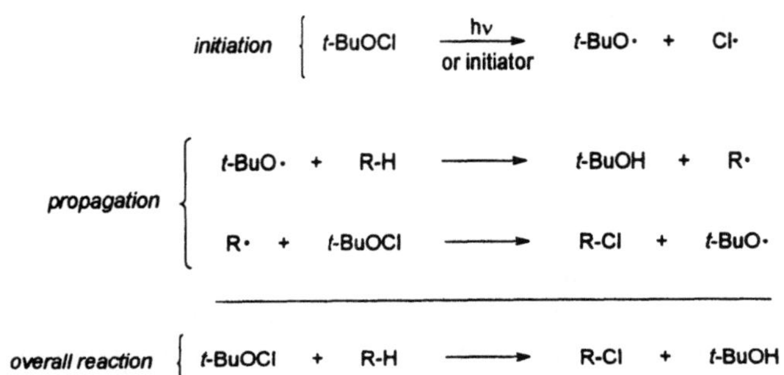

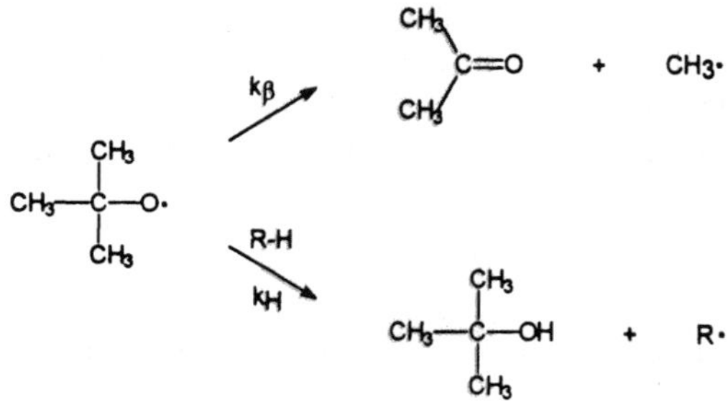

SCHEME 8.4

solvent on the partitioning of *t*-butoxyl radical between hydrogen abstraction and β-scission pathways (Scheme 8.4). Experimentally, the rate constant ratio k_H/k_β is readily obtained by GC analysis of the product mixture and application of equation (8.11).

$$\frac{k_H}{k_\beta} = \frac{\text{Yield } t\text{-BuOH}}{\text{Yield } CH_3C(O)CH_3} \frac{1}{[RH]} \quad (8.11)$$

The relative yields of β-scission and hydrogen abstraction products varied as a function of solvent with k_H/k_β ranging from 5.5 in CF_2ClCCl_2F to 1.78 in CH_3CN to 0.70 in CH_3CO_2H (with cyclohexane as the hydrogen atom source) [16]. Although in principle this decrease in k_H/k_β, with increasing solvent polarity could be linked to a variation in the rate constants of either (or both) of the processes, the hydrogen abstraction reaction (k_H) was thought to be relatively insensitive to solvent effects, and the observed trend was ascribed to an increase in k_β [16]. This conclusion was based on the premise that the C=O of the product is highly polar and much of this polarity is expected to be present in the transition state of the β-scission process (equation (8.12)). Thus, an increase in solvent polarity would cause an increase in solvent stabilization of the developing carbonyl in the transition state, resulting in an enhancement of k_β in polar solvents.

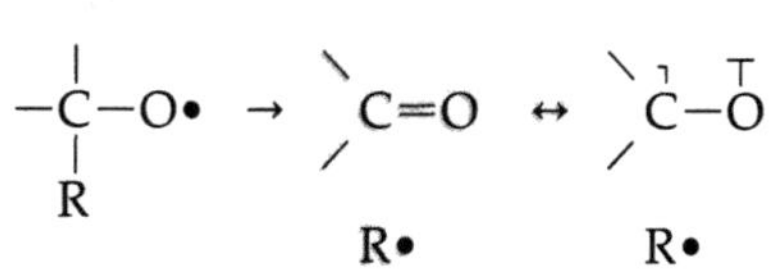

(8.12)

This hypothesis advanced by Walling has recently been substantiated by Lusztyk *et al.* via laser flash photolysis (LFP) and product studies on the cumyloxyl radical (5). Cumyloxyl radical was generated by the photolysis of dicumylperoxide (**4**) at 308 nm (equation (8.13)). (Cumyloxyl radical shows absorbances at 485 and 320 nm.) This study clearly demonstrated that, for cumyloxyl radical, the absolute rate constant for hydrogen abstraction from alkanes (k_H) is independent of solvent (Table 8.1) [17].

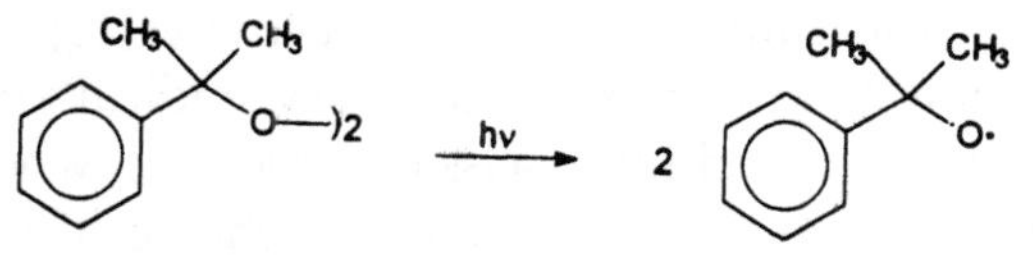

(8.13)

(**4**) (**5**)

In contrast, the absolute rate constant for β-scission of cumyloxyl radical varied as a function of solvent polarity. A good correlation between $\log(k_\beta)$ and the hyperfine coupling constant (a_N) for 4-amino-2,2,6,6-tetramethylpiperdinyl-*N*-oxyl radical (4-NH_2-TEMPO, (**6**) is found (Figure 8.4) [17]. (As noted in the introduction, a_N is a sensitive 'molecular level' probe of solvent polarity). An analogous plot of $\log(k_\beta)$ versus the Dimroth–Reichardt E_T [30] solvent polarity parameter was also linear [17].

These results are consistent with Walling's suggestion that for β-scission of alkoxyl radicals, there is an increase in the dipole moment of the C—O bond in the progression from reactant to transition state. (The increase rate in polar solvents can be attributed to increased solvent stabilization of the developing carbonyl moiety in the transition state.)

Recently, an alternative interpretation involving an equilibrium between solvated and unsolvated radicals was proposed to explain the

TABLE 8.1 Effect of solvent on the rate constant for hydrogen abstraction from cyclohexane by cumyloxyl radical at 30°C[a]

Solvent	$k_H \times 10^{-6}$, $M^{-1}s^{-1}$
CCl_4	1.1
C_6H_6	1.2
C_6H_5Cl	1.1
$(CH_3)_3COH$	1.3
CH_3CN	1.2
CH_3CO_2H	1.3

[a] All data from reference 17.

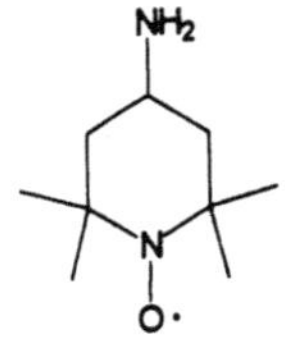

(6)

observed trends in cumyloxyl radical reactivity [18]. According to this proposition, only the unsolvated cumyloxyl radical undergoes both the hydrogen abstraction and the β-scission processes, whereas the solvated species reacts exclusively via β-scission (Scheme 8.5). The equilibrium between these two radicals would thus be affected by solvent leading to 'important solvent effects on the hydrogen abstraction reactions' [18].

This proposition was based upon the observation that the substituent effect for benzyl-substituted cumyloxyl radicals was unaffected by solvent. Specifically, the rate constant ratio k_β^X/k_β^H for the competitive β-scission of benzyl methyl substituted-benzyl carbinyloxyl radicals

FIGURE 8.4 Correlation between log(k_β) for cumyloxy radical and a_N for 4-amino-2,2,6,6-tetramethylpiperidinyl-*N*-oxide (data from reference 17).

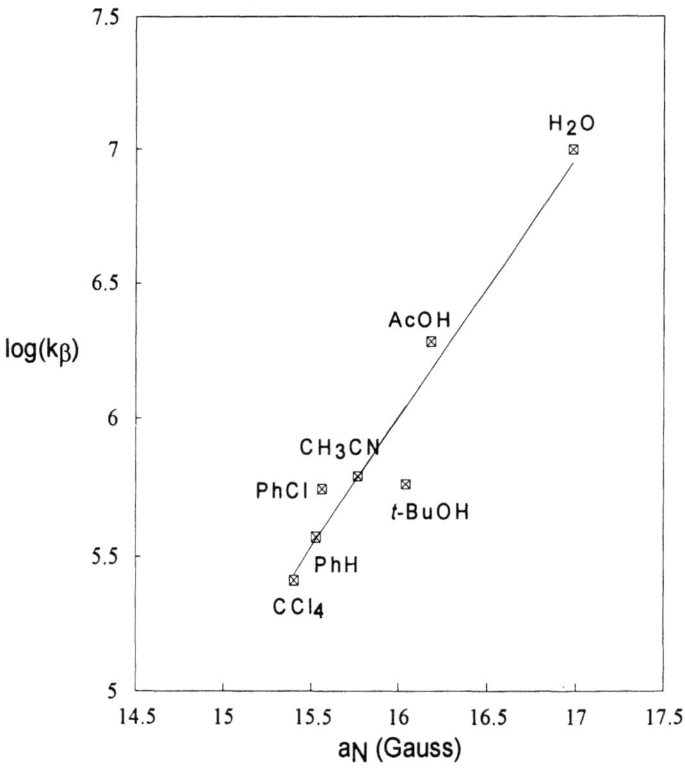

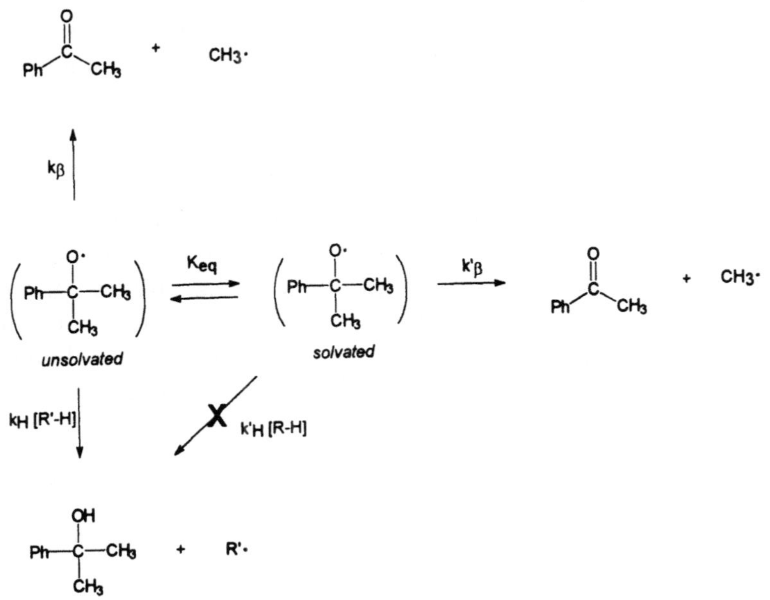

SCHEME 8.5

(Scheme 8.6) could be correlated in accordance with the Hammett equation to the σ^+ substituent constant ($\rho^+ = -0.89$) [18]. The intermediacy of two alkoxyl radicals (solvated and unsolvated) was suggested because it was found that ρ^+ did not vary with solvent.

In our view, however, the solvent polarity effect on k_β, which results primarily from dipolar stabilization of the developing carbonyl, should affect both k_β^X and k_β^H to the same extent because the substituent X cannot

SCHEME 8.6

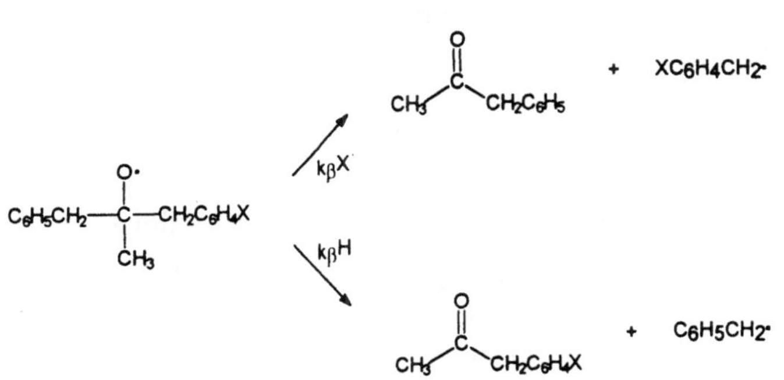

interact directly with developing the C=O bond (i.e. the substituent is electronically insulated from the reaction center). Consequently, there does not appear to be any compelling evidence to support this 'two intermediate' proposal.

Decarbonylation of Acyl Radicals

Acyl radicals ((**8**), R = alkyl or hydrogen) can be conveniently generated by the photolysis of aldehydes and ketones, which absorb light in the 230 to 330 nm region. An $n \rightarrow \pi^*$ transition of the carbonyl group results from the absorption of light in this region, and the excited molecules then produce acyl radicals by undergoing a unimolecular bond scission, known as a Norrish type I cleavage (equation (8.14)) [19].

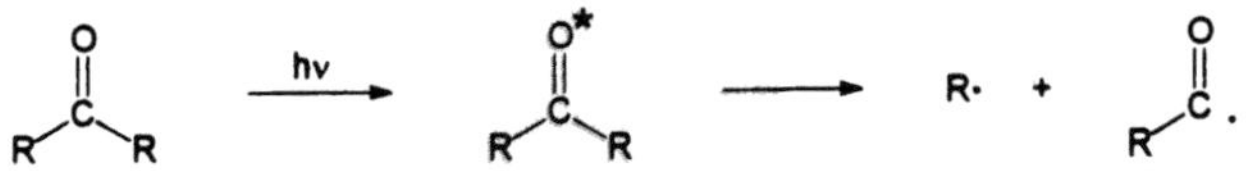

(8.14)

(**7**) (**8**)

Acyl radicals are also useful synthetic intermediates for carbon–carbon bond formation (e.g., Scheme 8.7) [20].

Recently, using laser flash photolysis, Tsentalovich and Fischer examined the effect of solvent on the decarbonylation rate of phenylacetyl radical (**9**) and pivaloyl radical ((**10**), Scheme 8.8). Unlike β-scission of alkoxyl radicals (wherein the dipole moment increases in the progression from reactant → transition state → products), for the decarbonylation of acyl radicals there is a net *decrease* in dipole moment along the reaction coordinate. (Alkyl radicals are non-polar; the dipole moment of carbon monoxide is only 0.1 Debye.) Consequently, the rate constant for decarbonylation decreases with increasing solvent polarity. (The rate constant decreased by a factor of approximately four in going from hexane to acetonitrile) [21,22].

These results were treated in accordance with the Kirkwood formula (equation (8.3)) using the rate constants found for alkane solvents ($\varepsilon = 2$) and acetonitrile ($\varepsilon = 36$). The experimentally calculated value of the quantity $\mu_R^2/r_R^3 - \mu_{ts}^2/r_{ts}^3$ agreed well with that estimated using semi-empirical molecular orbital theory (AM1).

Nitroxide Radical Trapping Reactions

Dialkyl nitroxides have the general structure (RR′NO•). Di-*t*-butyl nitroxide (t-Bu_2NO•) is an especially stable member of this class of free radicals, and has found extensive applications in chemistry. A popular

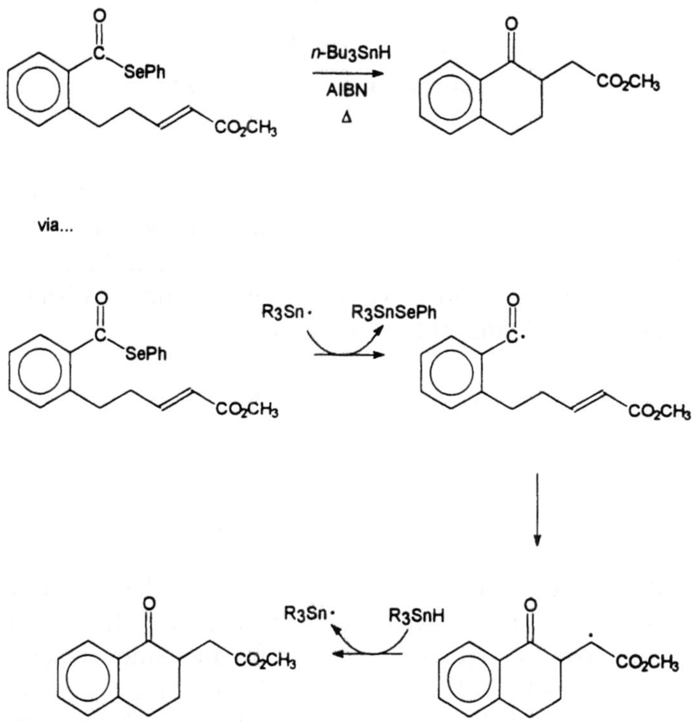

SCHEME 8.7

use of nitroxides is as 'spin labels' in electron spin resonance spectroscopy. In this technique, the stable nitroxide moiety is incorporated into the molecule or system of interest. The persistent ESR signal from the nitroxide then provides information on its local environment by means of the hyperfine splitting imparted by the interaction of the unpaired electron and the nearby nitrogen (and possibly other nuclei).

Another common use of nitroxides is as a 'free radical trap'. In this case, the nitroxide combines rapidly with any other free radicals in solution, thereby reducing or eliminating entirely reactions that may

SCHEME 8.8

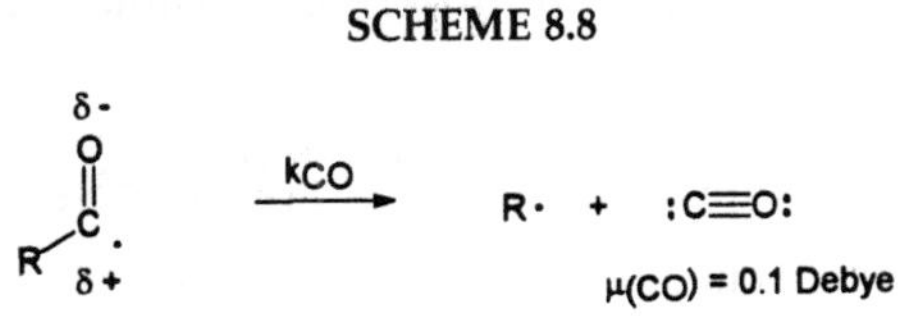

9, R = $PhCH_2$
10, R = $(CH_3)_3C$

proceed via a free radical pathway (equation (8.15)).

$$\begin{matrix} R \\ & \diagdown \\ & & N{-}O\bullet + R'\bullet \xrightarrow{k_T} \\ & \diagup \\ R \end{matrix} \quad \begin{matrix} R \\ & \diagdown \\ & & N{-}OR' \\ & \diagup \\ R \end{matrix} \qquad (8.15)$$

A recent study by Beckwith and co-workers investigated the effects of solvent on the rate constant (k_T) for reaction between carbon-centered and nitroxide radicals [23]. The kinetics of alkyl radical trapping of 2,2,6,6-tetramethylpiperidin-1-oxyl(TEMPO), (**11**) and 1,1,3,3-tetramethylisoindoline-2-oxyl (TMIO, (**12**) were examined.

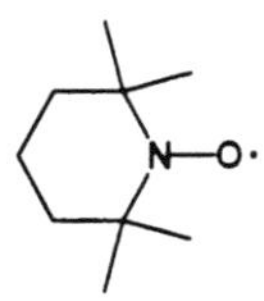

(**11**)

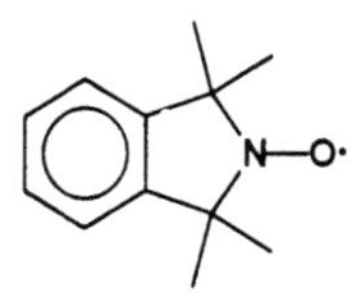

(**12**)

The authors used a combination of the 'radical clock' and laser flash photolysis techniques. The 'radical clock' method [24] is an approach where a unimolecular rearrangement or fragmentation reaction of the carbon-centered radical (for which the rate constant $k_{'clock'} \equiv k_C$ is well established) is allowed to compete with the bimolecular process of interest, in this instance the cross-coupling reaction involving the nitroxide. (Figure 8.5 shows two typical free radical clocks, and their application to the nitroxide trapping reaction). Because k_C is known, the desired rate constant k_T is simply obtained by determination of the product yields in accordance with equation (8.16).

$$\frac{k_T}{k_C} = \frac{\text{Yield } UT}{\text{Yield } RT} \frac{1}{[R_2NO\bullet]} \qquad (8.16)$$

The rate constants for nitroxide radical trapping were examined in 32 solvents, ranging from alkanes to aqueous methanol. Values of k_T/k_C for reaction of primary radical (**13**) with 1,1,3,3-tetramethylisoindoline-2-oxyl (TMIO) in several solvents are summarized in Table 8.2.

(**13**) (**14**)

The limitation of the radical clock approach is that it yields only the

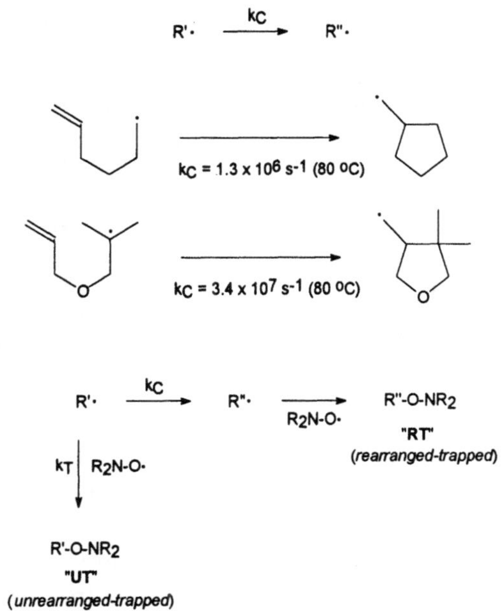

FIGURE 8.5 Illustration of the application of the radical clock method to determine K_T.

relative rate constant ratio k_T/k_C. Thus, it is unclear whether the apparent decrease in k_T/k_C with increasing solvent polarity (Table 8.2) has its origins in the radical coupling reaction, or in the clocking reaction, or both. This issue was resolved with laser flash photolysis experiments, which provided the *absolute* rate constants for the trapping reaction of the nitroxide radical TEMPO with benzyl, *n*-nonyl and neopentyl radicals in

TABLE 8.2 Effect of solvent on k_T/k_C using **13** as a radical clock and 1,1,3,3-tetramethylisoindoline-2-oxyl (TMIO) as a radical trap at 80°C[a]

Solvent	k_T/k_C (M^{-1})
n-pentane	33.1
cyclohexane	28.5
benzene	16.0
chlorobenzene	14.4
tetrahydrofuran	15.1
acetone	14.5
acetonitrile	8.1

[a] All data from reference 23.

TABLE 8.3 Effect of representative solvents on the absolute rate constant (k_T) for TEMPO trapping of benzyl radical measured by laser flash photolysis at 18°C[a]

Solvent	$k_T \times 10^{-7}$, $M^{-1}\,s^{-1}$
n-pentane	50 ± 1.5
Cyclohexane	41 ± 2
Benzene	18 ± 1
Chlorobenzene	17 ± 2
Tetrahydrofuran	23 ± 3
Acetonitrile	9.5 ± 0.7

[a] All data from reference 23.

a series of solvents. The results of these experiments are summarized in Table 8.3.

The observed trends in the variation of k_T from the laser flash photolysis studies mimic the variation in k_T/k_C values from the radical clock studies. These observations were taken as evidence that the changes seen in k_T/k_C result mainly from the solvent effect in the nitroxide radical termination reaction, k_T.

The importance of solvent polarity on k_T was demonstrated by observed correlation between the log of the rate constant for reaction of benzyl radical with TEMPO and solvent polarity expressed as a_N, the ESR hyperfine coupling constant for 4-NH_2-TEMPO (Figure 8.6).

The termination rate constants for reactions of nitroxides with alkyl and benzyl radicals are strongly influenced by solvent because of the varying degrees of solvation that are afforded the nitroxide (i.e. in a medium where the solvent interacts strongly with the nitroxide, the value of k_T decreases because it becomes more difficult for the attacking radical to displace the solvent molecules surrounding the nitroxide). As such, the termination rate constants of representative alkyl and benzyl radicals with nitroxides such as TEMPO fall substantially below the diffusion-controlled limit in a wide range of non-viscous solvents.

An important corollary of this work is that *the rate constant for the unimolecular rearrangement of the radical (i.e. the 'clocking' reaction) remains relatively constant in different solvents.* Additional evidence for the relative invariance of the clocking reaction rate constant (k_C) with solvent was based upon the observed trends in k_T/k_C for several clock/solvent combinations. Although the absolute values of k_C for the different clocks varied by over an order of magnitude, the k_T/k_C ratios varied identically with solvent.

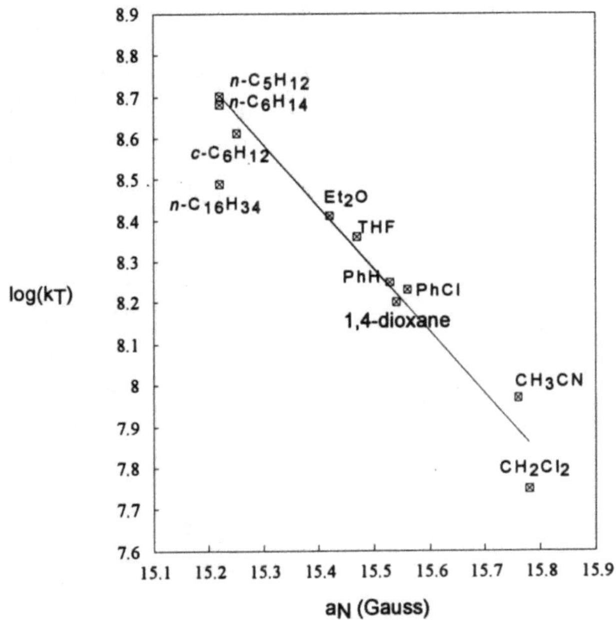

FIGURE 8.6 Plot of $\log(k_T)$ for benzyl radical trapping by TEMPO (at 18°C) vs. a_N for 4-NH_2-TEMPO in various solvents (data from reference 23).

This point is best illustrated graphically by plotting $\log(k_T/k_C)$ for the most and least reactive 'clocks' (**15** and **13**, respectively) versus a_N for the different solvents (Figure 8.7). The nearly identical value of the slope of the two lines in Figure 8.7 suggests that it is k_T that is solvent-dependent, not k_C. (Similar trends in k_T/k_C with solvent were observed with the other clocks used in this study.) The rationale for this proposition is derived from the reactivity-selectivity principle. Since all the radical clocks used in these experiments are primary radicals, the value of k_T is identical for all these radicals. However, the absolute rate constants for their rearrangement are markedly different, and it is unlikely that the solvent effect would be identical for each, unless of course, there were no solvent effect on the rate of rearrangement.

The observed insensitivity of the rearrangement rate constants in different solvents is an important observation since it means that, using the radical clock approach, it should be possible to measure solvent effects on a competing bimolecular rate constant without ambiguity, thereby foregoing the need for more elaborate and expensive techniques such as LFP.

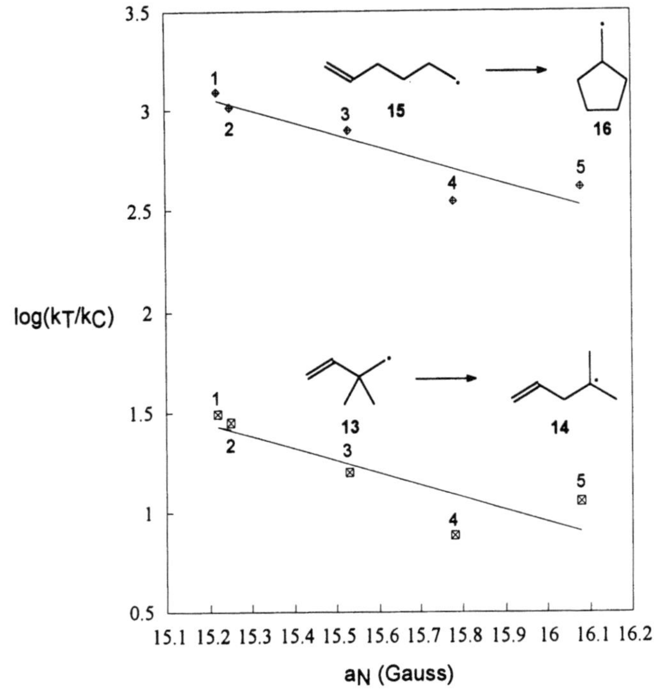

FIGURE 8.7 Plot of $\log(k_T/k_C)$ for (**13**) and (**15**) reacting with 1,1,3,3-tetramethylisoindoline-2-oxyl (TMIO) vs. a_N for 4-NH_2-TEMPO (data from reference 23). 1, *n*-hexane; 2, cyclohexane; 3, benzene; 4, CH_2Cl_2; 5, EtOH.

Thiyl Radical Additions

Thiyl radicals (RS•) can be readily generated by photolysis of the corresponding disulphides, or by hydrogen atom abstraction from a thiol, equations (8.18) and (8.19), respectively.

$$R{-}S{-}S{-}R \rightarrow 2\,R{-}\dot{-}S\bullet \tag{8.18}$$

$$R{-}S{-}H + R'\bullet \rightarrow R{-}S\bullet + R'{-}H \tag{8.19}$$

Interest in the reactivity of these species arises from their intermediacy in a variety of chemical transformations, such as the free radical addition of thiols, disulfides and sulfenyl chlorides to alkenes (Scheme 8.9) [25]. Thiyl radicals are also expected to be intermediates in the chemistry of living systems due to the action of light on the widely occurring disulfide linkage in biomolecules.

An excellent example of a solvent polarity effect on thiyl radical reactivity was provided by Ito and co-workers [26]. In these investigations,

R-S-H + CH_2=CHR' ⟶ R-S-CH_2-CH_2R'

R-S-S-R + CH_2=CHR' ⟶ R-S-CH_2CHR'SR

R-S-Cl + CH_2=CHR' ⟶ R-S-CH_2-CHR'Cl

SCHEME 8.9

the *p*-aminobenzenethiyl radical (*p*-$NH_2C_6H_4S$•) was generated by flash photolysis of the corresponding disulfide (equation 8.20)).

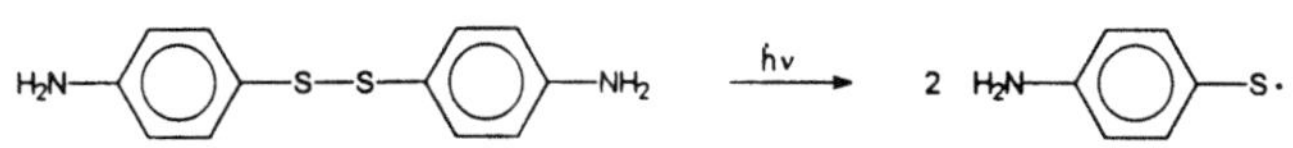

(8.20)

A large red shift was observed for λ_{max} of *p*-$NH_2C_6H_4S$• in polar solvents, suggesting greater stabilization of the free radical in these solvents. The reaction kinetics of *p*-$NH_2C_6H_4S$• were studied by monitoring the decay of the absorption attributable to the transient thiyl radicals [26].

The recombination rate constant (k_T: 2RS• → RSSR) was found to decrease with increasing solvent polarity. It was suggested that polar solvents cause a decrease in k_T due to the enhanced stabilization of *p*-$NH_2C_6H_4S$•, which can be explained on the basis of the increased contribution of dipolar resonance form (**18**) in these solvents [26].

(17) (18)

The rate constants for the reversible addition of the *p*-aminobenzenethiyl radical to styrene (equation (8.21)) were studied in 26 solvents. In general, the rate of addition of the thiyl radical to styrene was found to decrease with increasing solvent polarity, whereas the reverse rate constant was insensitive to solvent polarity. The authors found a reasonable linear relationship between the log of the addition rate constants k_1 and E_T [30], the Dimroth–Reichardt solvent polarity parameter (see 'main group solvents', Figure 8.8) [26].

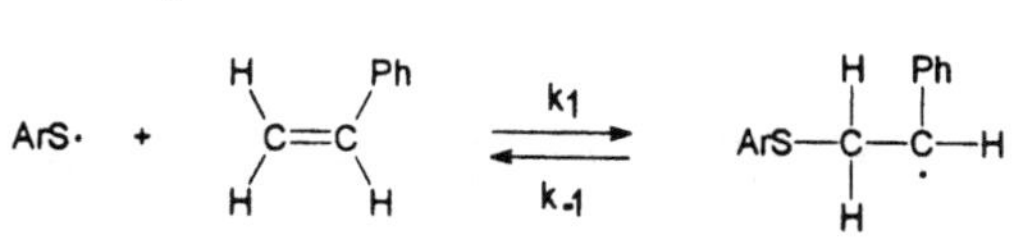

(8.21)

In Figure 8.8, deviations were observed from the main group of solvents (CCl_4, PhH, $PhCH_3$, $PhOCH_3$, THF, dioxane, pyridine, DMF, and DMSO)

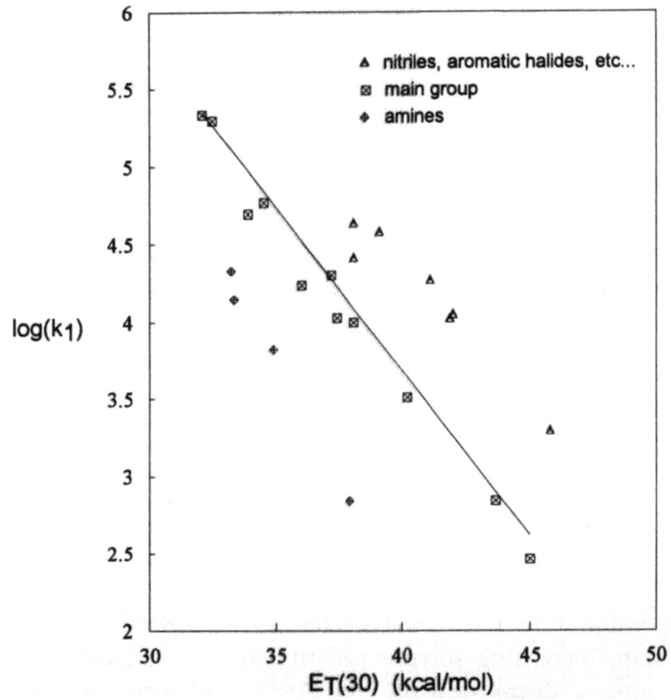

FIGURE 8.8 Plot of log (k_1) vs. E_T(30) for the addition of *p*-$NH_2C_6H_4S$• to styrene (data from reference 26).

by several classes of solvents (e.g., amines, nitriles and aromatic halides) [26]. Donor/acceptor interactions were suggested to account for these deviations. With the electron deficient solvents (e.g. nitriles, aromatic halides), *p*-$NH_2C_6H_4S$• is envisioned to donate electron density thereby decreasing the contribution of dipolar resonance form (**18**) to the resonance hybrid and resulting in the upward deviation in k_1 observed in these solvents. In contrast, for electron donating solvents (e.g. amines), *p*-$NH_2C_6H_5S$• functions as an electron acceptor and the contribution of the dipolar resonance form is enhanced resulting in downward deviation in these solvents.

SOLVENT (INTERNAL) PRESSURE EFFECTS

As noted in the introduction, variation of solvent can perturb the rate of a chemical reaction when the molar volumes of the reactant(s) and transtion state are different. This effect of solvent is attributable to a fundamental physical property of a liquid (internal pressure), and is analogous to the variation of rate associated with a variation of external pressure. The theoretical background for this effect has been developed by Hildebrand, and was discussed earlier.

For typical laboratory solvents, internal pressures vary from 2000 to 5000 atmospheres. For a reaction with a moderately high volume of activation (20 cm^3/mol), over the range of internal pressures available, the rate constant will vary by about one order of magnitude. When compared to the effect of solvent polarity on a polar process where rate constants may vary several orders of magnitude, the effect of internal pressure is very small indeed. Nonetheless, as the examples in this section will show, internal pressure can have a significant effect on the rate and selectivity of radical reactions (where solvent polarity effects are minimal).

cis-Diazene Inversion

Thermolysis of diazenes (commonly referred to as azo compounds, R—N=N—R) is an excellent methodology for the production of free radicals. Mechanistically, the process can be somewhat complicated because the process is not necessarily concerted (see below). (A non-concerted pathway would involve an intermediate diazenyl radical, R—N=N•). Moreover, radical production may be preceded by a thermal *trans* → *cis* isomerization of the diazene, with the *cis*-isomer being more reactive and hence the precursor which leads to radical formation.

As part of an extensive study of medium effects on *cis*-diazene thermolysis, the rate of $Z \rightarrow E$ isomerization of several diazenes (equation (8.22)) has been extensively studied as a function of both solvent and pressure by Neuman *et al.* [27]. For (R = R′ = adamantyl), the rate of isomerization has been examined in ten hydrocarbon and/or aromatic solvents and varies from $1.91 \times 10^{-4}\ s^{-1}$ in cumene to 2.6×10^{-4} in hexane [27].

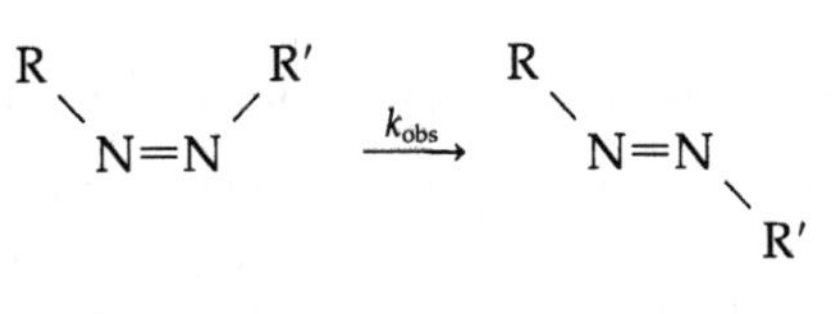

(8.22)

As shown in Figure 8.9, ln(k) correlates well with solvent internal pressure in accordance with equation (8.6). (Viscosity can be excluded as a possible factor because a plot of k vs $1/\eta$ is nonlinear.) Moreover, from the slope of the line in Figure 8.9 in accordance with equation 8.6 the

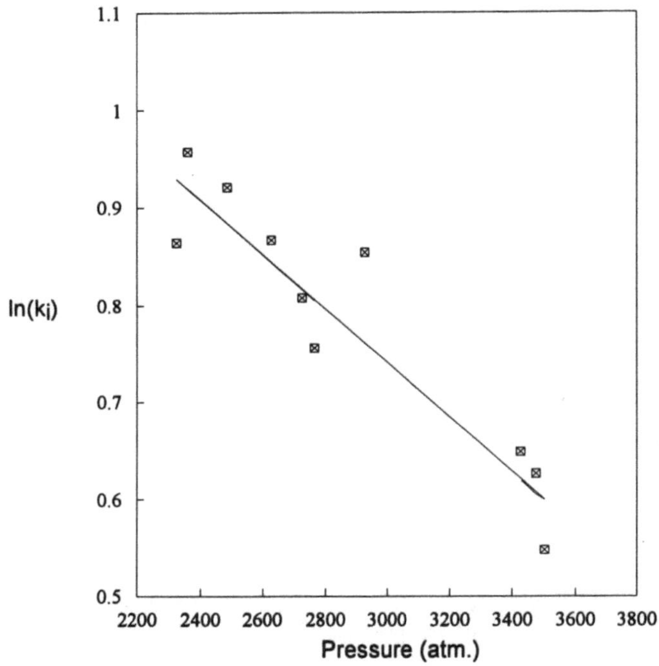

FIGURE 8.9 Variation of $\ln(k_i)$ for the $Z \rightarrow E$ isomerization of *cis*-diazene (R = R′ = adamantyl) with solvent internal pressure (data from reference 27).

derived volume of activation from this data (ΔV_{act} = + 7 cm^3/mol) [27] is identical to that obtained in hexane through the variation of *external* pressure [27]. Overall, these results are interpreted on the basis of a proposed mechanism for isomerization of diazo compounds via parallel radical and inversion (k_{inv}) pathways (Scheme 8.10) [27,28]. The observed ΔV_{act} is of the same order of magnitude as expected for a one-bond homolysis (*c.* +10 cm^3/mol) based upon 'typical' values reported by Asano and LeNoble in their 1978 compilation of $\Delta V'_{act}$s in solution [14].

Halogenation of Cyclopropane (and Derivatives)

Halogen atoms react with cyclopropane via two competing pathways: hydrogen abstraction and S_H2 ring opening (k_H and k_C, respectively; Scheme 8.11). The relative importance of the two pathways (i.e., the chemoselectivity) depends on the identity of the halogen. When X=Br, the ring opening pathway is generally preferred [29], while both ring-opened and hydrogen abstraction products are observed when X=Cl [30,31].

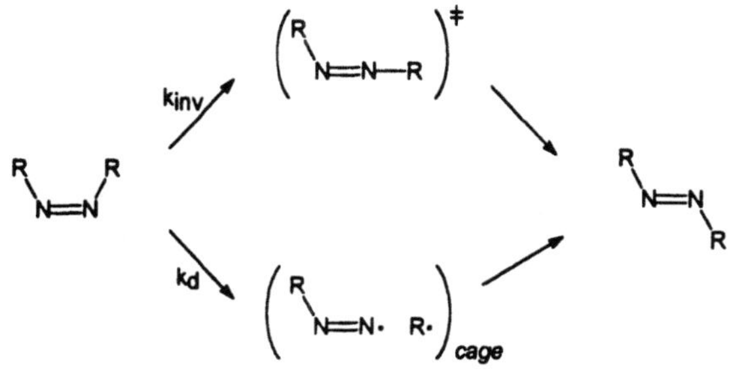

$k_i = k_{inv} + k_d$

SCHEME 8.10

As part of a study examining the role of stereoelectronic factors on the rate of ring opening of cyclopropylarenes by bromine atom [29,32], competition experiments were performed to assess the relative rate constant for ring opening of cyclopropylbenzene (k_C) vs. hydrogen abstraction from toluene (k_H, Scheme 8.12). The rate constant ratio k_C/k_H was found to vary by nearly a factor of 20 as a function of solvent. Solvent internal pressure was identified as the property responsible for the observed effect. The rate constant ratio k_C/k_H did not correlate well either to solvent polarity or viscosity, while plots of $\ln(k_C/k_H)$ vs. either P_i or δ^2 were linear in accordance with equation (8.6) (Figure 8.10). The derived

SCHEME 8.11

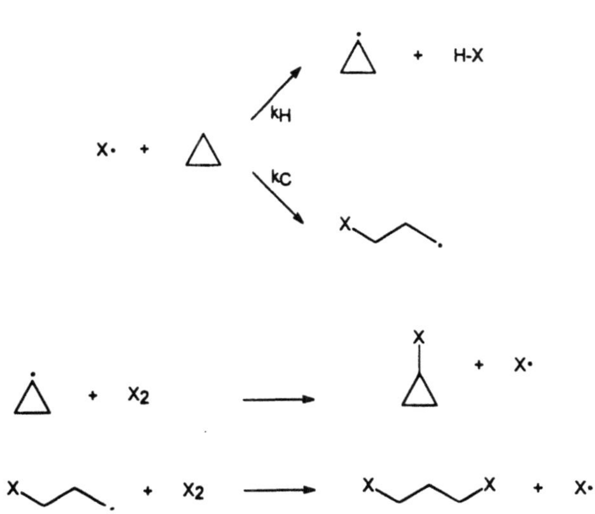

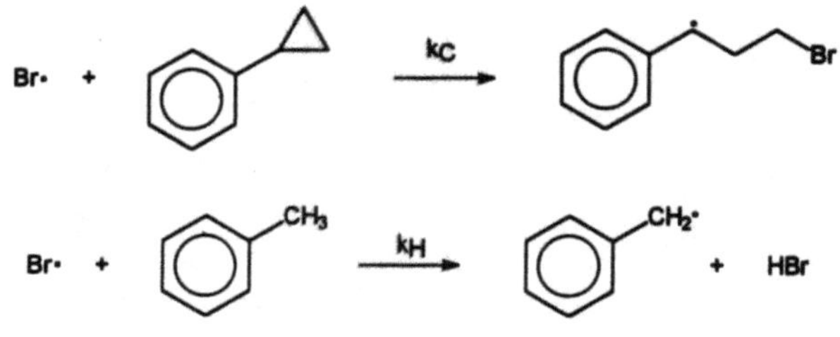

SCHEME 8.12

volume of activation from the slopes of these lines, $\Delta\Delta V_{act} = \Delta V_C^{\neq} - \Delta V^{\neq}_H$, was $-20\ cm^3/mol$.

Because the $S_H{}^2$ process (k_C) effectively involves the *addition* of $Br^{\bullet}$ to a cyclopropane ring (i.e. two reactants yield one product), it is reasonable to suspect that ΔV_{act} for this process would be more negative than hydrogen abstraction from toluene (k_H) wherein two reactants generate two products ($PhCH_2^{\bullet}$ + HBr). The magnitude of the observed $\Delta\Delta V_{act}$ is reasonable in light of published volumes of activation for a variety of radical additions and atom transfer reactions [14].

FIGURE 8.10 Variation of $\ln(k_C/k_H)$ for the competitive bromination of cyclopropylbenzene and toluene with the cohesive density (δ^2) of sovent (data from reference 22).

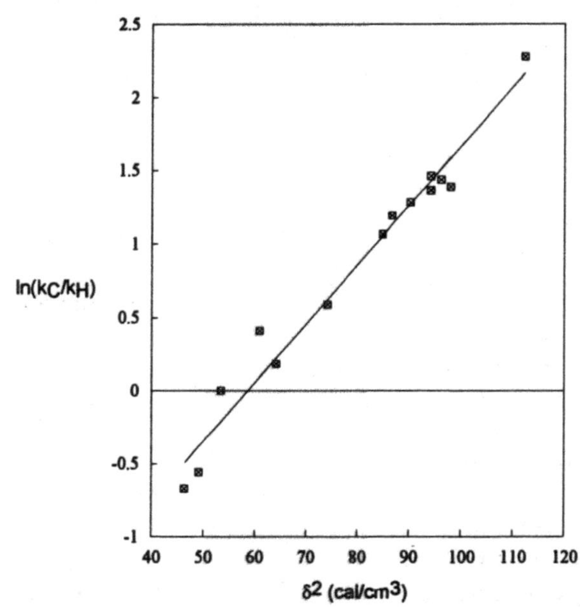

More recently it has been shown that solvent internal pressure is likely to be the factor responsible for some conflicting and contradictory results reported in the literature regarding the chlorination of cyclopropane [33]. In 1945, Roberts and Dirstine [30] reported that the *gas phase* chlorination of cyclopropane yielded primarily chlorocyclopropane and 1,1-dichlorocyclopropane (products resulting from hydrogen abstraction by Cl•). Only minor amounts of ring-opened (S_H2) products were observed implying $k_H > k_C$ (Scheme 8.11).

In contrast, Walling and Fredricks reported in 1962 that the same reaction carried out in the condensed phase (CCl_4 solvent) yielded comparable amounts of 1,3-dichloropropane (S_H2 product) and chlorocyclopropane [31], implying $k_H \approx k_C$.

This discrepency between the gas and condensed-phase results for this reaction remained unexplained until 1994 when Tanko and Suleman studied the reaction in several solvents and in the gas phase [33]. For the chlorination of cyclopropane, the rate constant ratio k_C/k_H was found to vary significantly with solvent. A plot of $\ln(k_C/k_H)$ vs. δ^2 in accordance with equation (8.6) was linear (Figure 8.11), implicating internal pressure as the factor responsible for the observed solvent effect. (This conclusion is reasonable in that this *intramolecular* S_H2 vs. hydrogen abstraction competion is analogous to the *intermolecular* cyclopropylbenzene/toluene competition for Br• which had been shown earlier to correlate to internal pressure). Moreover, extrapolation of the condensed phase results to zero internal pressure predicts the observed gas phase result (Figure 8.11) [33]. These observations provide compelling testimony for the potentially dramatic role internal pressure can play in affecting the outcome of a condensed phase reaction.

RADICAL–SOLVENT INTERACTIONS THAT PRODUCE 'RADICALOID' SPECIES

In this section, a unique type of solvent effect is examined that defies classification based upon the concepts of polarity, viscosity, and internal pressure around which this chapter is organized. In certain cases, a free radical can interact or react with specific functional groups on the solvent, which results in the formation of a transient species that retains the chemical characteristics of the 'free' radical (i.e. the same products are formed and no solvent molecules are incorporated in to the product). However, because this transient species is more stable than the 'free' radical, lower reactivity and enhanced selectivity is observed. This situation is directly analogous to carbene chemistry where the distinction is

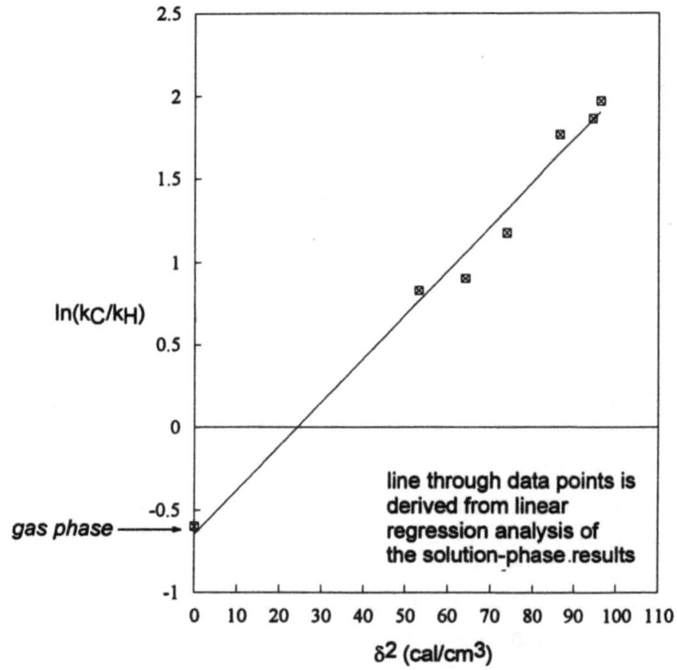

FIGURE 8.11 Chlorination of cyclopropane: Variation of log(k_C/k_H) with cohesive energy density (δ^2) of the solvent (data from reference 33).

made between a 'free carbene' and a 'carbenoid', the latter of which arises from complexation with another species (usually a metal). The same chemistry is observed with both species, but carbenoids are generally less reactive and more selective than the 'free' carbene.

By analogy, transient species resulting from a specific radical–solvent interaction are classified as 'radicaloid' for the organizational purposes of this chapter. To the best of our knowledge, all known examples of radicaloid formation via the interaction of a free radical with solvent involve the highly reactive chlorine atom.

Chlorine Atom/Arene Complexes

No discussion of solvent effects in radical reactions would be complete if it failed to include the effect of aromatic solvents on the selectivity observed in the free radical chlorination of alkanes. The free radical chlorination of alkanes is an important methodology for both the laboratory- and industrial-scale preparation of alkyl halides. This reaction also has the unique distinction of generally being one of the first

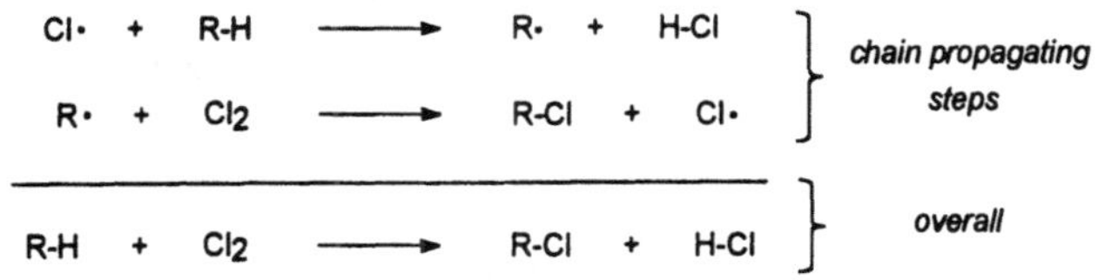

SCHEME 8.13

organic transformations discussed in most organic chemistry textbooks because it is an effective method for introducing functionality into alkanes, converting an unreactive C—H bond into a C—Cl bond. The mechanism of the reaction (Scheme 8.13) has been well-understood for at least six decades [15].

Overall, the reaction is a free radical chain process (chain lengths > 1000) involving the highly reactive and unselective chlorine atom [15]. (Absolute rate constants for Cl• + R—H → H—Cl + R• have been measured by laser flash photolysis and are at or near the diffusion-controlled limit in solution) [34]. Because of this high reactivity, Cl• exhibits nominal selectivity in hydrogen abstractions from alkanes, 3° (4.2) >2° (3.6) >1° (1), compared to the significantly less reactive but more selective bromine atom for which the reactivity order is 3° (>25,000) >2° (700) >1° (1) [35,36].

Early observations of enhanced selectivity in aromatic solvents. In a series of papers beginning in 1957, Russell demonstrated that the selectivity of alkane chlorinations could be dramatically increased by conducting the reaction in an aromatic solvent [37–40]. For example, chlorination of neat 2,3-dimethylbutane (23DMB) at 25°C resulted in formation of 1- and 2-chloro-2,3-dimethylbutane in yields of 60 and 40%, respectively. This relative yield indicates that the relative reactivities of the 3° and 1° hydrogens of 23DMB towards chlorine atom (designated as $r(3°/1°)$) is 4.2:1 (Note: $r(3°/1°) = (\text{Yield } 3° \text{ RCl}/\text{Yield } 1° \text{ RCl}) \times 6$). In contrast, when the same reaction is carried out in 8 M benzene, enhanced selectivity is observed: $r(3°/1°) = 59$.

To account for these observations, Russell proposed the formation of a *complex* between Cl• and benzene, and that this complex was the high-selectivity hydrogen abstractor. Russell's modified mechanism for free radical chlorination in aromatic solvents is summarized in Scheme 8.14 [37,38]. (Pertinent rate constants for all the kinetically significant steps are included.)

In order to understand Russell's observations and conclusions as well as those of subsequent workers in the field, it is instructive to consider the kinetic expression for the selectivity of the reaction as a function of the

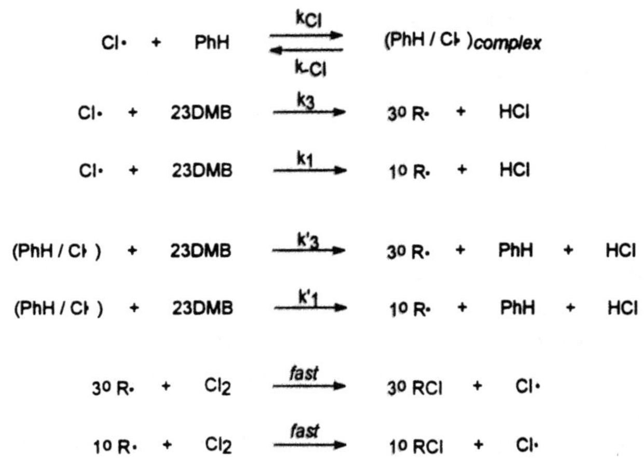

SCHEME 8.14

rate constants defined in Scheme 8.14 and the concentrations of all pertinent species (equation (8.23)).

$$r(3^\circ/1^\circ) = 6\left[\frac{\dfrac{k_3}{k_1}\left\{\left(1+\dfrac{k'_3+k'_1}{k_{-\mathrm{Cl}}}[\mathrm{23DMB}]\right)+\dfrac{k'_3}{k_{-\mathrm{Cl}}}\dfrac{k_{\mathrm{Cl}}}{k_1}[\mathrm{PhH}]\right\}}{\left(1+\dfrac{k'_3+k'_1}{k_{-\mathrm{Cl}}}[\mathrm{23DMB}]\right)+\dfrac{k'_1}{k_{-\mathrm{Cl}}}\dfrac{k_{\mathrm{Cl}}}{k_1}[\mathrm{PhH}]}\right] \tag{8.23}$$

Russell reported that while $r(3^\circ/1^\circ)$ did vary with benzene concentration, it was independent of 23DMB concentration [38]. In accordance with equation (8.23), this implies that the [23DMB] terms are unimportant ($((k'_3 + k'_1)/k_{\mathrm{Cl}})$ [23DMB] $\ll 1$) so that equation (8.23) can be rewritten as equation (8.24), where K_{Cl} is the equilibrium constant for the formation of the PhH/C1• complex. Thus, the invariance of selectivity with 23DMB concentration implies that free chlorine atom and the PhH/Cl• complex are at equilibrium. (In other words, this absense of a 23DMB concentration-dependance implies that $(k_3' + k_1')$ [23DMB] $\ll k_{-\mathrm{Cl}}$, i.e. the complex decomposes to free Cl• and thus equilibrates faster than it abstracts hydrogen.

$$r(3^\circ/1^\circ) = 6\left\{\frac{k_3 + k'_3 K_{\mathrm{Cl}}[\mathrm{PhH}]}{k_1 + k'_1[\mathrm{PhH}]}\right\} \tag{8.24}$$

Russell found that the selectivity also varied with the nature of the aromatic solvent. Specifically, selectivity was diminished when less electron rich arenes were used, and the variation of selectivity with solvent paralleled the basicity of the solvent [38,41].

Russell's observations and conclusions were subsequently confirmed by Walling and Mayahi in 1959 [42].

Proposed structure of the chlorine atom/benzene complex. Two possible structures for (PhH/Cl•) were considered by Russell, a hexahapto π-complex (**19**) and a 6-chlorocyclohexadienyl radical (σ-complex, **20**)). Russell assigned the π-complex structure to (PhH/Cl•) on the following basis: Cl• and PhH were believed to be in rapid equilibrium, and presumably formation of a weak π-complex would occur faster than σ-bond formation. In addition, the selectivity in different aromatic solvents correlated to the σ-donor ability of the solvent, but not to the rates of phenyl radical addition to the solvent (a process which produces a cyclohexadienyl radical) [38].

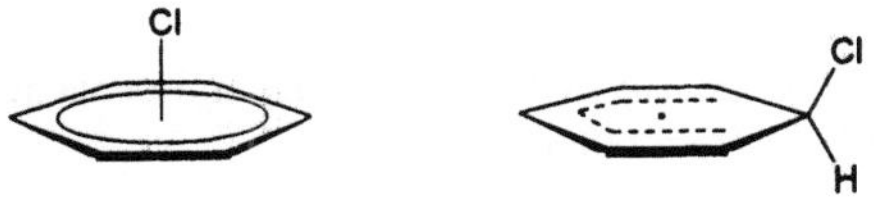

(**19**) (**20**)

In 1983, Skell reintroduced the suggestion that (PhH/Cl•) was actually the 6-chlorocyclohexadienyl radical (**20**) [43]. Indeed, (**20**) had traditionally been invoked to account for other chemistry in the chlorine atom/benzeme system, namely the formation of hexachlorocyclohexane from benzene (equation (8.25) and the formation of adducts with maleic anhydride (equation (8.26)) [44].

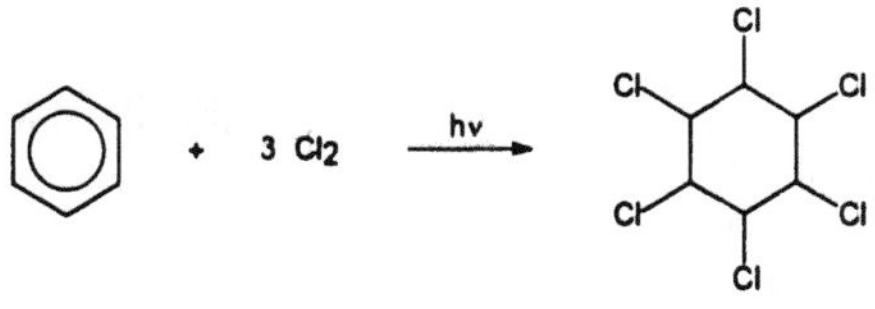

(8.25)

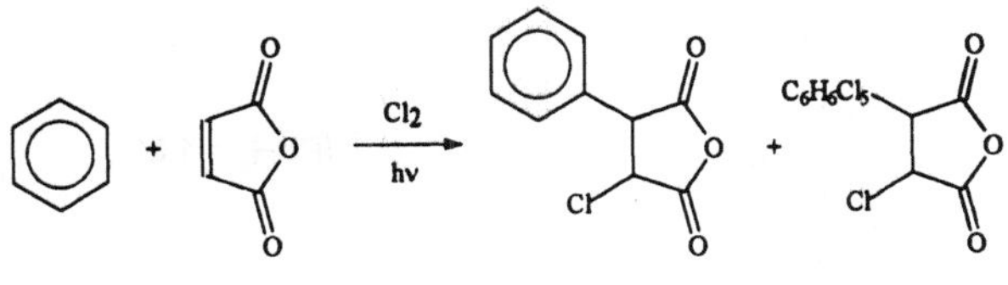

(8.26)

Skell's suggestion stirred up a hornet's nest of controversy, which continues to this day. However, before considering the precise structure of (PhH/Cl•) further, a review of some of the experimental data and conclusions subsequent to Russell's and Walling's reports (about which there is no disagreement) is appropriate.

Free Cl• and (PhH/Cl•) are not always at equilibrium. In 1983, Skell *et al.* found that in contrast to Russell's earlier report, $r(3°/1°)$ did in fact vary with the concentration of 23DMB [43]. Unlike the earlier reports of Russell and Walling however, these workers examined a much broader range of 23DMB concentrations. Specifically, it was found that at high 23DMB concentrations, the observed selectivity approached that of a free chlorine atom, while at very low 23DMB concentrations, the selectivity was nearly invariant with [23DMB]. A summary of Skell's results, using data from the 1986 full paper is presented in Figure 8.12 [44].

These observations are explicable since at very low 23DMB concentrations, the [23DMB] terms in equation (8.23) are negligible, and hence the selectivity is accurately expressed by equation (8.24) consistent with Russell's original hypothesis that Cl• and PhH/Cl• were at equilibrium, i.e. $(k_3' + k_1')[23DMB] \ll k_{Cl}$). However, at higher alkane concentrations,

FIGURE 8.12 Variation in selectivity with alkane and benzene concentrations in the chlorination of 23DMB (data from reference 44).

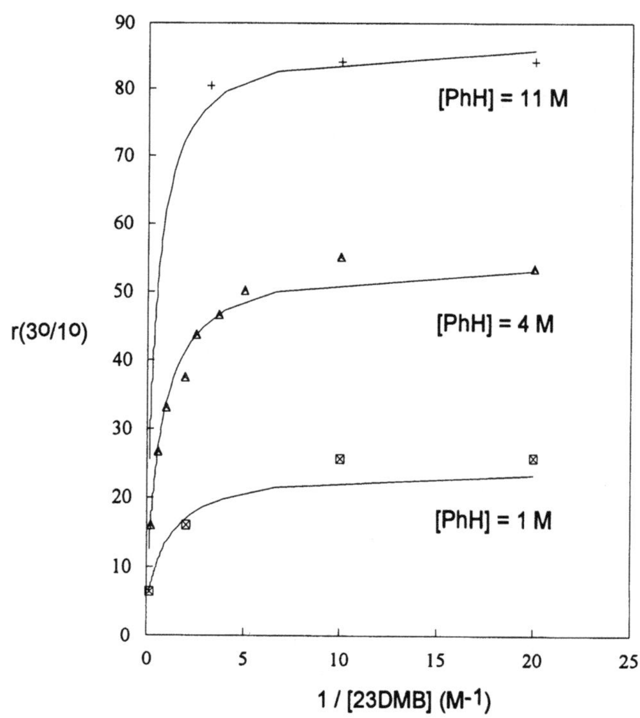

the 23DMB terms become increasingly important and in the limit, as [23DMB] $\rightarrow \infty$, $r(3°/1°) \rightarrow k_3/k_1$. (The selectivity approaches that of free Cl• because Cl• reacts with 23DMB before (PhH/Cl•) can be formed). The lines drawn in Figure 8.12 were obtained from a nonlinear, least-squares regression analysis of the data in accordance with equation (8.23), which yielded: $k_1/k_3 = 0.66$; $(k'_3 + k'_1)/k_{-Cl} = 1.4$; $k_{Cl}/k_1 = 3.3$, and $k'_3/k'_1 = 21$ (corresponding to $r(3°/1°) = 126$ for (PhH/Cl•)) [44]).

Direct observation of (PhH/Cl•) and important absolute rate constants. In 1985, Ingold *et al.* re-examined the issues raised by Skell, confirming his experimental observations and reporting for the first time absolute rate constants for all of the processes outlined in Scheme 14 [34]. Several of these are summarized in Table 8.4.

In addition, Ingold recorded the (PhH/Cl•) UV/VIS absorption spectrum ($\lambda_{max} = 320$ and 490 nm) and ascertained the equilibrium constant for its formation: $K_{Cl} = 200\ M^{-1}$. However, unlike Skell, Ingold concluded that the experimental data is more consistent with a π-complex structure for (PhH/Cl•). In the next section, the pros and cons of both positions are reviewed, the opinions of other 'observers' in the field are discussed, and finally one of the co-authors of this chapter (JMT) takes the perogative of offering his own views on the subject.

The structure of (PhH/Cl•) revisited. Too many chefs and not enough food in the kitchen? The original Russell suggestion that (PhH/Cl•) was a π-complex was based partly on the supposition that Cl• and (PhH/Cl•) were at equilibrium under all conditions. Presumably, σ-bond formation would require significant geometric changes (e.g. rehybridization) which would take longer than π-complexation where no alteration of the aromatic framework would be necessary. Skell's original 1983 communication showed that two hydrogen abstractors were not always at equilibrium, and on this basis (essentially echoing Russell's original argument) proposed that (PhH/Cl•) was the slower forming

TABLE 8.4 Absolute rate constants for several reactions in the chlorine atom/benzene system[a]

Reaction	Rate constant at 27°C
Cl• + PhH → (PhH/Cl•)	$k_{Cl} = 6 \times 10^9\ M^{-1}s^{-1}$
Cl• + 23DMB → HCl + R•	$k_3 + k_1 = 2.6 \times 10^9\ M^{-1}s^{-1}$
(PhH/Cl•) + 23DMB → R• + HCl + PhH	$k'_3 + k'_1 = 4.8 \times 10^7\ M^{-1}s^{-1}$
(PhH/Cl•) → PhH + Cl•	$k_{-Cl} = 3.0 \times 10^7 s^{-1}$

[a] All data from reference 34.

6-chlorocyclohexadienyl radical. In retrospect, since all the reactions of Cl• have been shown to be diffusion-controlled, arguments based upon the relative time frame for π- vs. σ-complex formation are no longer viable.

In their 1985 paper, Ingold *et al.* argued for the π-complex as the appropriate structure for (PhH/Cl•). While several arguments were advanced to justify this assignment, the most compelling of these was based upon spectroscopy: (PhH/Cl•) exhibits a charge-transfer band (equation (8.27)) at 490 nm in direct analogy to other atom/arene π-complexes [38]. (The UV/VIS spectrum of (PhH/Cl•) and other halogen atom/arene complexes had been reported by Bühler earlier, who assigned these spectra as π-complexes) [45].

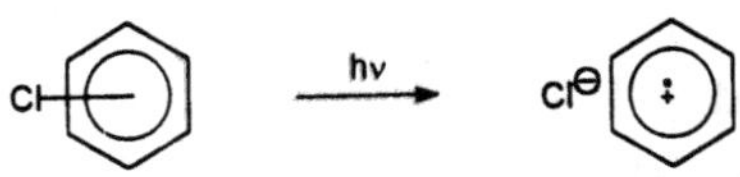

(8.27)

In his 1985 full paper, Skell counters that it is reasonable to suspect that the 6-chlorocyclohexadienyl radical would give a CT-band at 490 nm (as well as π-π^* band at 320 nm as is observed with cyclohexadienyl radical) because of substantial spin density at Cl arising from hyperconjugation (Scheme 15) [44]. New evidence was introduced, namely trapping experiments using maleic anhydride and Cl_2 that 'traditionally' had been assumed to react with the 6-chlorocyclohexadienyl radical. (The lower selectivities observed in the presence of these traps was interpreted on the basis of selective trapping of the 6-chlorocyclohexadienyl radical). Skell proposed a three intermediate scheme involving free Cl•, π-(PhH/Cl•), and the 6-chlorocyclohexadienyl radical, wherein only the latter species exhibits high selectivity [44].

In 1988, Walling reviewed these arguments, suggested that the hydrogen abstractors are free Cl• and π-(PhH/Cl•), and proposed that the trapping experiments could be explained by assuming that the π-complex

SCHEME 8.15

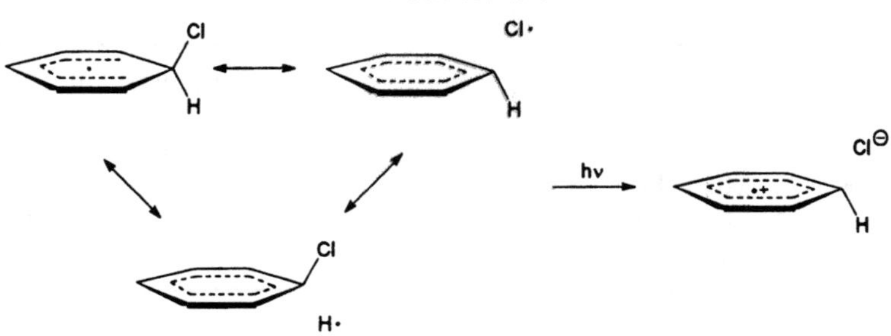

was in equilibrium with a small amount of the 6-chlorocyclohexadienyl radical [46].

A 1989 paper by Raner *et al.* examined the solution phase UV/VIS spectra of several X_2/arene and X•/arene complexes (X=Cl, Br, I) [47]. While λ_{max} for all the X_2/arene spectra examined correlate to each other and to those of I•/arene (i.e. λ_{max} correlates to the vertical ionization potential of the arene as expected for a charge-transfer band) [48], Cl•/arene and Br•/arene spectra exhibit what was described as a 'buck-shot pattern' [47]. The solid state structures of Cl_2/PhH and Br_2/PhH have been shown to be to hexahapto (**21**) by X-ray [49], hence on the basis of spectroscopic analogy, similar structures were proposed for the I_2/arene and I•/arene complexes by Ingold *et al.* [47]. However, because of the anamolous behavior of the Br•/arene and Cl•/arene complexes, the assignment of (PhH/Cl•) as a hexahapto π-complex was based more upon argument than experiment (i.e. the assumption that I•, Br•, and Cl• all form π-complexes of the same general structure).

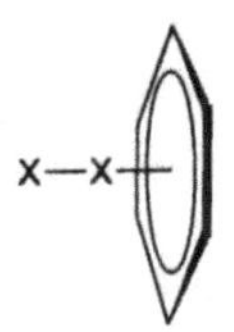

(21)

The difficulties in firmly establishing the structure of (PhH/Cl•) on the basis of the UV/VIS spectrum are (a) there is no reasonable model based upon which the UV/VIS characteristics of the 6-chlorocyclohexadienyl radical can be extrapolated (i.e. simple cyclohexadienyl radicals do not exhibit the type of intramolecular CT-interaction envisioned in Scheme 8.15), and (b) UV/VIS spectra do not provide enough detailed structural information to allow an unambiguous assignment.

ESR spectroscopy has been applied to the problem, but the conclusions eminating from this work do not appear to be generally accepted. A 1983 report in the Russian literature suggests that Cl• forms a structure intermediate between that of a pure hexahapto π-complex and a pure chlorocyclohexadienyl radical (e.g. **22**)) based upon the solid-state ESR spectrum [50]. Symons reports that the solid-state ESR spectra of I•/PhH

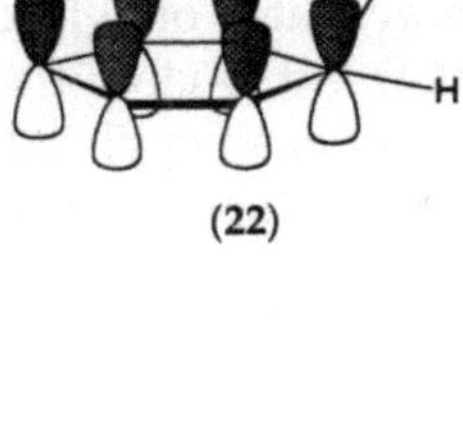

(22)

and Br•/PhH are consistent with a cyclohexadienyl radical structure (exhibiting large hyperfine splitting to the out of plane X and H) [51]. Unfortunately, it is difficult to assess the reliability of these reports, and to our knowledge these conclusions have not been verified by others.

Most recently, Benson has estimated ΔH° for Cl• + PhH → 6-chlorocyclohexadienyl radical via a thermodynamic cycle and the derived value ($\Delta H^\circ = -8.2 \pm 2$ kcal/mol) [52] was identical to that derived from the results of Ingold *et al.* for Cl• + PhH → PhH/Cl• ($\Delta H^\circ = -9.2 \pm 0.6$ kcal/mol) [34,52]. Benson thus argued that the transient observed by Ingold *et al.* was actually the 6-chlorocyclohexadienyl radical, but excluded this intermediate as the selective hydrogen abstractor on argumentative grounds (i.e. 'the uniqueness of these reactions makes them highly unlikely', (equation (8.28)) [52]. Instead it was suggested that a hitherto unobserved π-complex in rapid in-cage equilibrium with the 6-chlorocyclohexadienyl radical, is the high-selectivity hydrogen abstractor.

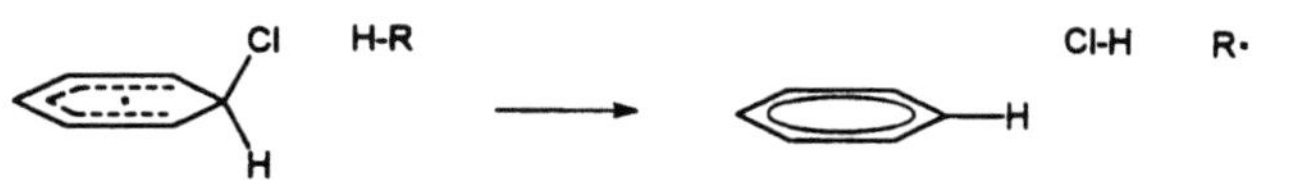

(8.28)

Finally, it is worth mentioning another possibility. The available data (ESR [53], IR [54]) unambiguously shows that F•/PhH is best represented as a cyclohexadienyl radical. Similarly, the concensus is that I•/PhH is a hexahapto (η_6) π-complex [47]. It is possible that for X•/PhH complexes, there is a continuum of structures in the series I → Br → Cl → F going from a pure π-complex to a pure σ-complex [44]. A clear distinction between an η_6, η_2, or η_1 complex for the structure of (PhH/Cl•) may not be possible, or for that matter meaningful because in the final analysis, the difference betwee these structures is relatively minor (i.e. the precise position of the chlorine atom and one of the hydrogens).

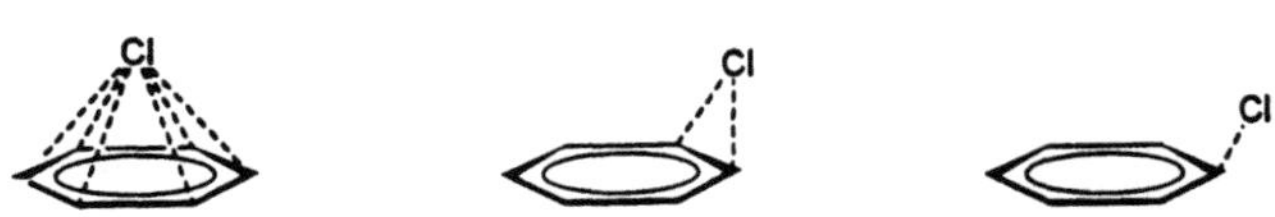

η_6 η_2 η_1

Where do we stand? In this author's opinion (JMT), there is not nearly enough unambiguous data available on which to base a defensible stand on the structure of (PhH/Cl•). It is clear that if the selective hydrogen abstractor is the 6-chlorocyclohexadienyl radical, then this radical must be different to other known cyclohexadienyl radicals (which does not seem unreasonable given the potential for hyperconjugative interactions involving the π-system and out of plane C—Cl bond, e.g π-σ^* interactions). Similarly, if the structure is truly the π-complex, then this structure is different from other X_2/arene and I•/arene π-complexes. The problems is that both positions lack the unambiguous experimental (or theoretical) basis needed to put the issue to rest.

With regard to the number of hydrogen abstractors, all the kinetic data can be accomodated with two hydrogen abstractors: free Cl• and something else (either π- or σ-(PhH/Cl•)). In retrospect, the three intermediate scheme proposed by Skell *et al.* in 1986, wherein the σ-complex was identified as the selective hydrogen abstractor and the π-complex was assumed to have a selectivity similar to free Cl• [44], is superfluous. Similarly, there is no experimental evidence to substantiate the recent hypothesis put forth by Benson [52] (i.e. that the spectroscopically observed transient is the 6-chlorocyclohexadienyl radical but an 'invisible' π-complex in rapid in-cage equilibrium is the hydrogen abstractor). Furthermore, there does not seem to be any way in which such a hypothesis can ever be proven (or disproven) experimentally. Benson's thermochemical arguments seem to indicate that the spectroscopically observed transient was the chlorocyclohexadienyl radical. Unfortunately, it is not evident as to how the thermodynamic data was obtained, specifically the heat of formation of the chlorocyclohexadienyl radical.

In short, the kinetic data are consistent with either proposition, the UV/VIS spectroscopic data is not definitive, and it is not clear as to whether the ESR results are reliable. Consequently, any conclusion regarding the structure of (PhH/Cl•) is highly speculative at this time.

While it is intriguing to argue and speculate regarding the true structure of (PhH/Cl•), it is important to bear in mind that the important issue in this system is the chemistry, which should not be obscured by disagreements over the true stucture of the intermediate. Indeed, a great deal of new and exciting observations pertaining to chlorine atom chemistry have been discovered since interest in the field was reignited by Skell's 1983 report. It would be unfortunate for the reader to view the disagreements regarding (PhH/Cl•) structure as the sole 'take-home message'.

The Chlorine Atom/CS_2 Complex

Concurrent with his initial report detailing enhanced selectivities when alkane chlorinations are conducted in aromatic solvents, Russell also found enhanced selectivities in carbon disulfide [38]. In fact, the selectivities in CS_2 were even greater than those found in benzene (e.g. for

TABLE 8.5 Absolute rate constants pertaining to the CS_2/chlorine atom complex[a]

Reaction	Rate constant at 21°C
$(CS_2/Cl\bullet) + 23DMB \rightarrow R\bullet + HCl + CS_2$	$1.7 \times 10^7\ M^{-1}\ s^{-1}$
$Cl\bullet + CS_2 \rightarrow (CS_2/Cl\bullet)$	$1.7 \times 10^{10}\ M^{-1}\ s^{-1}$
$(CS_2/Cl\bullet) \rightarrow Cl\bullet + CS_2$	$7.2 \times 10^6\ s^{-1}$

[a] All data from reference 34.

23DMB $r(3°/1°) = 49$ in 8 M PhH and 106 in 8 M CS_2 [38]. Skell *et al.* found that as with the chlorine atom/benzene system, the observed selectivity varied with alkane concentration (i.e. selectivity increased then levelled off with decreasing alkane concentration) [43]. The reversible formation of a highly selective $Cl\bullet/CS_2$ complex has been suggested, in direct analogy to $(PhH/Cl\bullet)$.

More recently, Chateauneuf has detected this complex spectroscopically ($\lambda_{max} = 370$ nm, shoulder at 490 nm) and ascertained several important rate constants for this system (Table 8.5), as well as the equilibrium constant for $Cl\bullet + CS_2 = (CS_2/Cl\bullet)$, $K_{eq} = 2.4 \times 10^4\ M^{-1}$ [55].

Compared to the chlorine atom/benzene system, the chlorine atom/CS_2 system has been remarkably devoid of speculation (and controversy) on the issue of structure (which is unknown at this time). Two plausible structures are a $CS_2/Cl\bullet$ donor/acceptor (π) complex, or a σ-complex (**23**) arising from bond formation between S and Cl. (An alternative σ-bonded structure, ClC(= S)S• is less likely because sulfur-centered radicals are generally poor hydrogen atom abstractors.)

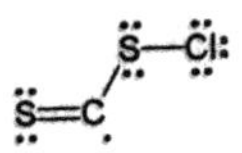

(23)

The Chlorine Atom/Pyridine Complex. Taking advantage of Russell's observations regarding enhanced selectivities attributable to a $PhH/Cl\bullet$ complex, Breslow *et al.* have utilized this technology to effect highly selective chlorinations of steroids. By incorporating an aromatic ring (generally an iodobenzene or pyridine) as a tether on the steroid, hydrogen abstraction could be accomplished intramolecularly and with an very high degree of selectivity using chlorinating agents such as $PhICl_2$ (equation (8.29)) [56].

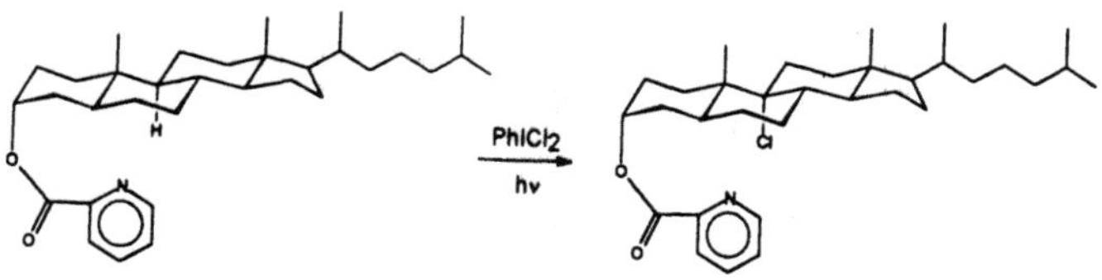

(8.29)

In their proposed mechanism, PhICl• delivers ('relays') a chlorine atom to the nitrogen atom of the pyridine ester template, which then 'reaches over' and selectively abstracts hydrogen intramolecularly. Reaction of the resulting alkyl radical with $PhICl_2$ regenerates PhICl• and yields the product alkyl chloride.

This report led to further study of the use of pyridine as a solvent for alkane chlorinations by Breslow *et al.* [57]. Indeed, it was found that the pyridine/Cl• complex exhibited even greater selectivity than (PhH/Cl•) in reactions with 2,3-dimethylbutane. (Thus, while chlorination of 0.1 M 23DMB in 4 M benzene yields $r(3°/1°) = 50$, in 4 M pyridine the selectivity is 200.) Using laser flash photolysis, absolute rate constants pertaining to this complex were obtained (Table 8.6). From this data, $K_{eq} = 3.4 \times 10^4$ M^{-1} for Cl• + Py = (Py/Cl•) [57]. (Chateauneuf has re-evaluated this equilibrium constant and places it more on the order of 1.2×10^5 M^{-1} [55]).

The UV/VIS spectrum of (Py/Cl•) (λ_{max} = 334 nm) was drastically different from (PhH/Cl•) (λ_{max} = 320, 490 nm). These results, in conjunction with *ab initio* calculations, suggest that Cl is complexed directly to the nitrogen of pyridine (e.g. σ^* structure (**24**)) [57].

Consistent with this assignment, the analogous complex formed between Cl• and 2,6-di-*t*-butylpyridine exhibits two absorption maxima (λ_{max} = 325, 420 nm) suggesting a structure more similar to (PhH/Cl•) [57]. It is suggested that the introduction of the bulky *t*-butyl groups effectively blocks complexation to the pyridine nitrogen. (Similar to his interpretation of the chlorine atom/benzene system, Benson has suggested that the high selectivity hydrogen abstractor in the chlorine atom/pyridine system is actually a π-complex, which has never been detected [52].)

TABLE 8.6 Absolute rate constants pertaining to the chlorine atom/pyridine complex.[a]

Reaction	Rate constant at room temperature
Cl• + Py → (Py/Cl•)	$1.5 \times 10^{10} M^{-1} s^{-1}$
(Py/Cl•) + 23DMB → R• + HCl + Py	$2.6 \times 10^5 M^{-1} s^{-1}$

[a] All data from reference 57.

(24)

Finally, Khanna *et al.* have provided evidence for a novel intramolecular hydrogen abstraction *via* a five-center transtion state in the chlorination of 2-methylquinolines (Scheme 8.16) [58].

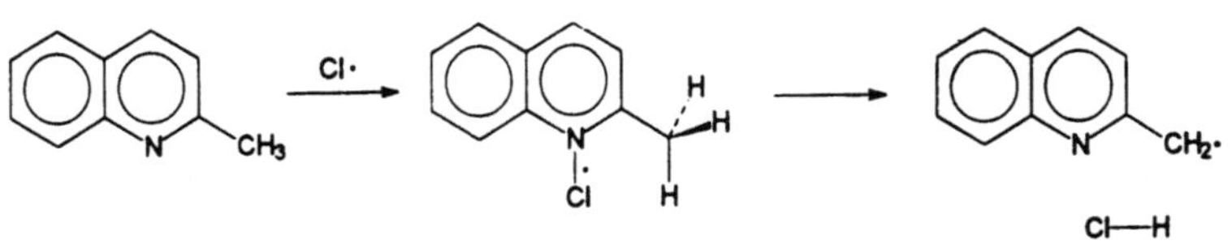

SCHEME 8.16

SOLVENT VISCOSITY AND CAGE EFFECTS

Diffusion-Controlled Reactions (Diffusive Radical/Radical Caged-Pairs)

An important reaction of neutral organic free radicals is bimolecular termination (i.e. dimerization: R• + R• → R—R or disproportionation: R• + R• → R_{-H} + R_{+H}, equation 8.30).

$$\mathrm{H{-}C{-}C\bullet + H{-}C{-}C\bullet \rightarrow H{-}C{-}C{-}H + {>}C{=}C{<}} \qquad (8.30)$$

It is envisioned that termination occurs via the two-step sequence outlined in Scheme 8.17. The two radicals diffuse together to form a *diffusive caged-pair* $(R\bullet/R\bullet)_{cage}$, which subsequently reacts to yield products. Because of the high reactivity of most sterically unhindered free radicals, the rate limiting step for termination is usually diffusion

SCHEME 8.17

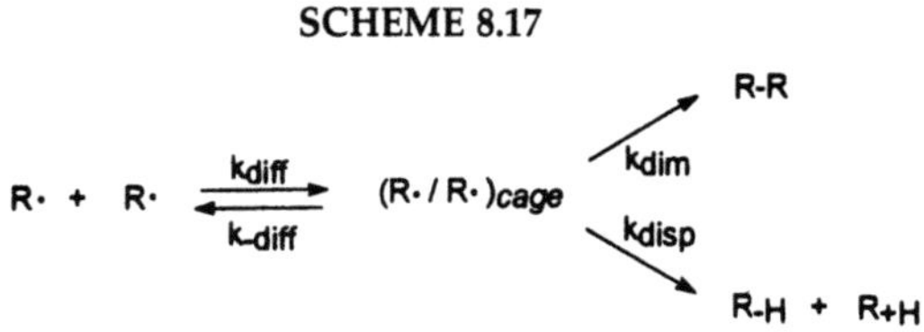

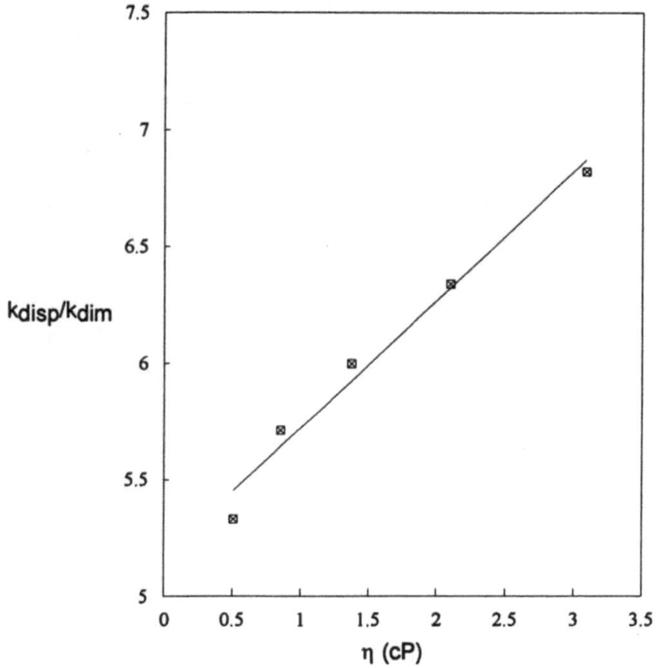

FIGURE 8.13 The disproportionation/dimerization ratio (k_{disp}/k_{dim}) for *t*-butyl radical as a function of solvent viscosity (data from references 60 and 61).

($k_T = k_{diff}$). These diffusion-controlled reactions are typically characterized by rate constants of the order 10^9–$10^{10}\,M^{-1}s^{-1}$ and by very low activation energies (*c.* 3 kcal/mol) in relatively non-viscous solvents [59].

The relative yields of disproportionation and dimerization products is solvent dependent in some cases. For *t*-butyl and 2-hydroxy-2-propyl radicals, k_{disp}/k_{dim} increases with viscosity (Figure 8.13) [60,61].

In contrast, for the 2-propyl radical, no such variation with viscosity is observed, and k_{disp}/k_{dim} remains invariant and approximately equal to unity [60]. This dichotomous behavior is thought to have its origins in the differing geometric requirements of the reactants in each reaction channel. *t*-Butyl and 2-hydroxy-2-propyl are radicals whose shapes roughly resemble an ellipsoid. For combination to occur, the collision must occur along the principal axes of the two colliding radicals. If the orientation is not correct upon first contact, a rotational reorientation of some element within the encounter complex (diffusive caged-pair) must occur. In highly viscous media, this reorientation becomes slower, and the disproportionation reaction (whose reorientation requirements for reaction are less stringent) becomes the more important reaction. For the 2-propyl radical, the reorientation requirements for the combination and disproportionation reactions are roughly equivalent, thus, the ratio k_{disp}/k_{dim} is relatively unaffected by changes in solvent viscosity [60].

The rate constant of a diffusion-controlled reaction is expected to exhibit a strong dependence on solvent viscosity since this solvent property mitigates the mobility of the two reacting species in solution. Quantitatively, the effect of solvent viscosity on the termination rate constant (k_T) of radicals involves a modification of the Smoluchowski equation (equation (8.9)) wherein a statistical factor is introduced. The statistical factor ($\sigma = \frac{1}{4}$) takes into account that of the four possible electronic states (one singlet and three triplet states for the radical pair), only radical pairs in the singlet state can react. The diffusion coefficient can be approximated by the Stokes–Einstein equation (see above). Thus, the termination rate constant $2k_T$ is expected to vary linearly as a function of temperature (T) and viscosity (η) according to equation (8.31) [62].

$$2k_T = \frac{RT}{1500\eta}\,\mathrm{M^{-1}\,s^{-1}} \qquad (8.31)$$

The reactions of *t*-butyl radical have been studied over a wide range of solvents and temperatures, and the evidence suggests complete diffusion control for the reaction, with $2k_T = 3.3 \times 10^9\ \mathrm{M^{-1}\,s^{-1}}$ at room temperature in tetradecane [62]. The activation energy (E_a) was found to be very close to the barrier for diffusion in that solvent. In addition, $2k_T$ varies in a linear manner with T/η as predicted by the classical model for diffusion controlled reactions (equation (8.31), Figure 8.14). All these observations can be taken as evidence for diffusion-control of this reaction [62].

Similarly, the dimerization of benzyl radicals is also essentially diffusion-controlled, with $2k_T = 4.1 \times 10^9\ \mathrm{M^{-1}\,s^{-1}}$ at room temperature in cyclohexane. Similar viscosity dependence for self-termination has been established for 2-propyl, 2,6-dichlorobenzyl, hydroxymethyl, 2-hydroxy-2-propyl and *t*-butyl acyl radicals [63].

As noted thus far, most radical termination reactions can be understood on the basis of their diffusion together to form a transient encounter complex (diffusive caged-pair). In this complex, the radicals are not bound, per se, but their mobility (e.g. their ability to separate) is obstructed by the surrounding solvent molecules.

A radical termination reaction will fall below the diffusion controlled limit if one of the radicals is stabilized by the solvent, as in the case of reaction of alkyl and nitroxide radicals in polar solvents (see above).

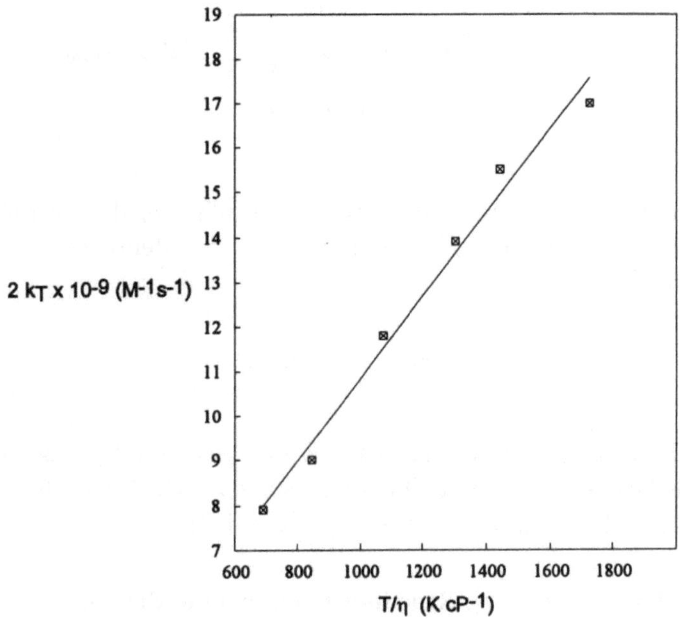

FIGURE 8.14 Variation in $2k_T$ for *t*-butyl radicals in heptane (data from reference 62).

Another phenomenon that may also result in k_T falling below the diffusion-controlled limit occurs when specific intermolecular interactions take place between the two radicals in the encounter complex (e.g. hydrogen bonding, π-complexation, etc.).

A good example of the latter effect is seen in the bimolecular termination reactions of phenoxyl radicals, which are considerably slower than would be predicted for diffusion-control. For 2,6-di-*t*-butylphenoxyl and 2,6-di-*t*-butyl-4-methylphenoxyl radicals, $2k_T = 2.6 \times 10^8$ and 1.8×10^8 M^{-1} s^{-1}, respectively (solvent *n*-heptane, at 254 K) [64]. The diminution of $2k_T$ by almost two orders of magnitude was shown not to derive from steric hinderance of the *t*-butyl groups, since the activation energy for the overall termination process was considerably below the barrier for diffusion. Furthermore, the similarly sterically hindered radical, 2,4,6-tri-*t*-butyl-benzyl was shown to terminate at diffusion controlled rates. An explanation was put forth that proposes a two-step process for the overall termination reaction (Scheme 8.18). In the first step, a diffusion controlled encounter between the two radicals produces a complex in which the two radicals are held together by some intermolecular attraction. This complex proceeds to form the dimeric product or decomposes back to the

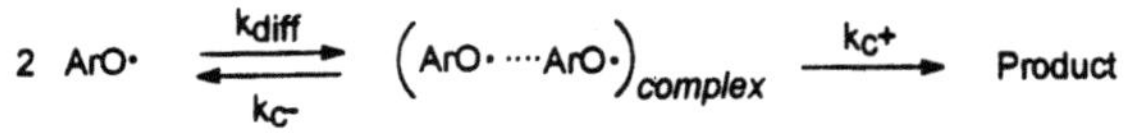

SCHEME 8.18

individual free radicals. According to this scheme, the overall rate is mitigated by the ratio k_{C^-}/k_{C^+}, giving rise to the attenuated value of $2k_T$ (equation 8.32) [64,65].

$$2k_T = \frac{2k_{\text{diff}}}{1 + k_{C^-}/k_{C^+}} \tag{8.32}$$

The stability and partitioning in k_{C^-}/k_{C^+} of this complex was shown to correlate best to the polarizability of the solvent (which reflects the ability of the solvent to respond to an electric field) [64].

Radical Forming Reactions (Geminate Radical/Radical Caged-Pairs)

Azo, peroxide and perester compounds have found widespread application as free radical initiators. The utility of these compounds as sources of free radicals stems from the fact that many of these compounds undergo thermal and or photochemical degradation to yield free radicals in a safe, reliable and predictable fashion (Figure 8.15). In the case of azo and perester compounds, the decomposition reaction proceeds with the concomitant production of molecular nitrogen and carbon dioxide, respectively. The rates of radical production are relatively insensitive to solvent polarity as illustrated by the rate constants for decomposition of *t*-butylperoxy-*α*-phenylisobutyrate ((**25**), Table 8.7) [63,66].

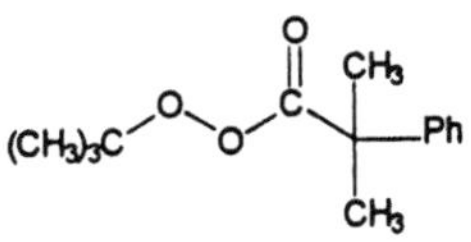

(25)

In contrast to the lack of a solvent polarity effect on unimolecular homolysis reactions, the effects of solvent viscosity are important and well-documented. In solvents of low viscosity, the initially formed (*geminate*) radical/radical caged-pair is relatively unencumbered by the surrounding solvent molecules and readily diffuses out of the solvent

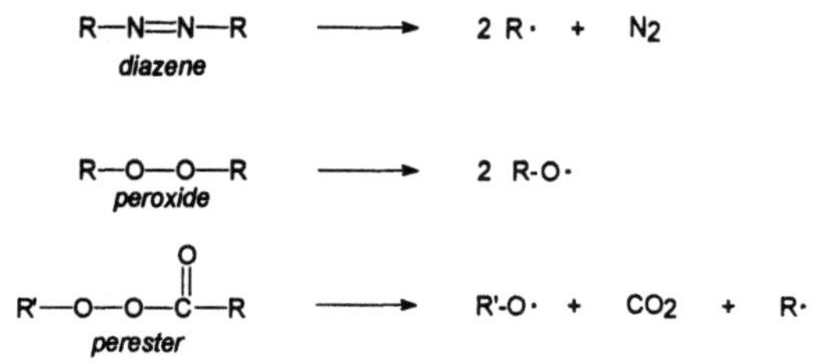

FIGURE 8.15 Typical classes of free radical initiators.

cage. These free radicals subsequently react *via* diffusive encounters in solution. In solvents of high viscosity, the lifetime of the geminate radical/radical caged-pair increases due to the reduction in translational mobility of the pair. Thus, in solvents of high viscosity a greater probability exists for an 'in-cage' termination reaction of the radical pair, which in some instances, regenerates the original radical precursor.

Quantitative relationships for the diffusive behavior of geminate radical/radical caged-pairs as a function of solvent viscosity have been developed. Generally, the effective rate constant for radical production can be correlated to $1/\eta$ or $\eta^{-1/2}$ [67,68].

The effect of solvent viscosity on the rate constant for the decomposition of *t*-butyl perbenzoate ((**26**), Figure 8.16) illustrates this effect [69].

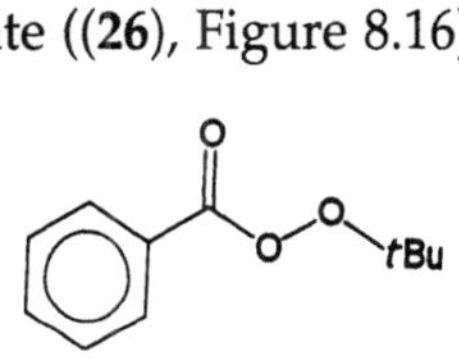

(**26**)

Another illustration of the effect of viscosity on the behavior of geminate radical/radical caged-pairs is provided by the photolysis of azomethane. In solution, photolysis of azomethane yields two products (Scheme 8.19): ethane (resulting from the in-cage reaction of the geminate methyl radical pair) and methane (resulting from cage escape and hydrogen abstraction from solvent). The relative yields of these two products varies linearly with $1/\eta$ (Figure 8.17) [70].

For certain free radical initiators, e.g. perester and azo compounds, there exists two possible fragmentation pathways: fragmentation can occur via sequential one-bond homolyses or alternatively, a simultaneous two-bond homolysis (Figure 8.18).

An elegant test to determine which of the these two pathways is operating in a particular radical initiator based upon the system's response to changes in solvent viscosity was devised by Pryor *et al.* [71].

TABLE 8.7 Rate constants for the decomposition of *t*-butylperoxy-*α*-phenylisobutyrate in several solvents at 60°C[a]

Reaction	$k \times 10^4$ (s^{-1})
Hexadecane	3.46
Eicosane	3.64
Dodecane	2.82
Cumene	5.22
Toluene	5.66
Chlorobenzene	6.98
Benzophenone	7.75
Methyl benzoate	7.55

[a] All data from reference 66.

This test is based on the premise that molecules that decompose by initially breaking one bond will show a retardation in rate for homolysis with increasing solvent viscosity because the initially formed radical pair

FIGURE 8.16 Effect of solvent viscosity on the rate constant for decomposition of *t*-butyl perbenzoate (26) at 130°C (data from reference 69).

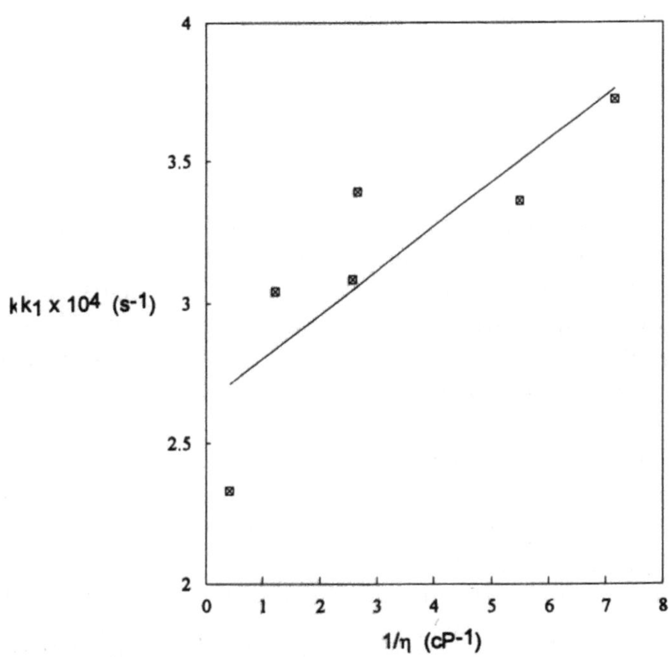

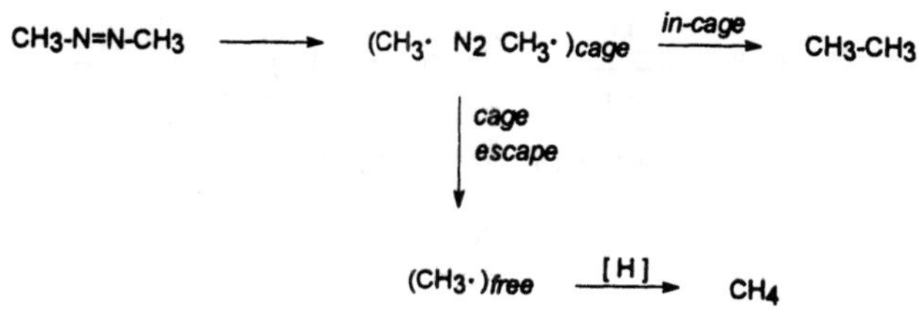

SCHEME 8.19

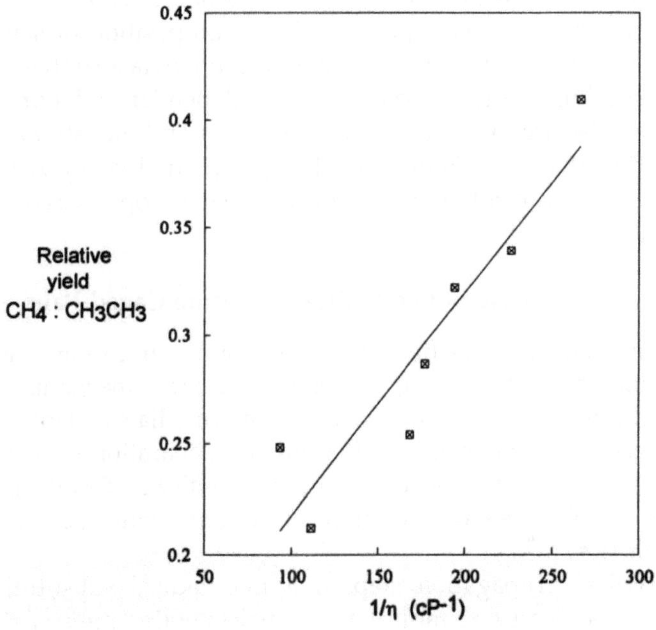

FIGURE 8.17 Ratio of methane:ethane in the photodecomposition of azomethane in various solvents 25°C (data from reference 70).

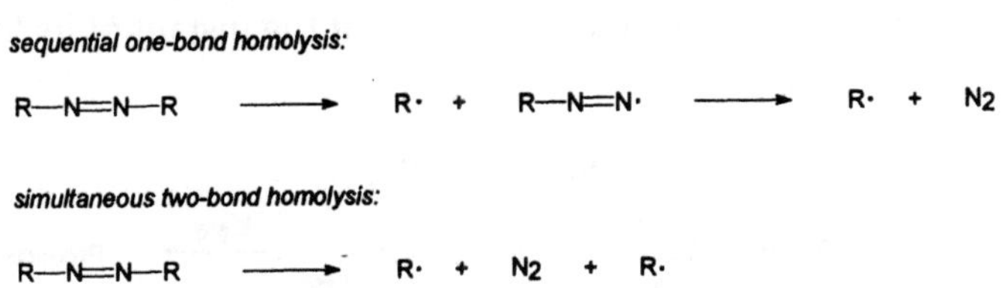

FIGURE 8.18 Possible fragmentation pathways for diazene initiators.

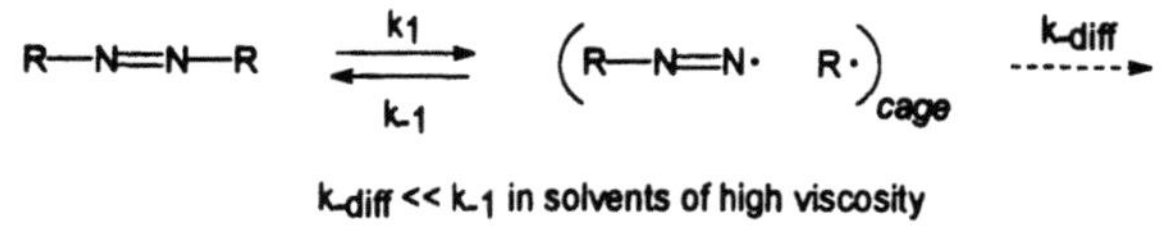

SCHEME 8.20

can undergo greater cage recombination in viscous media (Scheme 8.20).

In contrast, molecules that decompose *via* the simultaneous two-bond scission will be relatively insensitive to viscosity effects, since the probability of reforming two bonds is very low (Scheme 8.21).

The viscosity test was applied to the decomposition of several compounds, including azocumene and *p*-nitrophenylazotriphenylmethane (NAT). Based upon the observed viscosity dependence (Figure 8.19), the former was deduced to be a two-bond initiator and the latter a one-bond initiator [71]. (Note: in Figure 8.19, $1/k_{obs}$ is plotted vs. $(\eta/A_v)^{1/2}$, where A_v is the preexponential factor for the macroscopic viscosity of the solvent) [72].

Reactive Geminate Radical/Molecule Caged-Pairs

A unique type of cage-effect has been found in certain free radical substitution reactions that (a) proceed via a chain mechanism, and (b) involve highly reactive radicals as intermediates. Unlike typical cage-effects in radical chemistry that involve separation vs. reaction of a germinate or diffusive caged-pair of two radicals $(R\bullet/R'\bullet)_{cage}$, these reactions involve in-cage reaction of a germinate radical/molecule caged-pair $(R\bullet/M)_{cage}$.

In a typical propagation step of a free radical substitution chain process, a chain-carrying radical •X—Y reacts with a molecule A—B via atom abstraction (equation (8.33)):

$$\bullet X{-}Y + A{-}B \rightarrow A{-}X{-}Y + B\bullet \tag{8.33}$$

In instances where the radical produced in this step (B•) is *extremely reactive*, it may react with product A—X—Y at the instant of its 'birth' (as

SCHEME 8.21

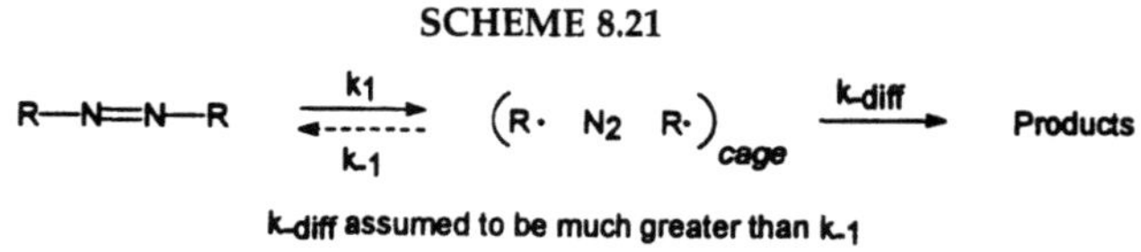

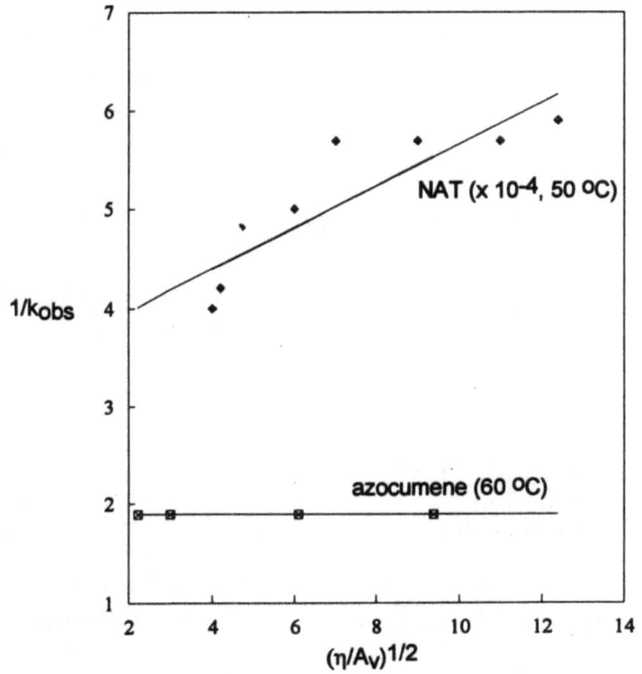

FIGURE 8.19 Rate constants for the decomposition of azocumene and NAT vs. viscosity (data from reference 71).

a geminate caged-pair) on a time scale competitive with cage escape (Scheme 8.22). This reaction of the geminate A—X—Y/B• caged-pair, in which B• abstracts Y from A—X—Y, leads to products that are different from those produced when A—X—Y and B• simply diffuse apart. As will become evident in this section, this novel type of cage-effect can only occur when the radical B• is extremely reactive. To put it more succinctly, the bimolecular rate constant for reaction of 'free' B• and 'free' A—X—Y

SCHEME 8.22

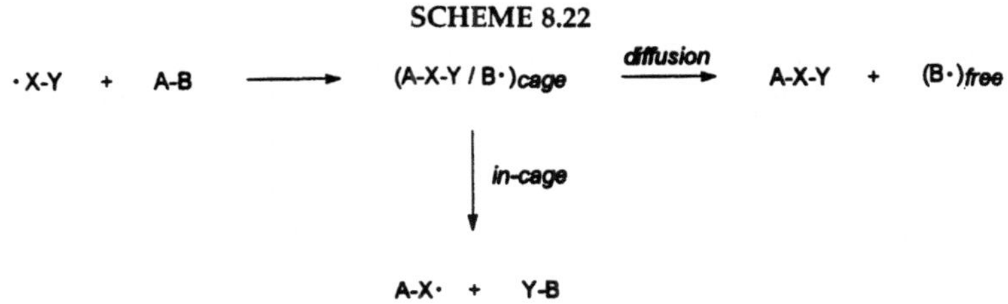

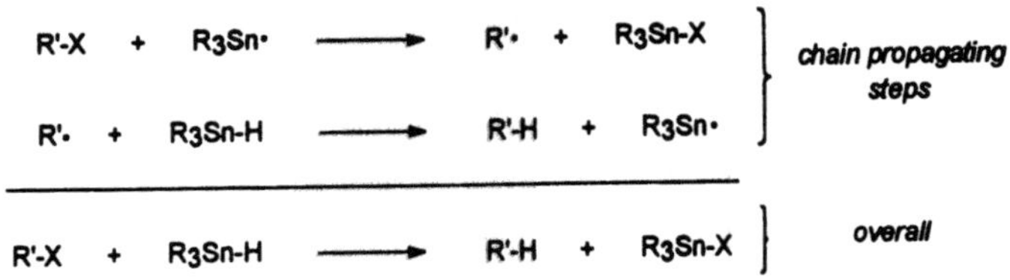

SCHEME 8.23

must be at or near the diffiusion-controlled limit. Clearly, solvent viscosity will be an important factor because of its effect on cage-lifetime. However, the chemical identity of the solvent may also be an important factor, especially if the solvent is reactive towards B• within the lifetime of the cage (see below).

Tin Hydride Reduction of Optically Active Cyclopropyl Halides. An early example where the course of a radical reaction was altered because of the involvement of a geminate radical/molecule caged-pair involves the reduction of cyclopropyl halides by tin hydrides. The chain-propagating steps for reduction of an alkyl halide by a tin hydride (R_3SnH) are depicted in Scheme 8.23. The reaction is generally initiated either photochemically, or thermally with a radical initiator. Because alkyl radicals (R′•) are essentially planar, optically active starting alkyl halide (R′-X) generally yields racemic product [73].

In 1970, Altman [74] reported that the reduction of optically active cyclopropyl bromide (**27**) with di-*n*-butyltindihydride (n-Bu_2SnH_2) resulted in a net retention of configuration. Moreover, the enantiomeric excess (%ee) of the product increases with decreasing n-Bu_2SnH_2 concentration and increases with increasing solution viscosity (Table 8.8).

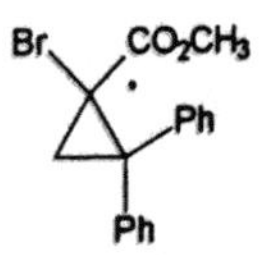

(**27**)

A mechanism consistent with these observations is presented in Scheme 8.24. Abstraction of bromine from (**27**) by •$SnBu_2H$ yields the cyclopropyl radical/$HSnBrBu_2$ caged-pair. Because it possesses a reactive hydrogen, $HSnBrBu_2$ is able to deliver a hydrogen to cyclopropyl radical (**28**) from the same face from which the bromine was removed yielding optically active product. Alternatively, diffusion apart yields 'free' (**28**), which can be attacked via diffusive encounter with H_2SnBu_2 at both faces with equal probability.

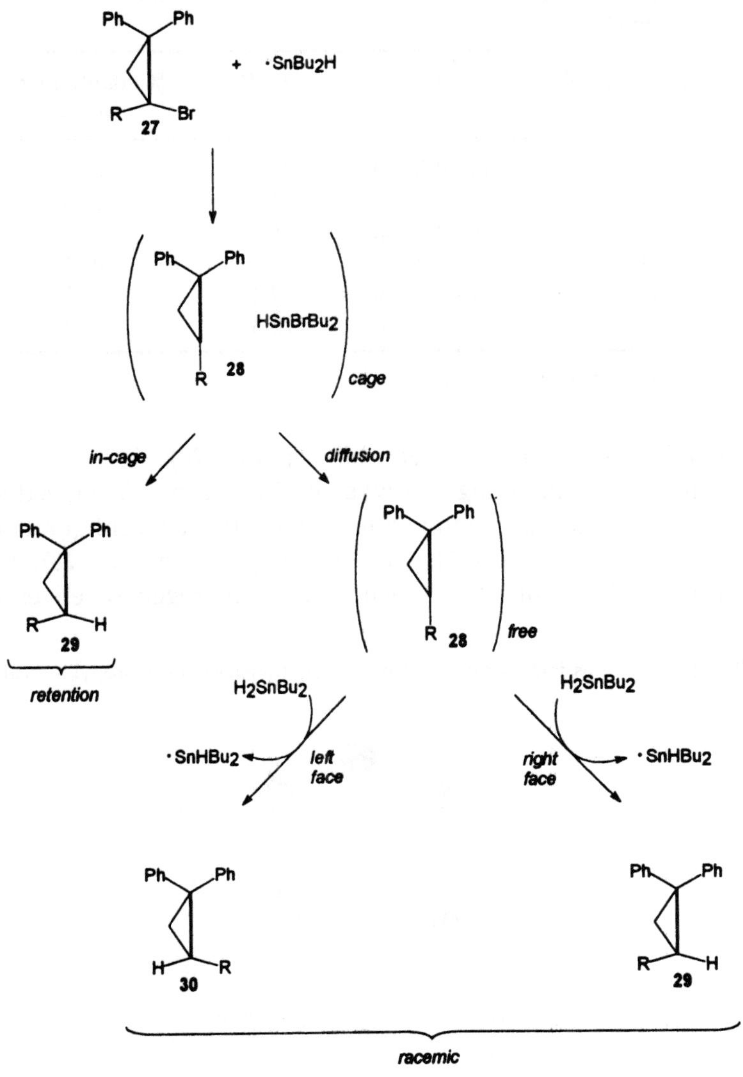

SCHEME 8.24

Thus, %ee increases with increasing viscosity because high solvent viscosity hinders both the translational and rotational diffusion of cyclopropyl radical (**28**), resulting in a longer cage lifetime and a greater amount of in-cage reaction. The dependance of %ee on [H_2SnBu_2] can be explained as follows: as the concentration of H_2SnBu_2 increases, this species becomes an increasingly important constituent of the cage-walls. Consequently, a hydrogen atom can be delivered to the opposite face by

TABLE 8.8 Reduction of cyclopropyl bromide **27** with n-Bu_2SnH_2[a]

n-Bu_2SnH_2, M	solvent	η, cP	% enantiomeric excess
0.71	paraffin oil	22.3	2.77
0.31	paraffin oil	51	4.03
0.173	paraffin oil	66	5.84
0.088	hexane	0.45	0.41
0.088	hexadecane	2.47	1.77
0.088	paraffin oil	75.8	5.87

[a] All data from reference 74.

the H_2SnBu_2 comprising the cage walls (Figure 8.20).

Finally, it is worth noting that when optically active (**27**) is reduced in neat Ph_3SnH, a 2.3%ee is found with a net inversion of configuration [75]. In this case, reaction of (**27**) with •$SnPh_3$ generates the (**28**)/$BrSnPh_3$ caged-pair, but now $BrSnPh_3$ does not have a hydrogen to deliver to (**28**),

FIGURE 8.20 Possible 'in-cage' hydrogen atom transfer to cyclopropyl radical (**28**).

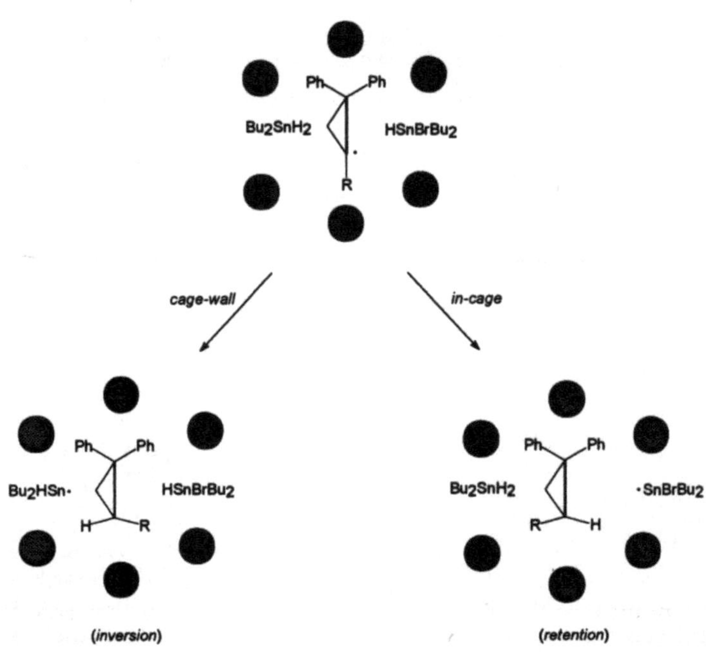

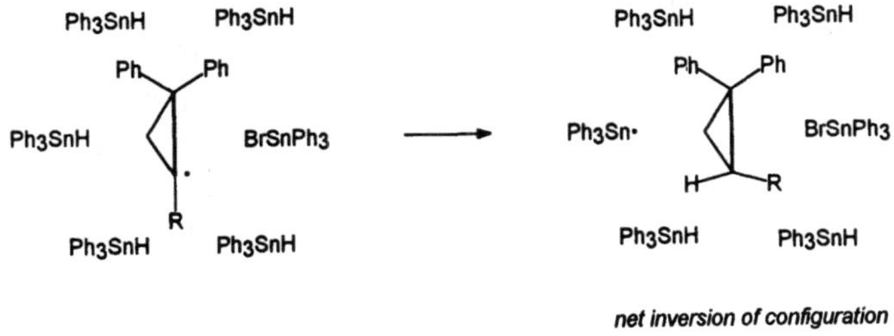

FIGURE 8.21 In-cage reaction of cyclopropyl radical (**28**) resulting in net inversion of configuration.

so the face from which the bromine was removed is effectively blocked. Consequently, the cage walls, composed solely of Ph_3SnH, deliver hydrogen to the opposite face resulting in net inversion of configuration (Figure 8.21).

The Chlorine Atom Cage-Effect. As noted earlier, the thermally- or photoinitiated reaction of molecular chlorine with an alkane yields the corresponding alkyl chloride and HCl via the chain process summarized in Scheme 8.13. Selectivity for this reaction is low (i.e. all possible monochlorides are produced in significant yields from alkanes that possess chemically non-equivalent hydrogens) because of the high reactivity and low selectivity of Cl• as a hydrogen abstractor. Despite the obvious maturity of the field, an important and intriguing aspect of free radical chlorinations remained unnoticed until 1985 when Skell and Baxter [76] found that chlorination of a low concentration of an alkane (in an inert solvent such as CCl_4) produced unexpectedly high yields of polychlorides. Thus while chlorination of cyclohexane (neat) yields only 6% of polychlorinated material (di- and tri-chlorocyclohexanes), chlorination of 0.03 M cyclohexane (in CCl_4) resulted in >50% polychlorinated material [76].

Undoubtedly, the reason that this facet of chlorine atom chemistry waited so long to be discovered is that typical, laboratory-scale procedures for alkane chlorination do not require the use of solvent, mostly for practical reasons. If a solvent were used, then it would have to be separated from the reaction products at the end of the reaction, which might not be trivial in cases where the desired product(s) were volatile. Moreover, the yields of monochlorinated products tended to be higher when the reaction was conducted neat, although it had been assumed up until that point that the polychlorides resulted from further chlorination of the monochlorinated product.

Skell and Baxter definitively ruled out the possibility that the polychlorides resulted from the further diffusive chlorination of the reaction products (i.e. $C_6H_{12} \rightarrow C_6H_{11}Cl \rightarrow C_6H_{10}Cl_2 \rightarrow$ etc.) on the following basis: in these experiments, the starting ratios of cyclohexane to Cl_2 were held constant at 10:1. Thus, at all times during the course of the reaction unreacted cyclohexane was present in nearly a ten-fold excess relative to any chlorinated cyclohexane. Furthermore, chlorocyclohexane, dichlorocyclohexane, etc. are all less reactive than cyclohexane towards chlorine atom [76].

To explain these results, it was suggested that at low alkane concentrations, polychloride formation was the result of a unique cage-effect. For simplicity, let RH_2 represent an alkane possessing two abstractable hydrogens. The proposed mechanism is depicted in Scheme 8.25. The first propagation step is the same as the 'normal' mechanism (Scheme 8.13), namely Cl• abstracts hydrogen from RH_2 yielding HCl and an alkyl radical. In the second step, encounter of alkyl radical (RH•) and molecular chlorine yields an alkyl chloride/chlorine atom geminate caged-pair $(RHCl/Cl\bullet)_{cage}$. This caged-pair partitions itself between three pathways, only two of which yield monochloride: (a) diffusion apart yielding monochloride RHCl and free Cl•, (b) reaction of Cl• with the alkane comprising the cage-walls, yielding HCl, RH•, and monochloride RHCl, and finally, (c) a second in-cage abstraction of hydrogen from RHCl by Cl• forming RCl• (which subsequently reacts with Cl_2 to yield dichloride RCl_2).

Scheme 8.25 completely explains the experimental observations. At low

SCHEME 8.25

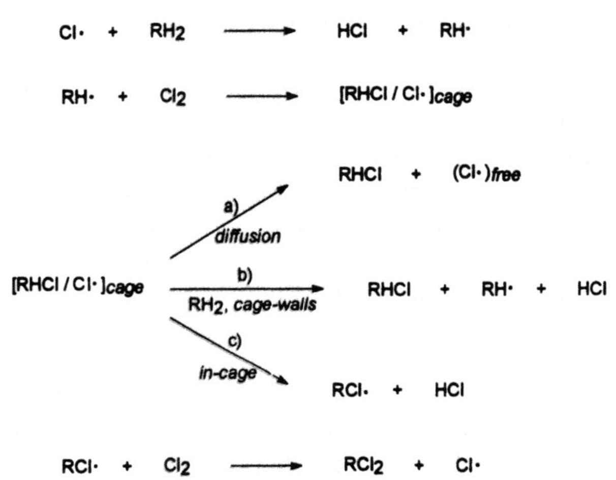

TABLE 8.9 Sites of Cl• attack arising from the diffusive and geminate cyclohexyl chloride/chlorine atom caged-pairs[a]

Diffusive c-$C_6H_{11}Cl$/Cl• caged-pair[b]	Geminate c-C_6H_{11} Cl/Cl• caged-pair[c]	Difference
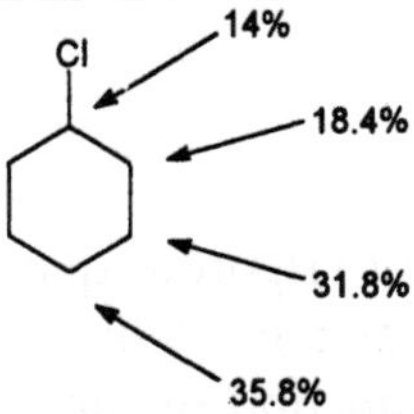	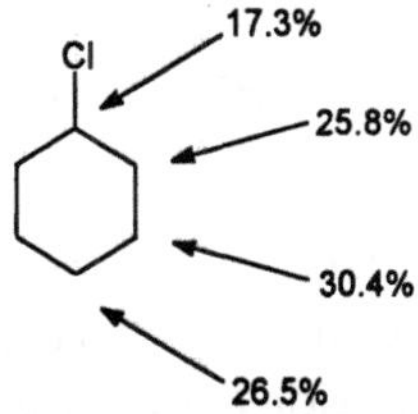	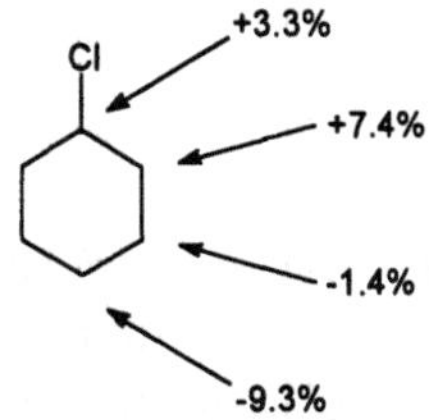

[a] All data from references 77 and 78.
[b] Obtained via chlorination of 0.5 M cyclohexyl chloride in CCl_4.
[c] Obtained via chlorination of 0.5 M cyclohexane in CCl_4.

alkane (RH_2) concentrations, the cage walls are comprised mostly of inert solvent molecules, consequently, path b (reaction of Cl• with RH_2 comprising the cage-walls) is inaccessible. The caged RHCl/Cl• pair partitions itself almost equally between paths (a) (which yields monochloride) and (c) (which yields polychloride). As the alkane concentration is increased, path (b) becomes increasingly important to the point where it becomes the dominant pathway at high alkane concentrations.

Other workers have confirmed and extended these observations. It has been shown that abstraction of the second hydrogen within the RHCl/Cl• caged-pair occurs at the site(s) in closest proximity to the initially formed chlorine atom. Raner *et al.* [77] studied the regiochemistry of the second hydrogen abstraction in the cyclohexyl chloride/chlorine atom geminate caged pair produced in the chlorination of cyclohexane. Their results (Table 8.9) clearly show an 'apparent' enhanced reactivity at positions 1 and 2 in the geminate c-$C_6H_{11}Cl$/Cl• caged pair. These workers suggest that the second in-cage hydrogen abstraction occurs on a time scale competitive with molecular rotation [77]. (Analogous results for cyclohexane chlorination were obtained by Tanner *et al.* [78].)

Similarly, Tanko and Anderson [79] found that while 1,3-dichloro-2,2-dimethylpropane and 1,1-dichloro-2,2-dimethylpropane resulted from the diffusive encounter of neopentyl chloride and chlorine atom in a ratio (1,3:1,1) of 2.2:1, the geminate caged pair (produced in the chlorination of neopentane) yielded a 1,3:1,1 ratio on the order of 4–5 [79]. The 'apparent' enhanced reactivity of the methyl hydrogens arises because they are the closest to the initially produced chlorine atom (Scheme 8.26).

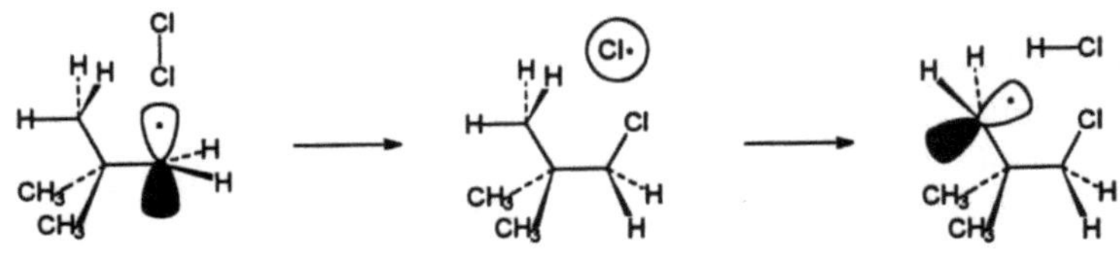

SCHEME 8.26

The chlorine atom cage-effect has been subject to rigorous quantitative treatment. A simple kinetic analysis reveals that the relative yields of monochlorides and polychlorides (M/P) is a function of the rate constants defined in Scheme 8.27, and the concentrations of the pertinent species comprising the cage walls (equation (8.34)).

In Scheme 8.27, an additional reaction has been added: Cl• + S → Cl—S•, where S is any species present in solution which reacts with chlorine atom. The reaction between Cl• and S is not necessarily limited to an addition process as depicted. The only 'chemical' requirement for this reaction is that it consume Cl• in-cage. It is worth noting that S will only perturb the M/P ratio if the bimolecular reaction S + Cl• → is diffusion-controlled. Also, it is assumed that $[RHCl]_{cage}$ does not vary with solvent, hence the in-cage hydrogen abstraction is treated as a pseudo first-order process ($k'_{RHCl} = k_{RHCl}\,[RHCl]_{cage}$).

SCHEME 8.27

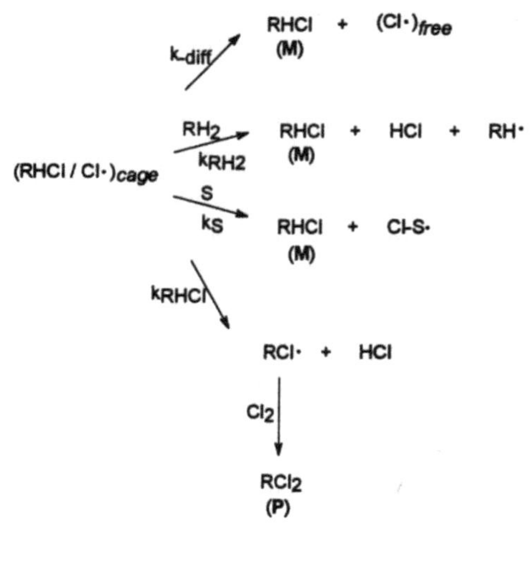

$$\frac{M}{P} = \frac{k_{RH_2}[RH_2]_{cage} + k_S[S]_{cage} + k_{-diff}}{k'_{RHCl}} \tag{8.34}$$

In the absense of any other reactive species in solution (i.e., $[S] = 0$), equation (8.34) predicts that M/P will vary linearly with alkane concentration. Indeed, linearity was observed in the chlorination of neopentane (Figure 8.22) [79] but for some substrates (e.g. 2,3-dimethylbutane and cyclohexane) upward curvature was noted at high alkane concentrations (Figure 8.23) [77,79].

Raner *et al.* [77] hypothesized that the upward curvature observed for 23DMB and cyclohexane suggested that the second hydrogen abstraction within the geminate RHCl/Cl• caged pair occurs on a time scale comparable to molecular rotation. In order for this second hydrogen abstraction to occur, RHCl must rotate in order to provide a hydrogen to Cl• for abstraction. At high alkane concentrations, the cage walls are extremely

FIGURE 8.22 Effect of neopentane concentration on the neopentane mono- to polychloride ratio (data from reference 79).

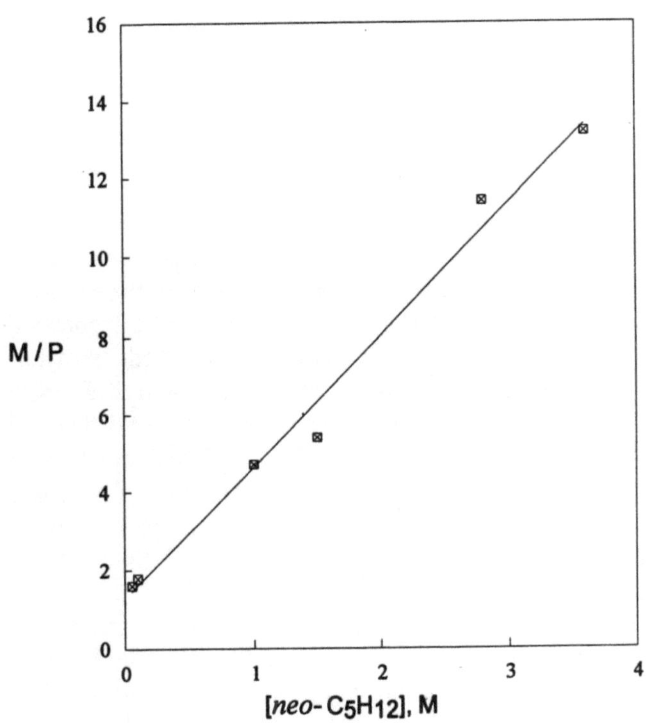

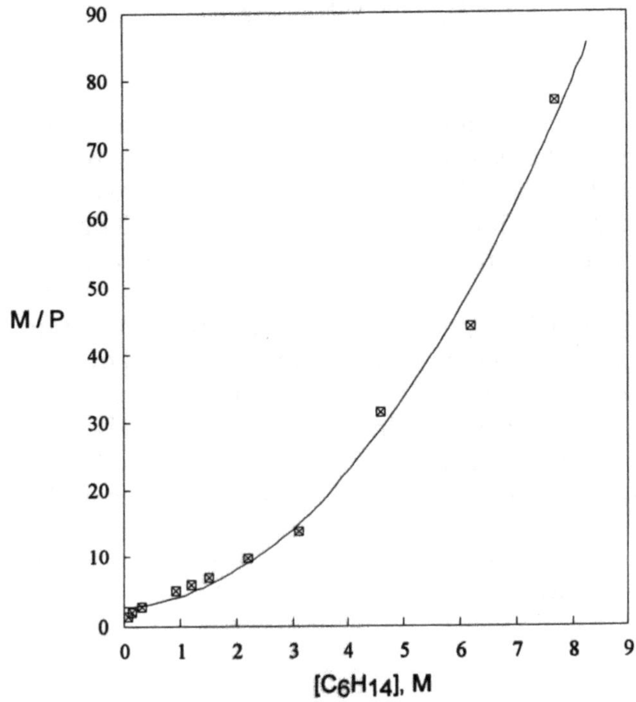

FIGURE 8.23 Effect of 2,3-dimethylbutane (23DMB) concentration on the 23DMB mono- to polychloride ratio (data from reference 79).

reactive and react with Cl• before this rotation can occur, resulting in a higher ratio of mono- to polychlorides than predicted on the basis of equation (8.34). Tanko and Anderson argued that the absence of curvature in the neopentane/Cl_2 system may simply be a consequence of the higher symmetry of neopentyl chloride (i.e. little rotation is needed because the one of the three methyls of neopentyl chloride is in close proximity to Cl• in the geminate caged-pair, Scheme 8.26) [79].

Tanner and coworkers [78] have recently challenged these interpretations by demonstrating that viscosity was not held constant in these experiments. Specifically, as the alkane concentration increases, the solution viscosity decreases. For example, the relative viscosity of 7.6 M 23DMB in CCl_4 is 0.361 versus that of neat CCl_4. (CCl_4 is often used as the inert diluent in these experiments.) Thus, as the concentration of alkane is increased, the diffusional rate constant k_{-diff} (Scheme 8.27) also increases ($k_{-diff} \propto 1/\varepsilon$, see above). Tanner suggests that 'curvature of these type of plots is not, therefore, diagnostic of the rotational requirement for the

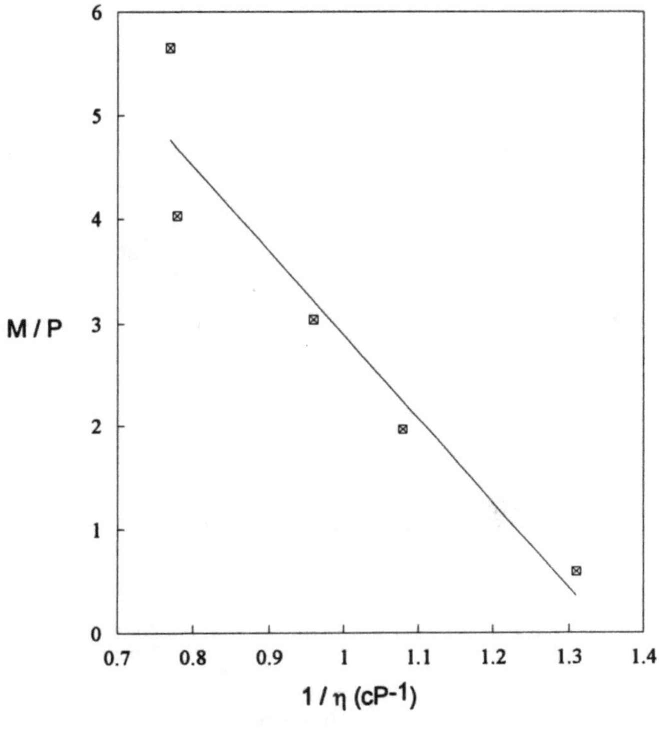

FIGURE 8.24 Mono- to polychloride ratio observed in the chlorination of 0.84 M neopentane as a function of solution viscosity (data from reference 78).

reaction of the caged alkyl halide, while linearity of these plots may only be fortuitous' [78].

Indeed, M/P is expected to vary with viscosity, as Tanner was able to confirm by studying the mono- to polychloride ratio at constant alkane concentration in solvents and solvents mixtures of varying viscosity. (For example, see Figure 8.24.)

Finally, equation (8.34) predicts that (at constant alkane concentration) any species S sufficiently reactive towards Cl• and present is high enough concentration to be a constituent of the cage-walls, will perturb the mono- to polychloride ratio. Moreover, a plot of the M/P ratio vs. [S] should yield a straight line whose slope is equal to the rate constant ratio k_S/k'_{RHCl}. (Note that these are intrinsic rate constants since the reacting species do not need to diffuse together in order to react.) In this manner, Tanko and Anderson were able to measure the relative rates of complexation of chlorine atom with several arenes. (Cl• reacts with benzene at a

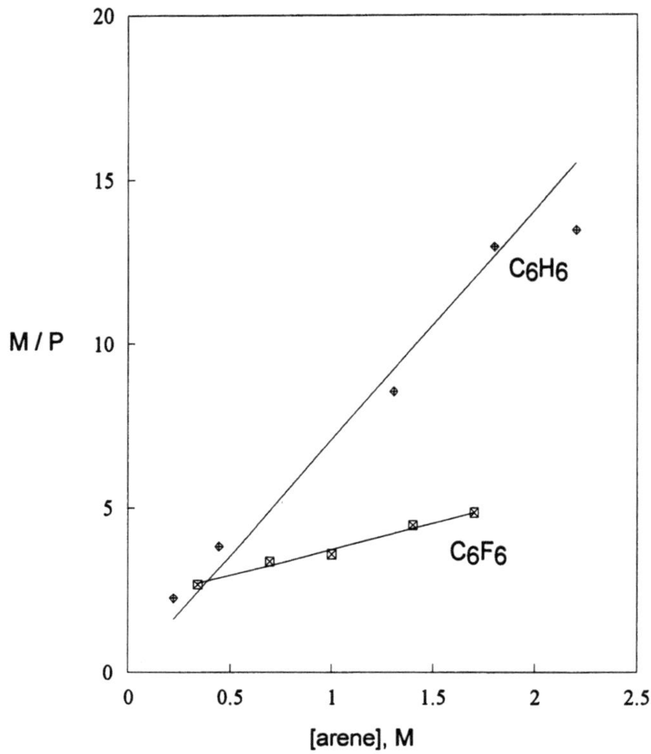

FIGURE 8.25 Effect of benzene and perfluorobenzene on the neopentane (0.1 M) M/P ratio (data from reference 79).

diffusion-controlled rate to yield a complex, see above). Representative data from this work is presented in Figure 8.25 [79].

FUTURE DIRECTIONS: FREE RADICAL REACTIONS IN SUPERCRITICAL FLUID (SCF) SOLVENT

A major limitation associated with any empirical study of the effect of solvents on reaction rate is the necessity to study the reaction in several different solvents. The problem is that each different solvent will present a unique molecular functionality with which the solute(s) may interact. Thus, while one is varying the solvent in order to study the effect of a particular liquid property, the molecular functionality is also changing.

Ideally, one would like to be able to vary important solvent parameters such as dielectric constant and viscosity without changing the solvent. An added bonus would be enjoyed if this 'magic' solvent were also non-toxic and environmentally bengin, unlike most of the solvents commonly used for radical reactions. In a sense, these dreams are becoming reality as the frontiers of radical chemistry move from conventional to supercritical fluid (SCF) solvents derived from natural compounds such as CO_2 and H_2O. This section focuses on the use of SCF media (with an emphasis on CO_2) as solvents for radical reactions.

The supercritical state is attained when a substance is taken above its critical temperature and pressure. Solvent properties of a supercritical fluid such as dielectric constant and viscosity tend to be intermediate between that of a conventional liquid solvent and the gas phase. Moreover, these (and other) solvent properties vary as a function of temperature and pressure. Thus, the possibility of studying solvent effects without varying the solvent is feasible in SCF media. The use of SC-H_2O and SC-CO_2 is especially appealing from an environmental perspective since these are non-toxic, naturally-occurring materials.

Free Radical Chain Reactions in SC-CO_2

The first report of a free radical chain reaction successfully conducted in SC-CO_2 was the oxidation of cumene (Scheme 8.28) by McHugh and co-workers in 1989 [80]. These workers found that CO_2 pressure and viscosity had little effect on the initiation rate constant (k_i) and the ratio of the propagation rate constant to the square root of the termination rate constant ($k_p/k_t^{1/2}$) [80].

DeSimone and co-workers [81] reported in 1992 the homogeneous free radical polymerization of 1,1-dihydroperfluorooctyl acrylate in SC-CO_2 (equation (8.35) as an alternative to the CFC (chlorofluorocarbon)) solvents typically used for the synthesis of fluoropolymers. Homogeneity

SCHEME 8.28

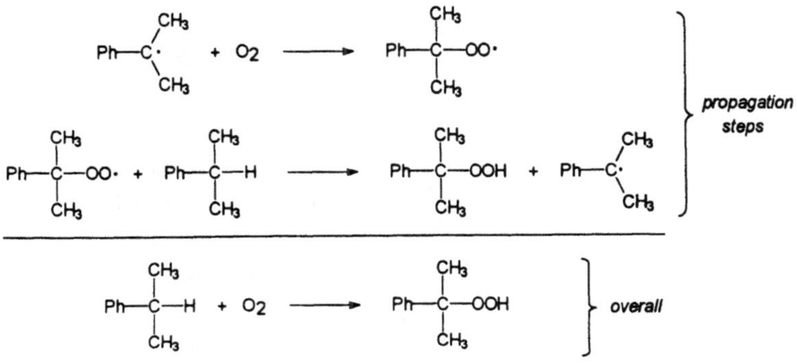

was maintained throughout the course of the reaction, the yield (65%) was good, and the resulting polymer was identical in all respects to that prepared in a Freon solvent [81].

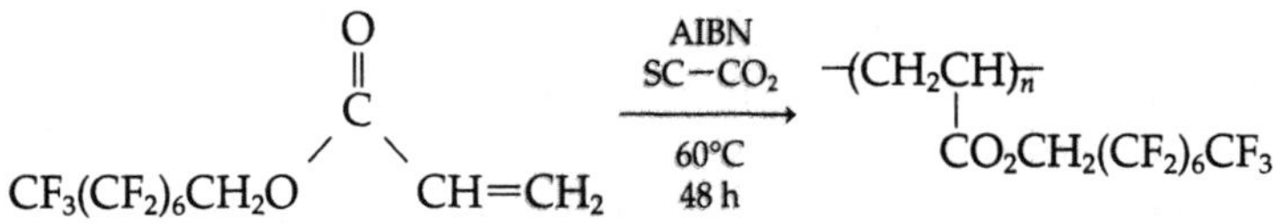

(8.35)

In 1994, Tanko and Blacker [82] reported that SC-CO_2 was a suitable solvent for the free radical brominations of alkylaromatics using molecular bromine. Furthermore, SC-CO_2 was shown to be an effective replacement for CCl_4 as a solvent for the Ziegler reaction, which is a classical method for allylic and benzylic bromination using *N*-bromosuccinimide (NBS, (equation (8.36)) [82].

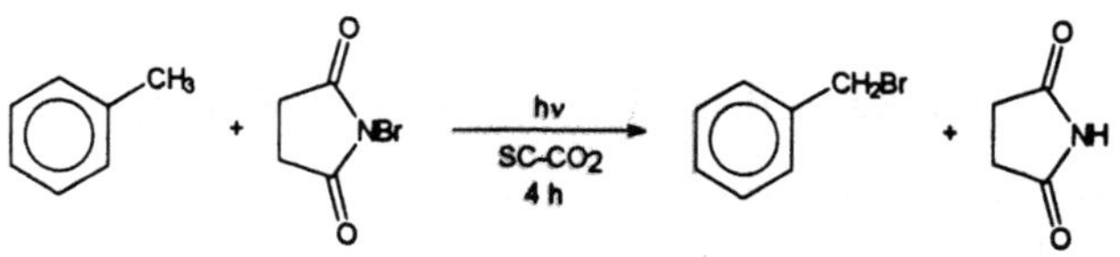

(8.36)

These reports helped establish the feasibility of SC-CO_2 as a medium for free radical chain reactions. Studies of SC-CO_2 as a reaction medium from a physical organic perspective are in their infancy, although it is likely that the field will mature in the next several years. In the remainder of this section, several examples where 'solvent effects' have been studied in SCF media are presented.

8.6.2 Pressure Effects in SCF Solvent

The effect of pressure on the rate constant of a reaction conducted in SCF solvent is similar to the effect of external pressure when the reaction is conducted in a conventional organic solvent (equation (8.7)). Generally, this behavior is observed for pressure-sensitive reactions such as the Diels–Alder reaction at moderate to high pressures. However, at pressures, just above the critical point, anamolous behavior is observed (i.e. the rate constant varies dramatically with pressure). This anamolous behavior in the vicinity of the critical pressure is the result of a phenomenon unique to SCF media call *clustering*, wherein solvent molecules congregate in close proximity about a solute molecule. (For example, in the Diels–Alder reaction of isoprene and maleic anhydride in SC-CO_2, Paulaitis and Alexander found [83] that at CO_2 pressures greater than 150 bar, the variation of ln(k) with pressure paralleled that observed in ethyl acetate solvent (corresponding to $\Delta V_{act} = -37\ cm^3/mol$). However, at pressures just above the critical pressure of CO_2, the slope of the ln(k) vs. P plot corresponds to $\Delta V_{act} \approx -500\ cm^3/mol$ [83].

An example of the apparent effect of solvent clustering on the rate of a radical reaction in SC-CO_2 was provided by Chateauneuf [84], who measured absolute rate constants for hydrogen atom abstraction by triplet benzophenone (BP^3) from isopropyl alcohol and 1,4-cyclohexadiene (equation (8.37)). The rate constant was observed to drop by about one order of magnitude as pressure was increased 10–15 bar above the critical pressure of CO_2. As pressure was increased further, however, the rate constant did not vary significantly (Figure 8.26) [84].

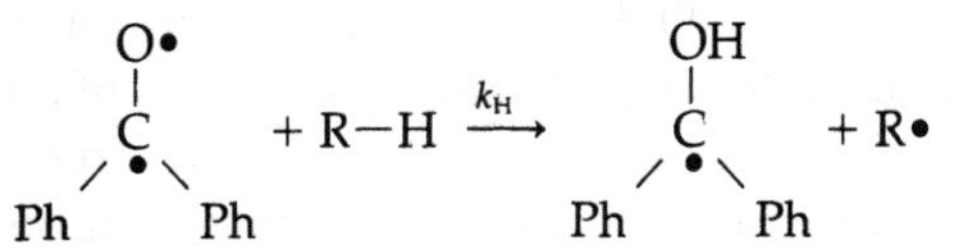

(BP^3) (8.37)

However, this anamalous pressure effect in SCF media does not always

FIGURE 8.26 Effect of CO_2 pressure on the rate constant for abstraction of hydrogen from cyclohexadiene by triplet benzophenone at 33°C (data from reference 84).

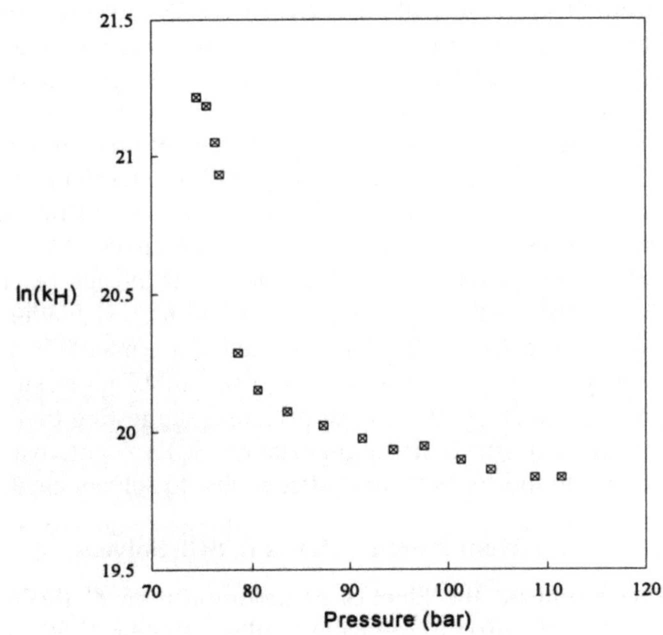

manifest itself in terms of altered reaction selectivities. For example, Tanko and Blackert [82] examined the free radical bromination of toluene and ethylbenzene in SC-CO_2. Within experimental error, the relative reactivity of ethylbenzene vs. toluene in SC-CO_2 was pressure insensitive (including at pressures in vicinity of the critical point), and nearly identical to that observed in conventional organic solvents [82]. This insensitivity to pressure may simply arise from the fact that the difference in the activation volumes for hydrogen abstraction from toluene and ethyl benzene is small (<5 cm^3/mol [14].

Viscosity Effects in SC-CO_2

Because the viscosity of a SCF is significantly less than that of a liquid solvent, it is reasonable to suspect that the lifetimes of both geminate and diffusive caged-pairs will be less in SC-CO_2. For free radical initiators (azo compounds, peroxides, etc.), in-cage recombination of the geminate radical/radical caged-pairs is less likely to occur in SC-CO_2 resulting in higher efficiency in the production of free radicals. Similarly, because of the diminished lifetimes of diffusive caged-pairs, it is likely that diffusion-controlled rate constants in SCF can exceed the $10^{10}\ M^{-1}\ s^{-1}$ limit imposed by conventional solvents.

Chateauneur [85] examined triplet–triplet annihilation of benzophenone in SC-CO_2 *via* laser flash photolysis. The observed rate constants varied from $2k_{TTA} = 1.4 \times 10^{11}$ to 3.4×10^{10} at 35°C at pressures ranging from 73 to 284 bar. The measured rate constants were in excellent agreement with those calculated from measured CO_2 viscosities and exceed the conventional 10^{10} limit imposed by laboratory solvents [85].

Fox and Johnston [86] studied the photolysis of unsymmetrical dibenzyl ketone (**31**) in SC-CO_2. In accordance with the mechanism presented in Scheme 8.29, bibenzyls A—A, A—B, and B—B will be formed in a statistical ratio of 1:2:1 if all bibenzyl formation occurs *via* diffusive encounter of benzyl radicals. To the extent that in-cage reaction of the geminate A•/B• benzyl radical pair is important, deviation from the ratio in favor of the A—B dimer is expected. Experimentally, a statistical distribution of bibenzyls was observed at several CO_2 pressures, including pressures just above the critical pressure, suggesting that because of low viscosity, cage effects are unimportant in SCF solvent. In addition, no evidence was found for any effect attributable to solvent clustering [86].

Solvent Polarity Effects in SCF Solvent

DeSimone examined the kinetics of decomposition of the free radical initiator 2,2′-azobis(isobutyronitrile) (AIBN, equation (8.38)) in benzene

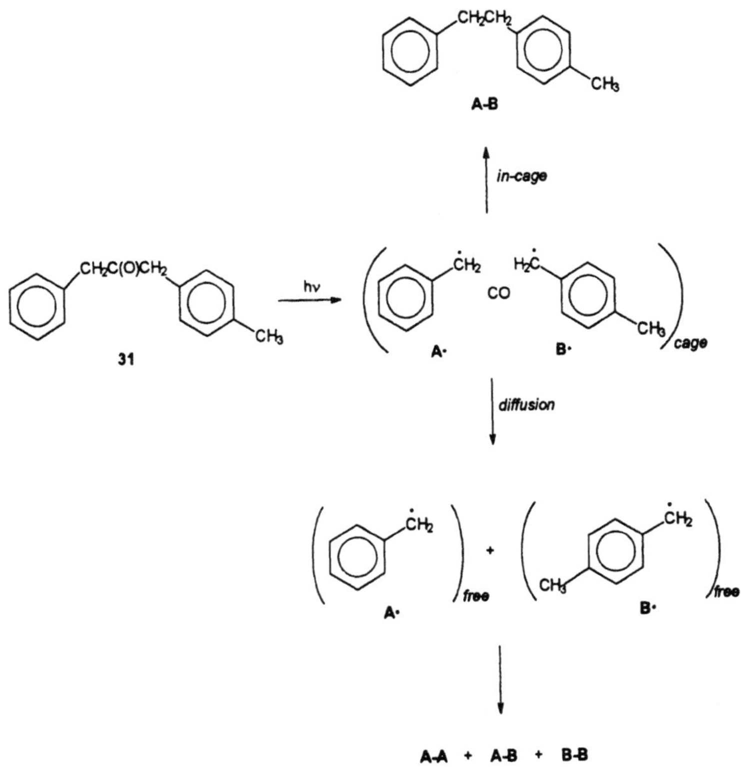

SCHEME 8.29

and SC-CO_2 solvent [81,87]. In conventional solvents, AIBN decomposition rates vary slightly (by a factor of 2–4) as a function of solvent polarity: in (k_d) is found to correlate with the Kirkwood function $(\varepsilon - 1)/(2\varepsilon + 1)$ with the reaction rate increasing with the dielectric constant of the solvent (87). These observations suggest an increase in dipole moment in the progression from reactants to transition state. It is envisioned that either (a) decomposition is preceded by a *trans* → *cis* isomerization, and that the *cis*-form has a greater dipole moment, or (b) an intermediate diazenyl radical (R—N=N•) is produced.

$$\mathrm{NC{-}\underset{CH_3}{\overset{CH_3}{C}}{-}N{=}N{-}\underset{CH_3}{\overset{CH_3}{C}}{-}CN} \xrightarrow{k_d} 2\ \mathrm{(CH_3)_2\dot{C}{-}CN} + \mathrm{N_2} \qquad (8.38)$$

In SC-CO_2, k_d is found to increase with increasing pressure, reaching a

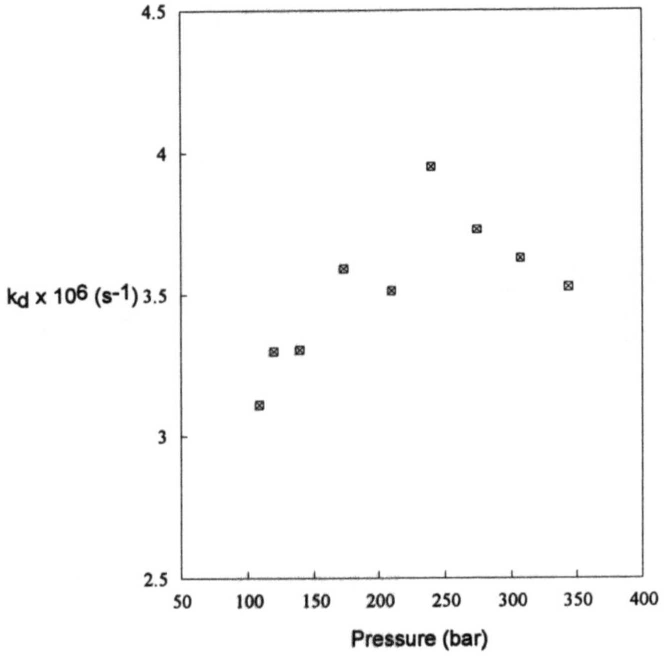

FIGURE 8.27 The effect of CO_2 pressure on the rate constant for the decomposition of AIBN in SC-CO_2 at 59°C (data from reference 87).

maximum at about 250 bar. Afterward, $\ln(k_d)$ decreases with a further increase in pressure (Figure 8.27) [87]. This curvature in the $\ln(k_d)$ vs. P plot is nicely explained by two competing effects: solvent polarity and pressure. As CO_2 pressure increases from 100 to 250 bar, the dielectric constant of CO_2 increases dramatically. Above 250 bar, the dielectric constant continues to increase with pressure, but much more gradually. The measured volume of activation for AIBN decomposition in benzene is $+24\ cm^3/mol$. Thus, in the region 100 to 250 bar, AIBN decomposition rate increases with increasing solvent polarity consistent with what is observed in conventional solvent. At pressures above 250 bar, the polarity increases only modestly with pressure, consequently, the effect of pressure dominates and the rate constant decreases [87].

CLOSING REMARKS

As with any virtually any other class of reactive intermediates, the reactions of neutral free radicals are subject to solvent effects. These solvent effects arise when solvent either induces a change in the Gibbs energy of activation for the reaction or restricts the mobility of the reactants.

Activated complex theory provides the basis for understanding changes in the Gibbs energy of activation due to differential solvation (stabilization) of the reactant(s) and transition state, as is the case when the dipole moment changes and solvent polarity effects are important, or in the more extreme instance where the radical interacts with specific functional groups on the solvent to form a complex ('radicaloid'). Activated complex theory also provides the theoretical base for understanding the effect of internal (and external) pressure on reaction rate, which arises when there is a volume change going from reactant(s) to transition state.

Mobility only becomes an issue when the chemical step involving the radicals is not rate-determining (i.e. reactions involving diffusive or germinate caged-pairs). In these cases viscosity is the important property of the solvent that mitigates the outcome of the reaction.

It is also important to realize that the magnitude of solvent effects in radical reactions is considerably smaller than those observed for polar processes. Thus, while a polar process such as the S_N1 reaction may experience a variation in rate greater than ten orders of magnitude with a change in solvent [88], even the most dramatic solvent effects in reactions of neutral free radicals are not greater than one to two orders of magnitude.

Although the effects are not as dramatic, they can be of practical significance. Radical reactions are becoming an increasingly important methodology in synthetic chemistry. For a radical partitioning itself between two pathways (e.g. cyclization vs. atom abstraction, addition vs. fragmentation, etc.), a one order of magnitude change in rate of one of the steps may mean the difference between a 50% yield and a 90% yield. Consequently, an appreciation for the role of solvent provides a means of manipulating the reactivity/selectivity, and hence the chemistry of neutral free radicals.

Acknowledgments

Financial support from the National Science Foundation (CHE-9113448) and the Thomas F. Jeffress and Kate Miller Jeffress Memorial Trust Fund is gratefully acknowledged. We also thank Dr Keith Ingold and Dr Janusz Lusztyk for helpful comments.

REFERENCES

1. KIRKWOOD, J. G. (1954) *J. Chem. Phys.* **2,** 351.
2. REICHARDT, C. (1988) *Solvents and Solvent Effects in Organic Chemistry, 2nd ed.,* VCH: New York, pp 190–203.
3. $(\varepsilon - 1)/(2\varepsilon + 1) = 1/2 - 3/(4\varepsilon) + 3/(8\varepsilon^2) - 3/(16\varepsilon^3) + \ldots$, see: ESPENSON, J. H. (1981) *Chemical Kinetics and Reaction Mechanisms,* McGraw Hill, New York.
4. C. REICHARDT (1964) *Angew. Chem. Int. Ed,* **3,** 30.
5. BENTLEY, T. W. and LLEWELLYN, G. (1990) In *Progress in Physical Organic Chemistry,* Taft, R. W., Ed., Wiley, New York, pp. 121–158.
6. (a) KNAUER, B. R. and NAPIER, J. J. (1976) *J. Am. Chem. Soc.* **98,** 4395. (b) REDDOCH, A. H. AND KONISHI, S. (1979) *J. Chem. Phys.* **70,** 2121.
7. BARTON, A. F. M. (1975) *J. Chem. Educ.* **48,** 156.
8. (a) BARTON, A. F. M. (1975) *Chem. Revs.,* **75,** 731. (b) BARTON, A. F. M. (1983) *CRC Handbook of Solubility Parameters and Other Cohesion Parameters;* CRC Press, Boca Raton.
9. ALLEN, G., GEE, G. and WILSON, G. J. (1960) *Polymer* **1,** 456.
10. DACK, M. R. J. (1974) *J. Chem. Educ.* **51,** 231.
11. HILDEBRAND, J. H. and SCOTT, R. L. (1964) *The Solubility of Nonelectrolytes,* Dover Publications, New York, pp 62–105, pp 419–434.
12. MOORE, J. W. and PEARSON, R. G. (1981) *Kinetics and Mechanism, 3rd ed.,* Wiley, New York, pp. 46–250.
13. AMIS, E. S. and HINTON, J. F. (1973) *Solvent Effects on Chemical Phenomena,* Academic Press, New York, pp 306–310.
14. ASANO and T. LENOBLE, W. J. (1978) *Chem. Rev.* **78,** 407.
15. POUTSMA, M. L. (1969) In *Methods in Free Radical Chemistry,* Vol. 1, E. S. Huyser, ed., Marcel Dekker: New York, pp 79–193. POUTSMA, M. L. (1973) In *Free Radicals, Vol. 2,* Kochi, J. K., ed., Wiley, New York, pp 159–229.
16. WALLING, C. and WAGNER, P. J. (1964) *J. Am. Chem. Soc.* **86,** 3368.
17. AVILA, D. V., BROWN, C. E., INGOLD, K. U. and LUSZTYK, J. (1993) *J. Am. Chem. Soc.* **115,** 466.
18. KIM, S. S., KIM, H. R., KIM, H. B., YOUN, S. J. and KIM, C. J. (1994), *J. Am. Chem. Soc.* **116,** 2754.
19. ISAACS, N. S. (1987) *Physical Organic Chemistry,* Wiley: New York, pp 791–793.
20. See for example: BOGER, D. L. and MATHVINK, R. J. (1992) *J. Org. Chem.* **57,** 1429 and references therein.
21. TSENTALOVICH, Y. P. and FISCHER, H. J. (1994) *J. Chem. Soc. Perkin Trans 2* **4,** 729.
22. A similar effect is observed in the analagous decarboxylation of the

4-methoxybenzoyloxyl radical (p-$CH_3OC_6H_4CO_2 \rightarrow p$-$CH_3OC_6H_4^{\cdot} + CO_2$) where the rate constant for this process decreases by over an order of magnitude when the solvent is changed from CCl_4 to CH_3CN. See: CHATEAUNEUF, J., LUSZTYK, J. and INGOLD, K. U. (1988) *J. Am. Chem. Soc.* **110,** 2877.

23. BECKWITH, A. L. J., BOWRY, V. W. and INGOLD, K. U. (1992) *J. Am. Chem. Soc.* **114,** 4983.
24. GRILLER, D. and INGOLD, K. U. (1980) *Acc. Chem. Res.* **13,** 317.
25. ABELL, P. I. (1973) In *Free Radicals, Vol. 2,* Kochi, J. K., ed., Wiley, New York, pp. 63–112.
26. ITO, O. and MATSUDA, M. (1982) *J. Am. Chem. Soc.* **104,** 568.
27. NEUMAN, R. C., JR. and GUNDERSON, H. J. (1992) *J. Org. Chem.* **57,** 1641.
28. NEUMAN, R. C., JR. and BINEGAR, G. A. (1983) *J. Am. Chem. Soc.* **105,** 134. NEUMAN, R. C., JR., BERGE, C. T., BINEGAR, G. A., ADAM, W. and NISHIZAWA, Y. (1990) *J. Org. Chem.* **55,** 4564.
29. TANKO, J. M., MAS, R. H. and SULEMAN, N. K. (1990) *J. Am. Chem. Soc.* **1990, 112,** 5557 and references therein.
30. ROBERTS, J. D. and DIRSTINE, P. H. (1945) *J. Am. Chem. Soc.* **67,** 1281.
31. WALLING, C. and FREDRICKS, P. S. (1962) *J. Am. Chem. Soc.* **84,** 3326.
32. TANKO, J. M., SULEMAN, N. K., HULVEY, G. A., PARK, A. and POWERS, J. E. (1993) *J. Am. Chem. Soc.* **115,** 4520.
33. TANKO, J. M. and SULEMAN, N. K. (1994) *J. Am. Chem. Soc.* **116,** 5162.
34. BUNCE, N. J., INGOLD, K. U., LANDERS, J. P., LUSZTYK, J. and SCAIANO, J. C. (1985) *J. Am. Chem. Soc.* **107,** 5464.
35. RUSSELL, G. A. (1973) In *Free Radicals, Vol. 1,* Kochi, J. K., ed., Wiley, New York, pp. 275–331.
36. LÜNING, U. and SKELL, P. S. (1985) *Tetrahedron* **41,** 4289.
37. RUSSELL, G. A. (1957) *J. Am. Chem. Soc.* **79,** 2977.
38. RUSSELL, G. A. (1958) *J. Am. Chem. Soc.* **80,** 4987.
39. RUSSELL, G. A. (1958) *J. Am. Chem. Soc.* **80,** 4997.
40. RUSSELL, G. A. (1958) *J. Am. Chem. Soc.* **80,** 5002.
41. These observations regarding the effect of arene structure on selectivity have been confirmed and extended. See: BUNCE, N. J., JOY, R. B., LANDERS, J. P. and NAKAI, J. S. (1987) *J. Org. Chem.* **52,** 1155. RANER, K. D., LUSZTYK, J. and INGOLD, K. U. (1989) *J. Am. Chem. Soc.* **111,** 3652.
42. WALLING, C. and MAYAHI, M. F. (1959) *J. Am. Chem. Soc.* **81,** 1845.
43. SKELL, P. S., BAXTER, H. N., III AND TAYLOR, C. K. (1983) *J. Am. Chem. Soc.* **105,** 120.
44. SKELL, P. S., BAXTER, H. N., III, TANKO, J. M. and CHEBOLU, V. (1986) *J. Am. Chem. Soc.* **108,** 6300.

45. BÜHLER, R. E. and EBERT, M. (1972) *Nature* **214**, 1220.
BÜHLER, R. E. (1972) *Radiation Res. Rev.* **4**, 233.

46. WALLING, C. (1988) *J. Org. Chem.* **53**, 305.

47. RANER, K. D., LUSZTYK, J. and INGOLD, K. U. (1989) *J. Phys. Chem.* **93**, 564.

48. A common criteria for assignment of charge transfer complexes is linearity in the plot of the charge transfer energy of several complexes with the same acceptor against the vertical ionization potential of the donor, see reference 45.

49. HASSEL, O. and STROMME, K. O. (1958) *Acta Chem. Scand.* **12**, 1146.
HASSEL, O. and STROMME, K. O. (1959) *Acta Chem. Scand.* **13**, 1781.

50. SERGEEV, G. B., PUKHOVSKII, A. V. and SMIRNOV, V. V. (1983) *Russ. J. Phys. Chem.* **57**, 589 (translated from *Zhurnal Fizicheskoi Khimii*, **57**, 977).

51. EDWARDS, J., HILLS, D. J., MISHRA, S. and SYMONS, M. C. R. (1974) *J. Chem. Soc. Chem. Commun.* **1974**, 556. MISHRA, S. P. and SYMONS, M. C. R. (1981) *J. Chem. Soc. Perkin Trans* 2, 185

52. BENSON, S. W. (1993) *J. Am. Chem. Soc.* **115**, 6969.

53. COCHRAN, E. L.; ADRIAN, F. J. and BOWERS, V. A. (1970) *J. Phys. Chem.* **74**, 2083.
YIM, M. B. and WOOD, D. E. (1975) *J. Am. Chem. Soc.* **97**, 1004.

54. JACOX, M. E. (1982) *J. Phys. Chem.* **86**, 670.

55. CHATEAUNEUF, J. E. (1993) *J. Am. Chem. Soc.* **115**, 1915.

56. BRESLOW, R., BRANDL, M., HUNGER, J. and ADAMS, A. D. (1987) *J. Am. Chem. Soc.* **109**, 3799.

57. BRESLOW, R., BRANDL, M., HUNGER, J., TURRO, N., CASSIDY, K., KROGH-JESPERSEN, K. and WESTBROOK, J. D. (1989) *J. Am. Chem. Soc.* **109**, 7204.

58. KHANNA, R. K., ARMSTRONG, B., CUI, H. and TANKO, J. M. (1992) *J. Am. Chem. Soc.* **114**, 6003.

59. LEFFLER, J. E. (1993) *An Introduction to Free Radicals*, Wiley, New York, pp 56–76.

60. SCHUCH, H.-H. and FISCHER, H. (1978) *Helv. Chim. Acta* **61**, 2463.

61. Neuman was able to definitively exclude a possible correlation of this data to solvent internal pressure by examining the effect of externally applied pressure on k_{disp}/k_{dim} for *t*-butyl radical. See:
NEUMAN, R. C. and FRINK, M. E. (1983) *J. Org. Chem.* **48**, 2430.

62. SCHUH, H. and FISCHER, H. (1976) *Int. J. Chem. Kinet.* **8**, 341.
SHUCH, H.-H. and FISCHER, H. (1978) *Helv. Chim. Acta* **61**, 2130.

63. WALLING, C. (1985) *Tetrahedron* **41**, 3887.

64. RUEGGE, D. and FISCHER, H. (1988) *J. Chem. Soc., Faraday Trans. 1* **84**, 3187.

65. See also FETI, M., INGOLD, K. U. and LUSZTYK, J. (1994) *J. Am. Chem. Soc.* **116**, 9440.

66. HERKES, F., FRIEDMAN, J. and BARTLETT, P. D. (1969) *Int. J. Chem. Kinet.* **1**, 193.

67. NOYES, R. M. (1961) *Progr. React. Kinet.* **1,** 129. See also:
KOENIG, T. and FISCHER, H. (1973) In *Free Radicals, Vol 1,* Kochi, J. K., ed., Wiley: New York, pp. 157–189.

68. KOENIG, T. and DEINZER, M. (1968) *J. Am. Chem. Soc.* **90,** 7014.
KOENIG, T. (1969) *J. Am. Chem. Soc.* **91,** 2558.

69. KOENIG, T., BEINZER, M. and HOOBLER, J. A. (1971) *J. Am. Chem. Soc.* **93,** 938.

70. KODAMA, S. (1962) *Bull. Chem. Soc. Jpn.* **35,** 827. See also:
NODELMAN, N. and MARTIN, J. C. (1976) *J. Am. Chem. Soc.* **98,** 6597.

71. PRYOR, W. A. and SMITH, K. (1970) *J. Am. Chem. Soc.* **92,** 5403.

72. The viscosity of a solvent can be written in the form of the Arrhenius equation: $\eta = A_v \exp(-E_v/RT)$, where E_v is the energy barrier that must be surmounted before flow can occur, see reference 71 and references therein.

73. NEUMANN, W. P. (1987) *Synthesis* 665. CURRAN, D. P. (1988) *Synthesis* 417.

74. ALTMAN, L. J. and ERDMAN, T. R. (1970) *Tetrahedron Letters* 4891.

75. ALTMAN, L. J. and NELSON, B. W. (1969) *J. Am. Chem. Soc.,* **91,** 5163.

76. SKELL, P. S. and BAXTER, H. N., III (1985) *J. Am. Chem. Soc.* **107,** 2823.

77. RANER, K. D., LUSZTYK, J. and INGOLD, K. U. (1988) *J. Am. Chem. Soc.* **110,** 3519.

78. TANNER, D. D., OUMAR-MAHAMAT, H., MEINTZER, C. P., TSAI, E. C., LU, T. T. and YANG, D. (1991) *J. Am. Chem. Soc.* **113,** 5397.

79. TANKO, J. M. and ANDERSON, F. E., III (1988) *J. Am. Chem. Soc.* **110,** 3525.

80. SUPPES, G. J., OCCHIOGROSSO, R. N. and MCHUGH, M. A. (1989) *Ind. Eng. Chem. Res.* **28,** 1152.

81. DESIMONE, J. M., GUAN, Z. and ELSBERND, C. S. (1992) *Science* **257,** 945.

82. TANKO, J. M. and BLACKERT, J. F. (1994) *Science* **263,** 203.

83. PAULAITIS, M. E. and ALEXANDER, G. C. (1987) *Pure & Appl. Chem.* **59,** 61.

84. ROBERTS, C. B., CHATEAUNEUF, J. E. and BRENNECKE, J. F. (1992) *J. Am. Chem. Soc.* **114,** 8455.

85. ROBERTS, C. B., ZHANG, J., BRENNECKE, J. F. and CHATEAUNEUF, J. E. (1993) *J. Phys. Chem.* **97,** 5618.
ROBERTS, C. B., ZHANG, J., CHATEAUNEUF, J. E. and BRENNECKE, J. F. (1993) *J. Am. Chem. Soc.* **115,** 9576.

86. O'SHEA, K. E., COMBES, J. R., FOX, M. A. and JOHNSTON, K. P. (1991) *Photochemistry and Photobiology* **54,** 571.

87. GUAN, Z., COMBES, J. R., MENCELOGLU, Y. Z. and DESIMONE, J. M. (1993) *Macromolecules* **26,** 2663.

88. $Y_{OTs} = 5.29$ and -4.99 for 60% H_2SO_4/H_2O and $CH_3CON(CH_3)_2$, respectively (reference 5), meaning that solvolysis of 2-adamantyl tosylate is more than 10^{10} times faster in the former solvent at 25°C.

Index

D

E

F

G

N

O

P

T

U

V

FSC
www.fsc.org
MIX
Papier aus verantwortungsvollen Quellen
Paper from responsible sources
FSC® C105338

If you have any concerns about our products,
you can contact us on
ProductSafety@springernature.com

In case Publisher is established outside the EU,
the EU authorized representative is:
Springer Nature Customer Service Center GmbH
Europaplatz 3, 69115 Heidelberg, Germany

Printed by Libri Plureos GmbH
in Hamburg, Germany